Climate Change and the Economic Importance and Damages of Insects

Ahmed Karmaoui
Moulay Ismail University of Meknes, Morocco & Moroccan Center for Culture and Science, Morocco

A volume in the Advances in Environmental Engineering and Green Technologies (AEEGT) Book Series

IGI Global
PUBLISHER of TIMELY KNOWLEDGE

Published in the United States of America by
IGI Global
Engineering Science Reference (an imprint of IGI Global)
701 E. Chocolate Avenue
Hershey PA, USA 17033
Tel: 717-533-8845
Fax: 717-533-8661
E-mail: cust@igi-global.com
Web site: http://www.igi-global.com

Library of Congress Cataloging-in-Publication Data

Names: Karmaoui, Ahmed, 1982- editor.
Title: Climate change and the economic importance and damages of insects / edited by Ahmed Karmaoui.
Description: Hershey, PA : Engineering Science Reference, [2022] | Includes bibliographical references and index. | Summary: "The edited book includes a set of chapters about Climate Change and the Economic Importance and Damages of Insects"-- Provided by publisher.
Identifiers: LCCN 2022041668 (print) | LCCN 2022041669 (ebook) | ISBN 9781668448243 (hardcover) | ISBN 9781668448250 (paperback) | ISBN 9781668448267 (ebook)
Subjects: LCSH: Beneficial insects--Climatic factors. | Insect pests--Climatic factors. | Honeybee--Climatic factors. | Insect-plant relationships. | Beneficial insects--Ecology. | Insect pests--Ecology. | Honeybee--Ecology.
Classification: LCC SB931 .C63 2022 (print) | LCC SB931 (ebook) | DDC 632/.799--dc23/eng/20220928
LC record available at https://lccn.loc.gov/2022041668
LC ebook record available at https://lccn.loc.gov/2022041669

This book is published in the IGI Global book series Advances in Environmental Engineering and Green Technologies (AEEGT) (ISSN: 2326-9162; eISSN: 2326-9170)

British Cataloguing in Publication Data
A Cataloguing in Publication record for this book is available from the British Library.

For electronic access to this publication, please contact: eresources@igi-global.com.

Advances in Environmental Engineering and Green Technologies (AEEGT) Book Series

ISSN:2326-9162
EISSN:2326-9170

Editor-in-Chief: Sang-Bing Tsai, Zhongshan Institute, University of Electronic Science and Technology of China, China & Wuyi University, China; Ming-Lang Tseng, Lunghwa University of Science and Technology, Taiwan; Yuchi Wang, University of Electronic Science and Technology of China Zhongshan Institute, China

MISSION

Growing awareness and an increased focus on environmental issues such as climate change, energy use, and loss of non-renewable resources have brought about a greater need for research that provides potential solutions to these problems. Research in environmental science and engineering continues to play a vital role in uncovering new opportunities for a "green" future.

The **Advances in Environmental Engineering and Green Technologies (AEEGT)** book series is a mouthpiece for research in all aspects of environmental science, earth science, and green initiatives. This series supports the ongoing research in this field through publishing books that discuss topics within environmental engineering or that deal with the interdisciplinary field of green technologies.

COVERAGE

- Alternative Power Sources
- Green Transportation
- Sustainable Communities
- Biofilters and Biofiltration
- Electric Vehicles
- Green Technology
- Policies Involving Green Technologies and Environmental Engineering
- Contaminated Site Remediation
- Cleantech
- Air Quality

IGI Global is currently accepting manuscripts for publication within this series. To submit a proposal for a volume in this series, please contact our Acquisition Editors at Acquisitions@igi-global.com or visit: http://www.igi-global.com/publish/.

The Advances in Environmental Engineering and Green Technologies (AEEGT) Book Series (ISSN 2326-9162) is published by IGI Global, 701 E. Chocolate Avenue, Hershey, PA 17033-1240, USA, www.igi-global.com. This series is composed of titles available for purchase individually; each title is edited to be contextually exclusive from any other title within the series. For pricing and ordering information please visit http://www.igi-global.com/book-series/advances-environmental-engineering-green-technologies/73679. Postmaster: Send all address changes to above address.

Titles in this Series

For a list of additional titles in this series, please visit:
http://www.igi-global.com/book-series/advances-environmental-engineering-green-technologies/73679

Applying Drone Technologies and Robotics for Agricultural Sustainability
Pethuru Raj (Reliance Jio Platforms Ltd., Bangalore, India) Kavita Saini (Galgotias University, India) and Vinicius Pacheco (Office of the Prosecutor General, Brazil)
Engineering Science Reference • ©2023 • 318pp • H/C (ISBN: 9781668464137) • US $250.00

Intelligent Solutions for Optimizing Agriculture and Tackling Climate Change Current and Future Dimensions
Ahmed Karmaoui (Moulay Ismail University, Morocco & Moroccan Center for Culture and Science, Morocco)
Engineering Science Reference • ©2023 • 269pp • H/C (ISBN: 9781668446492) • US $215.00

Recent Advances in Global Meliponiculture
Shamsul Bahri Abd Razak (Universiti Malaysia Terengganu, Malaysia) Tuan Zainazor Tuan Chilek (Universiti Malaysia Terengganu, Malaysia) and Jumadil Saputra (Universiti Malaysia Terengganu, Malaysia)
Engineering Science Reference • ©2023 • 250pp • H/C (ISBN: 9781668462652) • US $215.00

Biomass and Bioenergy Solutions for Climate Change Mitigation and Sustainability
Ashok Kumar Rathoure (M/s Akone Services, India) and Shankar Mukundrao Khade (Ajeenkya D.Y. Patil University, India)

For an entire list of titles in this series, please visit:
http://www.igi-global.com/book-series/advances-environmental-engineering-green-technologies/73679

IGI Global
PUBLISHER of TIMELY KNOWLEDGE

701 East Chocolate Avenue, Hershey, PA 17033, USA
Tel: 717-533-8845 x100 • Fax: 717-533-8661
E-Mail: cust@igi-global.com • www.igi-global.com

Table of Contents

Detailed Table of Contents

Growing evidence of the deleterious impacts of climate change on the agricultural sector and human health requires research for understanding the underlying mechanisms of damage. Insects being ectothermic organisms are vulnerable to temperature changes, precipitation, and carbon dioxide concentrations, which may alter their phenology. Seasonal perturbations also affect plant phenology. Consequently, there is a loss of synchrony among them. Though the effect varies between species, the overall impact is a decline in the pollinator population and a rise of pest infestations of plants. Such mismatches between insect pests and their natural enemies can render biological control ineffective. Since the environmental requirements of insect ecotypes vary across the globe, climatic alterations have triggered mass migrations resulting in transmission of parasites and diseases of both plants (via pests and pollinators) and animals (via vectors) across continental barriers. Thus, climate change can affect both the beneficial and harmful insects that may threaten the future food security and biodiversity.

Section 2
Climate Change, Insects, and Crops

the species concerned, with serious demographic consequences. This results in an alteration of the synchronization between plants and insects.

Honey bees (Apis sp.) have gained research attention lately because of their role in pollination, food production, and biodiversity conservation. Additionally, they can serve as excellent environmental biomonitors because they can concentrate pollutants in their products like honey and beeswax. Interestingly, honey bee colonies are resilient to stressors while the individuals are sensitive to them. Consequently, the negative impacts of stressors on their population tends to go unnoticed. However, research shows their population is declining rapidly. Pollution, intensification of agriculture, habitat loss, fragmented landscapes, reduction of floral heterogeneity, and climate change have been reasoned as the major factors for this observation. Their ectothermic nature makes them sensitive to climatic conditions which may have long-term effects on the environment. However, this field has been poorly investigated. Further research endeavours can provide sufficient information to plan conservation strategies for honey bee populations and ensure future food security.

Foreword

Considering our current era, the anthropogenic era, the fact that climate change represents the greatest threat to the Human future on earth is no longer a myth. Imagine a world without food, without trees, without agriculture and yields, terrifying isn't it?. Unfortunately, we have already started witnessing the beginning of a global scale crisis. Extreme weather events, intensified drought, and the spread and outbreaks of illnesses, pests, and infectious diseases are the most evident clues of such situations. Furthermore, worldwide economic trade is the biggest loser from climate change, as was evidenced by the recent coronavirus pandemic.

Indeed, *Climate Change and the Economic Importance and Damages of Insects* is another valuable and comprehensible book edited by Prof. Ahmed Karmaoui and published to reinforce our understanding of the relationship between climate change and the economic losses raising from insects damages. This book is an authentic and promising release, written by authors from many backgrounds, including from several leading research institutions and universities. Although the book has a particular emphasis on the economic importance and damages of insects, some chapters focus on critical issues such as agriculture and natural resources management. The book illustrates the large and diverse challenges that climate change poses to several aspects of Human lifestyle. In addition, the editor and the authors collectively bring new insights into how we can address and manage our current and most urgent challenges related to climate change.

The current book distinguishes itself from most available books by squeezing a considerable amount of knowledge into approximately 300 pages. Given the intensive involvement of Prof. Karmaoui and the authors, we cannot escape such a book without being fascinated by its rich and accurate scientific content. Finally, and most importantly, the book could be considered as the baseline for future contributions aiming at building reliable and strong knowledge.

Jalal Kassout
National Institute of Agronomic Research (INRA), CRRA, Marrakech, Morocco

Preface

Insects are considered as the largest group of animals on earth. They represent more than one million documented species and about half of all species at global scale. This makes this group of animals essential for global functioning and survival. It is well known that the importance of pollination ecosystem service provided by many groups of insects, such as the honeybees, in the support of global food security. In fact, the insects provide many important commercial products including honey, silk, dyes, lac, and wax, etc.

Regarding the ecological importance, it is difficult to list them all. Generally, the insects are abundant in all terrestrial ecosystems and present different sizes, colors, number of wings and legs with oval or elongated body. The high diversity of insects has significant ecological interactions. They offer valuable ecosystem services including soil fertilization, bioturbation, nutrient cycling, parasite suppression (pest control), seed dispersal, plant growth and play a role of food for many other predator species. Many other species are used as indicators to monitor the status and health of environment due to their sensitivity to habitat alteration and pollution of water, soil, and air.

Despite their ecologic and economic importance, pest insects are considered as a real threat of the agriculture and public health. For instance, diptera is among the most injurious pests that causes diseases such as leishmaniasis, trachoma, dengue fever, sleeping sickness, and malaria for human health. Termites for example, frequent insects in the soil ecosystem, cause material damage for wood constructions and farming equipment. They cause annually billions of economic losses.

Agricultural entomologists study food, textile, and timber pests to limit the resulting damage. Entomologists make great efforts to reduce the risk of insects that cause disease to humans and animals through a number of measures, including combating temporary water where mosquitoes and flies breed or using insecticides, and using animals that feed on insect pests.

In contrast, there are a large number of insects that are very beneficial to humans. For example, the silkworm produces precious threads, and the bees produce valuable liquid, give wax and pollinate many types of plants and agricultural crops. Some insects, such as mantis, ladybird, and others feed on insect pests. This prompted specialists in the field to pay attention to the production of these insects.

Honeybees is the most significant example of economic use. Many tribes across the world depend on honey as a food and healing source due to its medicinal properties. A large part is produced in beekeeping farms. Otherwise, honey bees pollinate different types of crops to increase production by visiting flowers to collect pollen and nectar or one of them… Without honey bees, the crops that need to pollinate are reduced in terms of quantity and variety. In addition to pollination and honey production, there are many other valuable products such as the queen bee breeding. Beekeepers breed pure or hybrid queens to produce packages. To build these tablets, bees secrete beeswax (a by-product) by special glands in the abdomen. Beeswax is also used in the manufacture of cosmetics and luminescent candles. Honey bees also produce royal jelly, which is used in the treatment of many diseases, which made it a great medical value that exceeds its profits twice the production of honey. It is secreted by young worker bees by the salivary glands to feed the larvae and the larvae of queens and queens. Pollen is one of the most important honey bee products because it contains many important and vital compounds, collected by placing special traps. The propolis production is another kind of valuable products. It is a resinous substance used by honey bees to seal cracks in their cells. Propolis is considered a successful treatment for wounds and some skin tumors. Another strange product is the bee venom used in the treatment of some rheumatic diseases.

Climate change is disrupting the distribution and abundance of this interesting group of animals. In fact, temperature and rainfall changes may affect the survival of many insects that will increase in some areas and decrease in others. On the other hand, the anthropogenic pressure is exacerbating this climatic effect. This may have without a doubt, serious repercussions on human well-being.

The edited book includes a set of chapters about *Climate Change and the Economic Importance and Damages of Insects*. It was divided into an introduction and four main sections.

The introduction provides a background of topic summarizing what is known in the field of climate change and Importance of Insects. This introductive chapter highlights that Growing evidences of deleterious impacts of climate change on the agricultural sector and human health requires research for understanding the underlying mechanisms of damage. Insects being ectothermic organisms are vulnerable to temperature changes, precipitation and carbon dioxide concentrations which may alter their phenology. Seasonal perturbations also affect plant phenology. Consequently, there is a loss of synchrony among them. Though the effect varies between species, the overall impact is decline in pollinator population and rise of pest infestations of plants. Such mismatches between insect pests and their natural enemies can render biological control ineffective. Since the environmental requirements of insect ecotypes vary across the globe, climatic alterations have triggered mass migrations resulting in transmission of parasites and diseases of both plants (via pests and pollinators) and animals (via vectors) across continental barriers. Thus, climate change can affect both the beneficial and harmful insects that may threaten the future food security and biodiversity.

Section 2 encompasses the topic "Climate Change, Insects, and Crops," including three chapters:

Chapter 2 is titled "Climate Change and Insect Pests: Economic Damages and Adaptive Strategies in Mediterranean Olive Groves". In this chapter, the authors showed that the climate change is a growing global issue and the most imminent concern threatening life forms. This applies in particular to the Mediterranean region where rates of most climate indicators are above global trends. Climate change could have implications for many facets of agricultural production systems. Although the olive tree (Olea europaea L.) is one of the most adapted species to the Mediterranean climate, its co-adaptation with pests may change with the considerable warming and drying trends in future climate projections. In fact, climate change will result in significant changes in diversity, distribution of insect pests, pest-plant interactions, natural enemy abundance, and the effectiveness of crop protection systems for pest control. Hence, a changed pattern in climatic factors has the potential to drastically affect the pest situation in olive groves and eventually lead to yield losses. This situation will create new challenges for olive growers who will have to protect their production and maximize their profit while adopting sustainable solutions.

Chapter 3 reviews the impact of climate change on the insect and mite pests of Moroccan citrus. Climate change parameters have greatly affected citrus production and pests. They affect agriculture by expanding the geographical range of insect and mite pests, and by modifying the lifecycle inducing an increase in the number of generations and survival during the wintering period. These changes in climatic parameters altered interspecific and intraspecific interactions between plants and pests, and increased the risk of invasive species, and reduced efficacy of biological control, especially those used by natural enemies, causing serious risks as well as crop damages. To mitigate the negative effects of climate change in plant agriculture, management strategies must be taken to deal with the new status of agricultural pests.

Chapter 4 reviews the impact of climate change on some insect pests damaging crops. Climate change affects the physiology of insects by accelerating their metabolism, growth, and reproduction, which increases their density, and ultimately leads to the consumption and infestation of crops damaged, this action, contributes to economic losses, especially in countries where agriculture plays an important role in their economic sector. This chapter aims to identify how climate parameters, especially temperature rise, insect pests, and crops, interact with each other so that insect pests develop in terms of their growth, and geographical expansion and fight against their natural enemies to damage crops. The evolution of these parameters is likely to develop or limit the pest species, in most cases, the increase in temperature promotes their development, also causes the action to expand their distribution, and further reduces the biological control, especially natural enemies. The control of agricultural pests can only be established through in-depth studies on the impact of climate change on the interaction between agricultural pests and crops and the contribution of strategies that incorporate the environment and physical resources, and social and financial benefits for farmers. Moreover, the application of adaptation practices, such as improved pest management, monitoring of climate and pest populations, and development of integrated models for predicting climate change, agriculture, and pests.

Section 3 explored the insect, human, and environmental health nexus through two chapters:

Chapter 5 attempted to link climate variables to the abundance of *P. sergenti*, the principal insect vector of leishmaniasis disease caused by *L. tropica*. Climate change is expected to influence health trough a change in the distribution and activity of vector borne diseases such as the Phlebotomine Sandflies that transmit leishmaniasis disease. In this chapter, we attempted to highlight the associations between climatic parameters and the activity of *Phlebotomus sergenti* that causes the anthroponotic cutaneous leishmaniasis (ACL). To this end, a correlation between the monthly data of the vector density and some climatic parameters in three selected sites from North Africa, South Europe, and Western Asia.

Chapter 6 explored the diversity of coleopteran species in arid environments highlighting the ecological importance and damages in the pre-Sahara of Morocco. Draa oases are declared as a part of the Biosphere reserve of Moroccan oases. These oases are agro-ecosystem dominated by the palm and fed by Draa Wadi, the longest in Morocco. Outside the oasis, the acacia trees and many pre-Saharan herbaceous species expand along the flooded areas and at the edges of temporary waterways. This desert vegetation creates a microclimate that favors the proliferation of a vast invertebrate diversity including Coleopteran species (Beetles) that support the soil and flora-fauna ecosystem services. These insects play a very important role in the health of this environment. Indeed, they can perform the functions of ecosystems such as pollination, purification, biological control, and soil fertilization. In this chapter, we provide new information and update the list of the Coleopteran species in the Draa Valley. A number of 31 species and 15 families were observed through several entomological trips.

Section 4 investigated the relationship between climate change and beneficial insects, focusing on the honeybees. This section includes two chapters:

Chapter 7 studied the climate change impact on honeybees (*apis spp.*) and their pollination services exploring the climate change and honeybees services.

Honeybee participates in the pollination of many crops and wild species and serves as a bioindicators of environmental health, and therefore plays an essential role in agricultural production, while it is currently facing serious threats worldwide, its disappearance will impact the balance of ecosystems. Global economic value of crop pollination services has been estimated at 153 billion euros per year, representing 9.5% of global agricultural food value. As bees are ectothermic species, the ambient temperature determines their activity and, therefore, climate change, with the increase in temperatures experienced in recent years, can influence their biology, distribution and behavior. Climate change affects negatively bees by impacting their food resources. Increased temperature induces a differential response of insects and plants that generate a temporal and spatial variation in the biological cycle of the species concerned, with serious demographic consequences. This results in an alteration of the synchronization between plants and insects.

Chapter 8 analyzed the honeybees as environmental biomonitors and reviewed the effects of climate change on their population. Honey bees (*Apis sp.*) have gained research attention lately because of their role in pollination, food production and biodiversity conservation. Additionally, they can serve as excellent environmental biomonitors because they can concentrate pollutants in their products like honey and beeswax. Interestingly, honey bee colonies are resilient to stressors while the individuals are sensitive to them. Consequently, the negative impacts of stressors on their population tends to go unnoticed. However, research shows their population is declining rapidly. Pollution, intensification of agriculture, habitat loss, fragmented landscapes, reduction of floral heterogeneity and climate change have been reasoned as the major factors for this observation. Their ectothermic nature makes them sensitive to climatic conditions, which may have long-term effects on the environment. However, this field has been poorly investigated. Further research endeavours can provide sufficient information to plan conservation strategies for honeybee populations and ensure the future food security.

Ahmed Karmaoui
Moulay Ismail University, Morocco & Moroccan Center for Culture and Science, Morocco

Acknowledgment

This book gathers a set of scientific studies from the field of economic entomology in the context of climate change which is, currently, one of the most important research areas concerning the well-being of humans. In this regard, the insect pests and vectors have gained much research attention in the recent years because of their negative economic impacts; particularly on agriculture, human and animal health. Strategies for regulating their population within the threshold limits have propelled such research endeavors. However, the beneficial insects need conservation to maintain their ecosystem services such as production of food and medicinal substances in the best interest of humans and animals.

This edited book presents an assortment of chapters from various fields highlighting the importance of insects and application of innovative techniques to curb their population or conserve them. This has been possible because of the collaborative contribution from researchers across Africa, Asia, and Europe. Sincere gratitude is due to all the authors, reviewers, and editorial advisory members for sharing their ideas, comments, reviews, and advices to produce and enhance this research effort. The support and editorial services provided by the IGI Global team is equally commendable in this regard.

Editorial Advisory Board

Section 1
Introduction

Chapter 1
Introduction

Debraj Biswal
https://orcid.org/0000-0002-7317-6774
Government General Degree College at Mangalkote, India

ABSTRACT

Growing evidence of the deleterious impacts of climate change on the agricultural sector and human health requires research for understanding the underlying mechanisms of damage. Insects being ectothermic organisms are vulnerable to temperature changes, precipitation, and carbon dioxide concentrations, which may alter their phenology. Seasonal perturbations also affect plant phenology. Consequently, there is a loss of synchrony among them. Though the effect varies between species, the overall impact is a decline in the pollinator population and a rise of pest infestations of plants. Such mismatches between insect pests and their natural enemies can render biological control ineffective. Since the environmental requirements of insect ecotypes vary across the globe, climatic alterations have triggered mass migrations resulting in transmission of parasites and diseases of both plants (via pests and pollinators) and animals (via vectors) across continental barriers. Thus, climate change can affect both the beneficial and harmful insects that may threaten the future food security and biodiversity.

Insects are the most diverse group of animals that provide ecosystem services as well as disservices; the ecosystem services include provisioning (e.g. production of food, fibre and other resources), support (e.g. pollination, decomposition and formation of soil), regulating (e.g. biological control and other feedback mechanisms essential for the maintenance of the supply of services) and cultural (e.g. non-material benefits like recreation, aesthetics and spiritual values). On the other hand, these animals may provoke damaging activity of pests and biological hazards. Each

DOI: 10.4018/978-1-6684-4824-3.ch001

aspect has been severely affected by climate change in recent years and the effects have been exacerbated by anthropogenic activities such as deforestation, agriculture intensification and land use changes (Noriega et al., 2018). Climate change and its effect on insects is a scientific issue, because these ectothermic animals are easily affected by climatic alterations, whatever their extent may be. In addition, the physiology and reproductive biology of insects are sensitive to environmental conditions such as light and temperature; however, these organisms may also be indirectly affected by the climate change impact on vegetation.

Insect pollinators are more valued because in addition to their contribution to food crop production, they help maintain biodiversity. In exchange for their pollination services, they obtain food from plants in the form of nectar and pollen. However, this mutualistic association is disrupted by monocultures, agricultural intensification, industrialization and other anthropogenic activities (Vasiliev and Greenwood, 2021). In addition, global warming also has deleterious effects on pollinators (Flores et al., 2019). The literature shows that negative effects on pollinators can be grouped into direct and indirect effects.

Directly, the health status of pollinators and their life cycles are affected by all the above factors. Poor health of the pollinators and their offspring make them vulnerable to several infectious diseases and parasites. Colonial species tend to be seriously affected because the infected individuals act as reservoirs for uninfected individuals, often resulting in the death of the entire colony (Vasiliev and Greenwood, 2021). The reduction in the pollinator population, in turn, affects plant productivity worldwide. This not only results in financial losses, but also threatens food security and ecosystem health. It is worth mentioning that warmer climates favour the survival and multiplication of parasites, which spread rapidly among wild and domesticated pollinators (Flores et al., 2019). According to reports, several new parasites have emerged with increased virulence in the last decade, even in places where they had not been reported before. This means that the parasites have been able to cross continental borders, survive, propagate and spread (Biswal, 2018). Researchers have come up with several theories to explain this phenomenon. Interestingly, the reasons are linked to the indirect effects of environmental changes and land use alterations on the pollinator population.

Plant phenology, which is closely related to the climatic conditions, especially light and temperature, has undergone dramatic changes in recent years due to the seasonal perturbations. Here, it should be kept in mind that the time of flowering should be appropriate to provide sufficient food for adult pollinators as well as their newly hatched offspring (Reddy et al., 2012). For colonial species like honey bees, there is the additional constraint of collecting and storing food that must be completed at an appropriate time because pollen loses its nutritional value if stored for a long period. The individuals feeding on them are eventually weak with poor immune systems exposing them to the risks of contracting diseases easily. Moreover, they are not physically fit to collect nectar and pollen by foraging through long distances. Unprecedented rainfall events and flash floods have been observed to be a very common feature of climate change. Such impacts during the flowering season can wash the pollen away from flowers leaving the pollinators famished. In addition, these weather conditions can also be a barrier to their foraging activities. The pollinators also avoid extreme heat because it leads to dehydration and exhaustion that can be fatal (Le Conte and Navajas, 2008). Similarly, prolonged harsh winters in temperate regions can deplete the food reserves in bee hives such that the honey bees emerging in the flowering season that follow are fragile. Interestingly, the reproductive structures of flowers and pollinators (as well as other insects) undergo structural and functional malformations under conditions of heat stress (Reddy et al., 2012). The former affects both the quantity and quality of the pollen and nectar produced while the latter causes pollinators to produce non-viable or unfit offspring.

As a result of all these effects, the pollinators have been observed to forage across long distances, often beyond their geographical distribution, in search of suitable environmental conditions and/or nutritious food sources. According to scientists, these migratory events are unwanted because they lead to genetic homogenization through hybrid formations and erosion of ecotypes. Though not visible, these genetic events may have adverse effects on biodiversity in the long run (Ellstrand and Reiseberg, 2016; Todesco et al., 2016). Here, it is worth mentioning that similar pollinator species (like several other insects) have evolved in different geographical areas of the world in an isolated manner and adapted themselves to the dominant climate and vegetation therein. These are referred to as 'ecotypes' which are essential for maintaining the indigenous plant species and thereby the global biodiversity (Reddy et al., 2012). However, introduction of invasive plant species for beautification or financial benefits through augmented agricultural and horticultural production or for any other reason have pushed several native plant species to the brink of extinction. This is another major reason that has deprived the pollinators from nutritional resources because of their lack of acquaintance with the newly introduced plant species (Potts et al., 2010). Here, the species-specific associations between the pollinators and their host plants influence such interactions. Monoculture, a common agricultural practice used currently, also affects the nutritional status of the pollinators negatively by restricting the variety of pollens from various floral species. This poses a vital problem, especially to those pollinators which forage on a wide range of plant species. These factors have also triggered mass migrations of pollinators that have caused the spread of parasites and other infectious agents to newer geographical regions. Apart from natural migratory events, global trade and introduction of domesticated pollinators to new areas can also have similar effects (Vasiliev and Greenwood, 2021).

Spreading of parasites and disease causing agents to newer locations have severe implications because the immune system of the pollinator ecotypes cannot cope up with the challenges from the newly invading parasites (Le Conte and Navajas, 2008). Eventually, these incidents can silently wipe out the population of important pollinator species or other beneficial insects much to the misfortune of the ecosystem. Reports from these fields are much more pronounced in case of vector insects because they directly affect the health of human beings (Kolimenakis et al., 2021). The impact of climate change on vectors, especially mosquitoes, have been much discussed and is an important research domain currently because they cause death of large proportions of humans worldwide. Moreover, their prominence in the temperate zones and the arid regions of the world, from where they were not previously recorded, clearly depict how far-reaching and serious the effects of climate change can be.

According to scientists; developments in transport, aviation and global trade have facilitated the worldwide distribution of mosquitoes (Wang et al., 2020). Milder winters (in temperate zones) and unprecedented rainfall events (in arid zones) have supported their survival and reproduction by generating favourable microclimates in previously unfavourable areas. Urbanization has aggravated the conditions by generating unique breeding habitats for mosquitoes that cannot be accessed by conventional vector control strategies (Biswal, 2021; Augustina et al., 2021). This has compelled the researchers to develop innovative techniques of mosquito control which are still being worked upon. Furthermore, studies have revealed that warmer climates stimulate the blood probing behaviour of gravid female mosquitoes which increases the risk of disease transmission (Kolimenakis et al., 2021). Though female mosquitoes are a source of ecosystem disservice, the males can actually pollinate some plant species thereby providing ecosystem service (Lutz et al., 2017). Moreover, their larvae form a part of the food chain in the aquatic ecosystem. Here lies the paradox! However, the importance of male mosquitoes as pollinators and that of mosquito larvae as components of the food chain have not been explored thoroughly. The obvious medical complications caused by the female mosquito vectors make them notorious elements for the human society. The root of the problem lies in their high adaptability and reproductive success even in the most adverse situations. The side effects of vector control strategies add another dimension to this.

Even though innovative vector control strategies like genetic manipulations appear promising, their applicability seem to be restricted to laboratories because they have not been successful in field conditions. Use of insecticides and pesticides continue to be the most common mosquito control technique. However, their negative impacts on beneficial insects and non-target organisms like fishes have been widely studied with reports being mostly negative (Davis et al., 2007; Russell et al., 2009; Rochlin et al., 2022). The same situation has been observed in case of agricultural insect pests' control. Research shows that global warming has widened the geographical range of several insect pests and enhanced their ability to overwinter (Skendžić et al., 2021). Their overall deleterious impact on agricultural production has been accelerated by their increased number of generations due to climatic alterations. Interestingly, their interaction with the host plants and their natural predators have also been reported to undergo alterations under the present scenario of climate change. The implications are grave since these conditions are rendering biological control of pests ineffective.

It is very well known that climate change is a major driver of insect pest population dynamics. Since insects are ectothermic organisms, temperature alterations exert the most obvious effects on their physiology and behaviour (Lawton et al., 2022). Experiments show that their metabolic rate increases with temperature rise augmenting their consumption rate along with their mobility and development. Increased consumption leads to increased agricultural losses while enhanced mobility leads to expansion of their geographical distribution (Deutsch et al., 2018). The latter is a serious threat because plants are likely to face additional pressures from migratory pests (Bebber, 2015). Here, one must understand that the problem is not just agricultural loss due to increased grazing by pests but also due to the increased risk of transmission of plant diseases (by pests) not prevalent in a particular region (Donatelli et al., 2017). Under these situations, the lack of preparedness of the plants as well as the farmers multiplies the agricultural loss several times. It has been predicted that sudden and frequent outbreaks of plant diseases transmitted by insect pests will be a common phenomenon in the near future (Lawton et al., 2022).

As far as the growth rates of insect pests are concerned, they have been hypothesized to vary in the temperate and tropical regions. Warmer climates in the temperate regions will increase their growth rates while it will decrease the same in the tropical regions (that already have warm climates). As such, agricultural losses due to herbivory by insect pests have been proposed to vary according to their geographical location (Skendžić et al., 2021). Interestingly, the soil dwelling insect pests of plants are buffered against the temperature extremes and, thereby, exhibit less deviations from their normal biology compared to those that live aboveground (Skendžić et al., 2021). Apart from temperature, humidity is also an important factor that regulates the growth of insect pest population. However, scientific data and climate models differ from the opinion that global warming will impact the insect pest populations uniformly. Differences in their levels of tolerance to ecological factors will largely dictate their responses to climatic alterations. Therefore, their phenology and life history traits will vary across the globe leading to mixed responses (Donatelli et al., 2017). Reduced physiological performance and contraction of distribution range of some insect pests could lower their deleterious effects on crop production. According to some researchers, certain insect pest species may become extinct in the next few years (Skendžić et al., 2021). Despite this, the dominance of invasive and migratory insect pests may increase, accelerating the problems all the more. These complicated and contradictory findings create room for further research.

Insects possess high plasticity and are therefore able to adapt themselves to changing climates. They have the capacity to alter their voltinism which gives them this advantage (Skendžić et al., 2021; Lawton et al., 2022). However, diapause disruption can have severe impacts on the ecosystem because its purpose is to slow down the metabolic activities of insects which, in turn, regulates pollination, consumption of host plants and interactions with other insect species (Gill et al., 2017). Research shows that temperature rise shortens the duration of diapause allowing the insect pests to remain active for longer periods during which they continue to exert their destructive actions. Nevertheless, the observations are not uniform because the effects are species-specific with some species being reported to decline in abundance due to diapause disruption (Gill et al., 2017). This arises from their inability to withstand the winter that follows. However, the most significant impact of diapause shortening is the resultant mismatch in the synchrony between host plant phenology and insect phenology. Such mismatches can be disastrous to the insect pests or the host plants depending on their species (Gill et al., 2017). No clear cut hypothesis can delineate these complicated interactions. These need thorough exploration to understand the consequences in the future.

Increases in ambient concentration of carbon dioxide (CO_2) is another important feature of global warming which can cause changes in the nutritional quality and the palatability of both the foliage and fruits of plants. High CO_2 concentration alters the carbon-nitrogen ratio (C:N) in plant parts often triggering greater consumption rates by insect pests since they require nitrogen (N) for their growth and development (Neelaboina et al., 2018). This phenomenon of voracious feeding by insect pests due to reduction of N content of plant parts is referred to as 'compensatory feeding'. Eventually, it amplifies the agricultural losses (DeLucia et al., 2008). However, augmented levels of atmospheric CO_2 serves to slow down their rates of growth and development and thereby their abundance. Consequently, no definite relationships have been established between the levels of atmospheric CO_2 and destruction caused by insect pests.

The effects of enhanced CO_2 concentrations in the atmosphere have been found to vary for C3 and C4 plants. The variations arise due to their differential ability to fix CO_2 (Hamim, 2005; Silva et al., 2020). In this respect, the C3 plants are not much efficient in utilizing augmented CO_2 levels because they possess the enzyme, ribulose bisphosphate carboxylase/oxygenase which has low substrate specificity. Interestingly, the productivity of the C3 crops has been hypothesized to increase under conditions of elevated levels of atmospheric CO_2 provided the plants are not acted upon by other stress factors which is impossible under the existing natural conditions (Hamim, 2005; Silva et al., 2020). As a result, the positive effects of high atmospheric CO_2 on plants are negated by changes in precipitation and temperature. Here, it is noteworthy that monophagous insects like silkworm (*Bombyx mori* larvae) are severely affected under these conditions, especially when their host plants like mulberry are C3 plants. Furthermore, steep rises in the vulnerability of both the host plants and silkworm larvae to pests, parasites and diseases under the climate change scenario are being reported currently (Neelaboina et al., 2018). Silkworms being domesticated species are more susceptible to such damaging factors. Consequently, the silk industry has been projected to suffer losses in the coming years. Similar results have also been obtained from studies with lac insects since alterations in climatic conditions have been found to exert negative impacts on their sexual maturity and thereby their performance and lac production (Mohanasundaram et al., 2014).

Coming back to the pest-plant interactions, they have also been documented to be affected by the alterations in the frequency and intensity of precipitation due to climate change (Tamura et al., 2022). Climate models predict rises in the frequency of intense precipitation events of shorter durations accompanied by storms in the coming years. Heavy rainfall and floods resulting therefrom can wash off the eggs and/or the developmental stages of insect pests lowering the pest population (Donatelli et al., 2017). On the contrary, events of drought can not only stimulate herbivory by the insect population but also lower the defence mechanisms of plants against them. Drought stressed plants are usually unhealthy and more prone to damage by insect pests causing huge agricultural losses (Skendžić et al., 2021). Insect pests which spend a part or whole of their life in the soil are more prone to be affected by irregular precipitation events and floods. Therefore, the effects of precipitation on insect pests will vary across the globe depending on the biogeography of the concerned species and the deviations from the dominant climatic conditions.

It is quite evident that climate change will be affecting the interactions between the insect pests and the host plants. The consequences, however, depend on the impact of climatic factors on the nutritional status of plants and reproductive biology of insect pests. Considering the available data, it may be said that the overall effect of climate change on the agricultural sector is destruction and financial loss due to the augmented activity of insect pests. The conditions have been aggravated further, in the recent years, because of the failure of the conventional pest control strategies and the biological control mechanisms being rendered ineffective by climatic alterations (Athanassiou et al., 2018). Current researches in this field focus on climate modelling and prediction of pest infestations that can largely help to keep the pest population within threshold limits. However, the environmental impacts of insecticides and pesticides (as in vector control) have drawn flak from the scientific community because they erode the population of several beneficial insects such as the pollinators and beetles (that serve as biological control agents) (Rochlin et al., 2022). The time of application of these agrochemicals has been found to be crucial in exerting their effects on the beneficial insects. Studies show if the pesticides are applied in the evening, when the pollinators and other beneficial insects are not active, their deleterious effects can be counteracted to a large extent. Additionally, avoiding rains is equally important to prevent them from being washed away. Current developments in the field of eco-friendly nanopesticides and phyto-products appear to give promising results and trials are on the way before they can be applied on large scales (Athanassiou et al., 2018). However, combining them with suitable climate models is equally important to determine the time of application of these substances.

The current book discusses all these aspects trying to focus on the latest research developments in these fields. It tries to summarize the effects of climate change on the beneficial insects (while discussing their importance) and the consequent financial losses in the agricultural sector and other industries dependent on them. It also aims to highlight the impact of climatic alterations on vectors and the associated human health complications. Furthermore, it documents the fluctuations in the dynamics of the insect pests that are being observed and have been predicted to occur due to climate change. Needless to mention, the destructive influence of these pests on agricultural and horticultural crops have been elaborated. At the same time, the sensitivity of their natural enemies to climatic factors causing the loss of effectiveness of natural biological control of pest population has been analysed. In this regard, the loss of synchrony in the phenology of the insect pests and their natural enemies under the climate change scenario have also been depicted. Similarly, the impacts of mismatches in the phenology of host plants and insects arising from climate change have also been elaborated. Additionally, the role of anthropogenic activities in aggravating the impacts of climate change on each of these aspects have been described thoroughly. Eventually, the research gaps in these fields have been highlighted to enlighten the scientific community regarding the direction of further exploration and formulation of mitigation strategies. This book is a small endeavour towards conservation of biodiversity and ensuring food security for the future generations.

DECLARATIONS

Funding: This research received no specific grant from any funding agency in the public, commercial, or not-for-profit sectors.

Competing Interests: The author has no competing interests to declare that are relevant to the content of this article.

REFERENCES

Athanassiou, C. G., Kavallieratos, N. G., Benelli, G., Losic, D., Usha Rani, P., & Desneux, N. (2018). Nanoparticles for pest control: Current status and future perspectives. *Journal of Pest Science*, *91*(1), 1–15. doi:10.100710340-017-0898-0

Augustina, I., Jabal, A. R., Permana, G. I., & Ratnasari, A. (2021). Distribution and ecology of mosquito larvae in Pahandut subdistrict, Palangkaraya city. *Journal of Physics: Conference Series*, *1918*(5), 052018. doi:10.1088/1742-6596/1918/5/052018

Bebber, D. P. (2015). Range-Expanding Pests and Pathogens in a Warming World. *Annual Review of Phytopathology*, *53*(1), 335–356. doi:10.1146/annurev-phyto-080614-120207 PMID:26047565

Biswal, D. (2018). *Contemporary Health Issues and Environmental Impact.* In M. Bose, S. A. Bandyopadyay, & S. Poddar (Eds.), *Impact of climate change on parasitic diseases: Do we need to bother?* (pp. 32–40). Lincoln University College. doi:10.31674/books.2018.chiei.ch05

Biswal, D. (2021). Recent Advancements in Eco-Friendly Mosquito Control Strategies. In D. Ghosh & S. Saha (Eds.), *Recent Trends in Environmental Science and Applied Ecology* (pp. 124–129). Bhumi Publishing.

Davis, R. S., Peterson, R. K., & Macedo, P. A. (2007). An ecological risk assessment for insecticides used in adult mosquito management. *Integrated Environmental Assessment and Management*, *3*(3), 373–382. doi:10.1002/ieam.5630030308 PMID:17695110

DeLucia, E. H., Casteel, C. L., Nabity, P. D., & O'Neill, B. F. (2008). Insects take a bigger bite out of plants in a warmer, higher carbon dioxide world. *Proceedings of the National Academy of Sciences of the United States of America*, *105*(6), 1781–1782. doi:10.1073/pnas.0712056105 PMID:18256185

Deutsch, C. A., Tewksbury, J. J., Tigchelaar, M., Battisti, D. S., Merrill, S. C., Huey, R. B., & Naylor, R. L. (2018). Increase in crop losses to insect pests in a warming climate. *Science*, *361*(6405), 916–919. doi:10.1126cience.aat3466 PMID:30166490

Donatelli, M., Magarey, R. D., Bregaglio, S., Willocquet, L., Whish, J. P. M., & Savary, S. (2017). Modelling the impacts of pests and diseases on agricultural systems. *Agricultural Systems*, *155*, 213–224. doi:10.1016/j.agsy.2017.01.019 PMID:28701814

Ellstrand, N. C., & Rieseberg, L. H. (2016). When gene flow really matters: Gene flow in applied evolutionary biology. *Evolutionary Applications*, *9*(7), 833–836. doi:10.1111/eva.12402 PMID:27468301

Flores, J. M., Gil-Lebrero, S., Gámiz, V., Rodríguez, M. I., Ortiz, M. A., & Quiles, F. J. (2019). Effect of the climate change on honey bee colonies in a temperate Mediterranean zone assessed through remote hive weight monitoring system in conjunction with exhaustive colonies assessment. *The Science of the Total Environment*, *653*, 1111–1119. doi:10.1016/j.scitotenv.2018.11.004 PMID:30759551

Gill, H. K., Goyal, G., & Chahil, G. (2017). Insect diapause: A review. *Journal of Agricultural Science and Technology A*, *7*(7), 454–473. doi:10.17265/2161-6256/2017.07.002

Hamim. (2005). Photosynthesis of C3 and C4 species in response to increased CO_2 concentration and drought stress. *HAYATI Journal of Biosciences, 12*(4), 131-138. doi:10.1016/S1978-3019(16)30340-0

Kolimenakis, A., Heinz, S., Wilson, M. L., Winkler, V., Yakob, L., Michaelakis, A., Papachristos, D., Richardson, C., & Horstick, O. (2021). The role of urbanisation in the spread of *Aedes* mosquitoes and the diseases they transmit—A systematic review. *PLoS Neglected Tropical Diseases*, *15*(9), e0009631. doi:10.1371/journal.pntd.0009631 PMID:34499653

Lawton, D., Huseth, A. S., Kennedy, G. G., Morey, A. C., Hutchison, W. D., Reisig, D. D., Dorman, S. J., Dillard, D., Venette, R. C., Groves, R. L., Adamczyk, J. J., Barbosa Dos Santos, I., Baute, T., Brown, S., Burkness, E., Dean, A., Dively, G. P., Doughty, H. B., Fleischer, S. J., ... Zuefle, M. (2022). Pest population dynamics are related to a continental overwintering gradient. *Proceedings of the National Academy of Sciences of the United States of America*, *119*(37), e2203230119. doi:10.1073/pnas.2203230119 PMID:36067290

Le Conte, Y., & Navajas, M. (2008). Climate change: Impact on honey bee populations and diseases. *Revue Scientifique et Technique (International Office of Epizootics)*, *27*(2), 499–510. PMID:18819674

Lutz, E. K., Lahondère, C., Vinauger, C., & Riffell, J. A. (2017). Olfactory learning and chemical ecology of olfaction in disease vector mosquitoes: A life history perspective. *Current Opinion in Insect Science*, *20*, 75–83. doi:10.1016/j.cois.2017.03.002 PMID:28602240

Mohanasundaram, A., Monobrullah, M. D., Sharma, K. K., Anees, K., Singh, R. K., Meena, S. C., & Verma, S. (2014). Climate change: Effect of weather parameters on production of summer season crop of rangeeni strain of Indian lac insect, Kerria lacca (Kerr) at Ranchi, Jharkhand. *Journal of Agrometeorology*, *16*(1), 108–113.

Neelaboina, B. K., Khan, G. A., Kumar, S., Gani, M., Ahmad, M. N., & Ghosh, M. K. (2018). Impact of climate change on agriculture and sericulture. *Journal of Entomology and Zoology Studies*, *6*(5), 426–429.

Noriega, J. A., Hortal, J., Azcárate, F. M., Berg, M. P., Bonada, N., Briones, M. J. I., Toro, I. D., Goulson, D., Ibanez, S., Landis, D. A., Moretti, M., Potts, S. G., Slade, E. M., Stout, J. C., Ulyshen, M. D., Wackers, F. L., Woodcock, B. A., & Santos, A. M. C. (2018). Research trends in ecosystem services provided by insects. *Basic and Applied Ecology*, *26*, 8–23. doi:10.1016/j.baae.2017.09.006

Potts, S. G., Biesmeijer, J. C., Kremen, C., Neumann, P., Schweiger, O., & Kunin, W. E. (2010). Global pollinator declines: Trends, impacts and drivers. *Trends in Ecology & Evolution*, *25*(6), 345–353. doi:10.1016/j.tree.2010.01.007 PMID:20188434

Reddy, P. V. R., Verghese, A., & Rajan, V. V. (2012). Potential impact of climate change on honeybees (*Apis* spp.) and their pollination services. *Pest Management in Horticultural Ecosystems*, *18*(2), 121–127.

Rochlin, I., White, G., Reissen, N., Martheswaran, T., & Faraji, A. (2022). Effects of aerial adulticiding for mosquito management on nontarget insects: A Bayesian and community ecology approach. *Ecosphere*, *13*(1), e3896. doi:10.1002/ecs2.3896

Russell, T. L., Kay, B. H., & Skilleter, G. A. (2009). Environmental effects of mosquito insecticides on saltmarsh invertebrate fauna. *Aquatic Biology*, *6*, 77–90. doi:10.3354/ab00156

Silva, R., Alves, R. C., & Zingaretti, S. M. (2020). Increased [CO2] Causes Changes in Physiological and Genetic Responses in C4 Crops: A Brief Review. *Plants*, *9*(11), 1567. doi:10.3390/plants9111567 PMID:33202833

Skendžić, S., Zovko, M., Živković, I. P., Lešić, V., & Lemić, D. (2021). The Impact of Climate Change on Agricultural Insect Pests. *Insects*, *12*(5), 440. doi:10.3390/insects12050440 PMID:34066138

Tamura, Y., Osawa, T., Tabuchi, K., Yamasaki, K., Niiyama, T., Sudo, S., Ishigooka, Y., Yoshioka, A., & Takada, M. B. (2022). Estimating plant–insect interactions under climate change with limited data. *Scientific Reports*, *12*(1), 10554. doi:10.103841598-022-14625-9 PMID:35794117

Todesco, M., Pascual, M. A., Owens, G. L., Ostevik, K. L., Moyers, B. T., Hübner, S., Heredia, S. M., Hahn, M. A., Caseys, C., Bock, D. G., & Rieseberg, L. H. (2016). Hybridization and extinction. *Evolutionary Applications*, *9*(7), 892–908. doi:10.1111/eva.12367 PMID:27468307

Vasiliev, D., & Greenwood, S. (2021). The role of climate change in pollinator decline across the Northern Hemisphere is underestimated. *The Science of the Total Environment*, *775*, 145788. doi:10.1016/j.scitotenv.2021.145788 PMID:33618305

Wang, Y., Yim, S., Yang, Y., & Morin, C. W. (2020). The effect of urbanization and climate change on the mosquito population in the Pearl River Delta region of China. *International Journal of Biometeorology*, *64*(3), 501–512. doi:10.100700484-019-01837-4 PMID:31811391

Section 2
Climate Change, Insects, and Crops

Chapter 2
Climate Change and Insect Pests:
Economic Damages and Adaptive Strategies in Mediterranean Olive Groves

Rachid Azenzem
National Institute of Agricultural Research, Rabat, Morocco

Al Feddy Mohamed Najib
National Institute of Agricultural Research, Rabat, Morocco

Jalal Kassout
National Institute of Agricultural Research, Rabat, Morocco

ABSTRACT

Climate change is a growing global issue and the most imminent concern threatening life forms. This applies in particular to the Mediterranean region where rates of most climate indicators are above global trends. Climate change could have implications for many facets of agricultural production systems. Although the olive tree (Olea europaea L.) is one of the most adapted species to the Mediterranean climate, its co-adaptation with pests may change with the considerable warming and drying trends in future climate projections. In fact, climate change will result in significant changes in diversity, distribution of insect pests, pest-plant interactions, natural enemy abundance, and the effectiveness of crop protection systems for pest control. Hence, a changed pattern in climatic factors has the potential to drastically affect the pest situation in olive groves and eventually lead to yield losses. This situation will create new challenges for olive growers who will have to protect their production and maximize their profit while adopting sustainable solutions.

DOI: 10.4018/978-1-6684-4824-3.ch002

INTRODUCTION

The olive (*Olea europaea* L.) tree is the most emblematic tree of the Mediterranean region with considerable economic, social, and ecological importance. The use of olive trees as a source of food, wood, and cattle feed explains the spread of olive groves alongside the development of human civilization (Rapoport et al., 2016). Nowadays, the olive is one of the most cultivated trees in the world and is found on all five continents for cultural (traditions, landscapes), economic (oil and olives), and ecological (landscape) reasons. In recent decades, the olive growing areas have expanded and the global production of olive oil has increased, though the effect of biennial or alternate production (high/low yield years) is still noticeable. Approximately 11.2 million ha are devoted to olive cultivation worldwide in 2019, and the Mediterranean basin accounts for more than 95% of this area, mainly in Spain (23%), Tunisia (19%), Italy (10%), Morocco (10%), Greece (8%) and Turkey (8%) (FAOSTAT, 2019) (Table 1). World olive oil production in 2019 is estimated at 3.1 million tons, and the main producing countries are Spain (36%), Italy (11%), Greece (9%), Tunisia (8%), Morocco (7%), and Turkey (7%) (FAOSTAT, 2019) (Table 1). The use of fast-growing, high-yielding varieties planted at higher densities, innovative orchard management practices such as pruning, tillage, and plant protection, as well as drip irrigation that allows for the application of fertilizer, are factors that have contributed to this significant increase in oil production. However, olive growers are now facing new difficulties and challenges, climate change being one of the most important. Indeed, global warming is predicted to result in increased temperatures, altered rainfall patterns, and other derived factors that may have unknown effects on the distribution, abundance, and impact of species in natural and agricultural systems (Gutierrez, et al., 2010). Such climatic variations have significant effects on insect pest populations and their dynamics (Woiwod, 1997). In fact, pest development is strongly influenced by environmental factors, including humidity and temperature. Therefore, climate change may affect the distribution and life cycle of pest species, their migratory behavior, and their interaction with natural enemies, and could result in an increase in their growth rate and the number of generations. Hence, a changed pattern in climatic factors has the potential to drastically affect the pest situation in olive groves and eventually lead to yield losses. More than 255 species make up the group of organisms that could be harmful to *O. europaea*, including insect pests, mites, nematodes, and phytopathogenic agents (Montiel-Bueno and Jones, 2002). According to estimates, annual losses due to insect pests account for approximately 15% of total production in the Mediterranean Basin (Haniotakis, 2005). In this region of the world qualified as a "hotspot" of global changes (Giorgi, 2006), these pests could be affected by climate change with direct impacts on their biology and indirect impacts related to their environment, host plants, and natural enemies. This situation will create new challenges for olive growers who will have to protect their production and maximize their profit while adopting economically viable solutions and environmentally friendly strategies in the framework of sustainable agriculture. This chapter provides an overview of the impacts of climate change on olive insect pests and strategies to mitigate their damage in olive groves.

Table 1. World's major countries regarding the olive area, olive production and olive oil production (FAOSTAT, 2019)

	Area (ha)	%	Prod (t)	%	Oil prod (t)	%
Algeria	431 506	4%	868 754	4%	92 200	3%
Egypt	103 000	1%	981 000	5%	41 300	1%
Spain	2 601 900	23%	5 965 080	28%	1 129 233	36%
Greece	903 080	8%	2 790 442	13%	290 476	9%
Italy	1 139 470	10%	2 194 110	10%	336 581	11%
Morocco	1 073 493	10%	1 912 238	9%	204 200	7%
Portugal	377 280	3%	925 950	4%	154 063	5%
Tunisia	2 159 004	19%	1 184 961	5%	239 500	8%
Turkey	879 177	8%	1 525 000	7%	217 800	7%
World	11 230 212		21 586 902		3 097 803	

CLIMATE CHANGE AND OLIVE PRODUCTION IN MEDITERRANEAN REGION

Climate change, a growing global issue, is the most imminent and urgent concern threatening human survival and livelihoods. In the Mediterranean basin, climate models and historical data indicate that the climate is changing faster than global trends due to anthropogenic greenhouse gas emissions. Recent studies indicate that the Mediterranean region has experienced significant warming over the past 40 years and that annual temperatures are now 1.5°C higher than in the pre-industrial era (1880-1899), which is well above current global warming trends (+1.1°C) (Levitus et al., 2017; Zittis et al., 2022). According to several climate projections, the climate of the Mediterranean region will experience generalized warming and a decrease in precipitation during the 21st century (Lionello and Scarascia, 2018; Zittis et al., 2022). Moreover, extreme weather occurrences are increasingly emerging as the new norm on a global scale (Singh and Zommers, 2014). This is particularly true in the Mediterranean region, where the frequency of severe weather events is rising. These occurrences include droughts, heatwaves, hydro-climate extremes, storm surges, windstorms, and dust storms (Lionello and Scarascia, 2018). These changes have implications for many facets of agricultural production systems throughout the world. In fact, several climate factors, including rising temperatures, increasing atmospheric CO_2 levels, and changing precipitation patterns, have varying interactions with plants through different mechanisms. Plants are affected by these alterations both directly, in terms of allocation of photosynthetic energy to certain tissues and organs, and indirectly, in terms of geographic distribution (Kassout et al., 2022) and population dynamics of pest species (Ponti et al., 2014). Among these plants, the olive tree is considered a sensitive indicator of the level of climate change in the region and subject to its impacts (Kassout et al., 2021). In this case, these predictions are expected to have serious negative effects on water relations (Fraga et al., 2020), oxidative pathways, and other physiological processes (Petridis et al., 2012; Biel et al., 2008), as well as on phenological schedules (Galán et al., 2005), final yield (Greven et al., 2009; Quiroga and Iglesias, 2009) and quality characteristics of the olive tree (Servili et al., 2009). Thus, high temperatures could increase the growing season's length and lead to a lack of chilling requirements for olive trees, resulting in problems with fruit bud growth and flowering (Pérez-López et al., 2008; Gutierrez et al., 2008; Abou-Saaid et al., 2022). The productivity of olive groves is also influenced by other climatic elements, including solar radiation, relative humidity and wind. For example, hot summer winds cause fruit drop, while dry summer winds tend to cause fruit to ripen too early and wilt (Therios, 2009). Water availability, in addition to the implications of global warming discussed above, poses a serious problem, especially in achieving reasonable yields (Iglesias and Garrote, 2015). Water stress can have a variety of adverse effects (Arampatzis et al., 2018), including reduced fruit set and flowering, small leaves, limited photosynthesis, flower abortion, and cluster abscission. In addition to traditional orchards, and although the olive is a relatively water-stress tolerant plant (Ponti et al., 2014), the expansion and intensification of olive cultivation has increased the growing share of intensive and super-intensive orchards adopting new management techniques, including the more frequent use of drip irrigation. To adapt to climate change, olive growers are exploring new techniques to reduce losses due to water stress, such as regulated deficit irrigation (Therios, 2009). On the other hand, the potential benefits of future increases in atmospheric CO_2 concentrations, as a key chemical component of photosynthesis, should also be considered as it could have a direct effect on plants by promoting their growth and development (Drake et al., 1997). To cope with climate change projections, industry players will need to adapt to a warmer, more water-poor future by developing climate-resilient varieties and adopting appropriate technical management strategies.

IMPACT OF CLIMATE CHANGE ON INSECT PESTS

In addition to its direct impact on plant yields, climate change may also have an indirect impact on pests and diseases. Indeed, an agroecosystem is mainly controlled by the interactions between its biotic and abiotic components. Pest pressure on agricultural systems is expected to intensify due to increased extreme weather events, changing precipitation patterns, rising temperatures, and increasing CO_2 concentrations. These factors have the potential to profoundly alter the distribution and intensity of insect infestation of crops, through direct impacts on the insect life cycle and indirect effects on hosts and natural enemies. Direct effects result from variations in population growth rates, wintering success, migration patterns, and habitat expansion while indirect effects result from trophic interactions with their environment.

Impact on Pest Population and Number of Generations

The increase in temperature due to climate change can have various effects on crop pest populations. Since insects are poikilothermic species, the temperature of their bodies is influenced by the ambient temperature. Therefore, the most significant environmental element influencing insect behavior, dispersal, growth, and multiplication is probably temperature (Kocmánková et al., 2010). An increase in temperature within a specific favorable range will allow insects to produce a greater number of generations per year, resulting in more crop damage. Yamamura and Kiritani (1998) found that insects could have one to five additional generations each year with a 2°C temperature increase. In addition, climate change projections indicate that rising temperatures will accelerate development, leading to a rapid increase in pest populations as the time to achieve reproductive maturity is reduced and life cycles are shortened (Bale et al., 2002). According to Shrestha (2019), each degree of temperature increase could result in an additional 10-25% increase in yield losses due to insect pests. Furthermore, Deutsch et al. (2019) concluded that as a result of climate change, insect populations in tropical regions are expected to grow at a slower rate, as the current temperature is already close to the ideal range for pest development and growth, while insect populations in temperate regions are expected to grow more rapidly. On the other hand, some pest species, such as thrips (Lewis et al., 1997), may have altered gender relationships, which could affect reproductive rates. Moreover, insect populations may increase as a result of lower winter mortality of insects brought on by warmer winter temperatures (Harrington et al., 2001). Nevertheless, Lehmann et al., (2020) reported that global warming might not always result in increased numbers of insect pests and corresponding commercial crop losses, due to the specificity of insect biological niche requirements, physiological tolerances, and the variable impact of temperature on their phenology and life cycle. Likewise, Thomas et al. (2004) suggest that some plant species may not be able to adapt to climate change, resulting in the extinction of host-specific pest species. Another key marker of climate change that could also influence insect pest populations is rainfall. For instance, in contrast to the ambient circumstances and drought, Staley et al (2007) found that increased summer rainfall events caused the rapid proliferation of wireworm populations in the upper soil. A further example involves bark beetles (Scolytidae) that can detect ultrasonic acoustic emissions created by the rupture or cavitation of water columns in the xylem of plants subjected to drought during the transpiration process (Yihdego et al., 2019).

Impact On Insect Pests Distribution and Overwintering Survival

As each species has unique climatic needs for development, survival, and reproduction, these requirements affect the distribution of insects and the plants that serve as their hosts. Hence, growers will face new pest problems as the pests expand into new areas and growing regions of their host plants. It is predicted that insect species will be forced to change their habitats by moving to new climatic regions and disappearing from unsuitable ones due to global warming (Hughes et al., 2003). Insect species that are limited by low temperatures at high latitudes may be better able to overwinter when temperatures rise, resulting in an expansion of their geographic range (EPA, 1989; Hill and Dymock, 1989). As temperatures rise, insect pests are expected to move from tropical and subtropical areas to temperate areas at higher elevations as a result of changes in the growing locations of their host plants (Shrestha, 2019). Moreover, unexpected outbreaks of insect pests can completely eradicate some crop species and facilitate invasion by exotic species (Kannan and James, 2009). It is estimated that a 1°C increase in temperature shifts the distribution of a species 200 km north or 140 km in elevation (Parry and Carter, 1989). In addition, some studies suggest that the geographic expansion of insect ranges toward the pole and high altitudes is likely to increase crop losses (Bjorkman et al., 2011; Wolf et al., 2008). There are many instances where pest ranges have been shown to shift in response to climatic factors, as in the case of *Ostrinia nubilalis* (Hubner), the European corn borer, which could shift its potential range northward by as much as 1,220 km with a temperature increase of 1 to 3°C (Porter et al., 1991).

Winter is the worst season for many insect pests because it greatly increases mortality and reduces populations for the following season. In order to survive in stressful environments, insects have developed a variety of behavioral responses, including diapauses (Bale and Hayward, 2010). This phenomenon corresponds to a period during which the development processes are interrupted and depends on environmental factors such as temperature, humidity and photoperiod (Taylor et al.,

1986). It is thought that insect diapause duration is shortened by increased metabolic activity at higher temperatures (Hahn and Denlinger, 2007). Insect populations may increase due to decreased winter insect mortality from higher winter temperatures (Hahn and Denlinger, 2007). Therefore, global warming would lead to increased insect population growth, early infestations, and subsequent agricultural damage from insect pests (Kocmánková et al., 2011). For example, Ouyanga et al. (2016) reported that as temperatures increase, *Helocoverpa armigera* pupae may emerge from overwintering diapause up to 7 days earlier.

Invasion By Alien Species And Outbreaks

Modern and rapid transportation systems, as well as globalization and liberalization of agricultural trade, have greatly increased the likelihood of introducing alien species. Along with these factors, climate change can also create favorable habitats for the establishment and spread of exotic species. The bioclimatic range of invasive species is generally wider than that of native insects, allowing alien insects to find a wider variety of acceptable environments (Walther et al., 2009). In response to rising temperatures, many insects are migrating north or towards higher altitudes. As a result, they can now expand to new locations that were previously unsuitable for their establishment. On the hand, extreme weather events, such as storms and hurricanes, can also move pests to new geographic areas where they can find favorable environmental circumstances for establishment (FAO, 2019). For instance, during the 2005 hurricane season, the cactus moth (*Cactoblastis cactorum*) was transported from the Caribbean islands to Mexico, where it posed a particularly serious ecological and financial threat to several species of prickly pear (Mafokoane et al., 2007; Burgiel and Muir, 2010). With a changing climate encouraging the introduction of insect-vulnerable cultivars or crops, the invasion of new insect pests will be the greatest threat to agricultural ecosystems (Gregory et al., 2009).

Pest outbreaks are caused by complicated conditions induced by a variety of variables, including host plant availability, enhanced metabolism, and changed voltinism. They may also result from unpredictable changes in the population of pests and their natural enemies. Their frequency and severity have increased due to changing climatic factors. Increased temperature has been shown to indirectly increase insect herbivory by accelerating photosynthesis in host plants and improving their nutritional quality, resulting in pest outbreaks (Visser and Both, 2005). Due to the low temperatures and frequent frosts, the Polar regions are mainly protected from pest outbreaks (Volney and Fleming, 2000). Global warming and more frequent droughts are expected to increase the frequency of insect outbreaks in temperate zones (Logan et al., 2003).

Impact on Crop-Pest Interactions

Interactions between crops and insect pests may be strongly influenced by the complex physiological effects of rising temperatures and CO_2 (Hare, 1992). An insect that feeds on plants must be able to adapt to its environment as well as to its host plant to complete its development. When host plants experience changes in their growth cycle, many insect species that depend on these plants may feel pressure to adapt. Hence, critical events in the plant life cycle, such as stem elongation, flowering and fruiting, have been found to be significantly influenced by temperature and photoperiod (Cleland et al., 2007). Subsequently, this can have a significant impact on the feeding and breeding behavior of related insect pests. Furthermore, plants growing in situations of high temperatures increased CO_2, and/or decreased precipitation may provide low-quality nutritional resources to herbivores (Han et al., 2019). Increased CO_2 generally results in increased C-N ratios (Ainsworth et al., 2005) and altered concentrations of carbohydrates, starch, and soluble proteins, which is the most common mechanism by which increased CO_2 affects insect behavior through changes in host plant quality (Coviella and Trumble, 1999). Herbivore responses to the rising carbon-to-nitrogen ratio are anticipated to include compensatory feeding, an increase in plant defense compounds, and pest species competition.

IMPACT OF CLIMATE CHANGE AND CLIMATIC FACTORS ON OLIVE MAIN INSECT PESTS

More than 255 species make up the group of organisms that could be harmful to *O. europaea*, including insect pests, mites, nematodes, and pathogenic microorganisms (Montiel-Bueno and Jones, 2002; Haniotakis, 2005). Olive fruit fly (*Bactrocera oleae*), olive psyllid (Euphyllura olivinea Costa), olive moth (*Prays oleae*), and black scale (*Saissetia oleae* Bern) are among the insect pests that have been identified as being of greatest concern (Daku et al., 2000) (Table 2). Other pests such as the bark beetle (*Hylesinus oleiperda*), and the otiorhynchus (*Otorhynchus cribricolis*), are considered secondary and can cause more or less important damages under certain conditions. Insects can also serve as vectors of infections that indirectly harm *O. europaea* (Table 2). As a recent example, the bacterium *Xylella fastidiosa*, which is the main cause of rapid olive decline syndrome, is spread between olive trees by the spittlebug *Philaenus spumarius* L. (Cornara et al., 2017). As stated above, climate change will have several impacts on the abundance and dynamics of insect pests and their related hosts. Despite the importance of olive production in the Mediterranean region, little research is currently available on the potential effects of climate change on the main olive pests.

Table 2. Main insect pests of the olive trees

Insect Pest Species		Organs attacked*	Geographical distribution*	Damages	References
Common Name	Scientific name				
Olive fly	*Bractocera oleae*	Olive fruits	Throughout the Mediterranean region, Canary Islands, Near East, India, North, East and South Africa.	Up to 80% of oil value and 100% of some table cultivars 5% to 30% of total olive production	Tzanakakis (2006); Mazomenos (1989); Katsoyannos (1992)
Olive moth	*Prays oleae*	Leaves, terminal buds, flowers and fruits.	Throughout the Mediterranean region, and as far as Russia (shores of the Black Sea: Crimea and Georgia)	49–63% of production, which equates to 8–11 kg per tree	Ramos et al. (1998); Patanita and Mexia (2004)
Olive black scale	*Saissetia olea*	Leaves, twigs, inflorescences	Throughout the Mediterranean region.	Leaf drop, reduced tree vigor and twig dieback Development of sooty mould	Bodenheimer (1951); Ben-Dov and Hodgson (1997)
Olive psyllid	*Euphyllura olivina*	Leaves, buds, young shoots, stems, inflorescences and fruiting shoots.	Throughout the Mediterranean olive-growing region.	Inflorescence and fruit drop Installation of ectoparasitic fungi	Chermiti(1983)
Olive weevil	*Otiorhynchus cribricollis*	Leaves	Throughout the Mediterranean region. Adventive species in California, Australia, New Zealand	Chewing the edge of the leaves Soil dwelling larvae can damage tree roots	Spooner-Hart et al. (2007)
Olive scale	*Parlatoria oleae*	Leaves, woody parts, fruits	Throughout the Mediterranean region, United States (California, Arizona, Maryland), Asia, Europe, North Africa, Middle East.	Attacks mainly olive fruits causing discoloration of the skin and deformation of the tissues	Brkić Bubola et al. (2014)
Olive borer	*Hylesinus oleiperda*	Trunk and branches.	Throughout the Mediterranean region as far as the Near and Middle East (Iran), northern Europe (Belgium, England, Denmark), Chile, Argentina.	Death of branches and twigs, caused by adults and larvae that feed on the wood	Holgado (2007)
Olive Bark Beetle	*Phloeotribus scarabeoides*	Twigs, flower clusters and above all fruiting clusters, pruning wood, trunk, branches and twigs of trees suffering from dieback.	Throughout the Mediterranean region, North Africa, Near and Middle East as far as Iran	Construction of feeding galleries in the young branches causing important losses in fruit and flowers Up to 60% decrease in yield in severely damaged trees	Golnzalez and Campos (1994); Jarraya (1986)

continued on following page

Table 2. continued

Insect Pest Species		Organs attacked*	Geographical distribution*	Damages	References
Common Name	Scientific name				
Pyralid moth	*Euzophera pinguis*	Branches and trunk.	Whole of the Mediterranean region, most of Europe: Denmark, France, central Europe, Portugal.	The larvae tunnel and block sap flow, reduce vegetative growth and dry out branches.	FAO/RNE (2010) ; Moussa et al. (2017)
Jasmine moth	*Palpita vitrealis*	Leaves, terminal buds and fruits.	Mediterranean, Near East, Canary Islands and Madeira, Japan, tropical countries of America.	The larvae devour leaves, tender shoots, buds and even the pulp of green olives	Badawi et al. (1976) ; Balachowsky (1972)
Olive thrips	*Liothripsoleae*	Leaves, young stems, terminal shoots, fruits	All the Mediterranean, olive-growing areas.	Necrosis and desiccation of shoots and fall of the drupes	Vono et al. (2020)
Leopard moth	*Zeuzera pyrina*	Leaf petioles, young twigs, twigs, branches, trunk.	Northern and southern Europe, North Africa, Near and Middle East, Iran, China, Japan	The larvae begin to burrow into the shoots and, as they develop, spread through the branches of the entire plant, eventually causing the tree to wither and die	Bonnemaison (1976)
Oleander scale	*Aspidiotus nerii*	Leaves, fruits.	Throughout the southern Palearctic region (countries bordering the Mediterranean).	It attacks leaves and olives, causing deformations in both parts of the tree, and discoloration of the invaded parts, due to the lack of chlorophyll in the places occupied by the pest	Molina de la Rosa et al. (2004)
Olive leaf gall midge	*Dasineura oleae*	Leaves, vegetative buds, flower stalks and stems.	Eastern Mediterranean region, Croatia, Italy	Larvae of *D. oleae* cause slight, indefinite, elongate swellings on olive leaves. The galls induced in these tissues interrupt the supply of nutrients to the growing fruit, and small fruits are produced.	Batta (2019)
Oystershell scale	*Lepidosaphes ulmi*	Leaves, twigs, fruits.	Vast geographical distribution: whole Palearctic region, introduced in America, reported in Asia, South Africa, Australia.	Discoloration and prematre deterioration of leaves Infested fruit is unsightly spotted and discolored and may drop prematurely It can damages branches and young trees and if left uncheckehd, mature plants will seriously weaken and stunt	Gill (1997)

* International Olive Council (2007)

Impacts of Climate Change on Olive Fruit Fly

The olive fruit fly (*Bactrocera oleae* Gmelin), a Diptera of the family Tephritidae, is one of the most important olive pests in the world. It can be responsible for losses of up to 80% of oil value and 100% of some table cultivars in areas of the world where it is established (Tzanakakis, 2006). It is estimated that it can damages 5% to 30% of the total olive production depending on the environmental conditions (Mazomenos, 1989; Katsoyannos, 1992), resulting in annual economic losses of more than one billion USD in the Mediterranean region (Van Asch et al., 2015). Economic damage is the result of adult oviposition punctures on the fruit surface of table olives and fruit drops caused by larval feeding. In addition to this damage, the direct destruction of the pulp by the larvae renders the fruit unusable for canning and decreases the quality and value of the oil pressed from these infested fruits due to the increased acidity resulting from the development of the microorganisms (Daane and Johnson, 2010). The development of the olive fly and the resulting number of annual generations are influenced by a variety of factors, including ambient temperature, fruit quantity and quality, humidity and microclimate within the canopy (Burrack and Zalom, 2008; Kounatidis et al., 2008). Hence, depending on temperature and local conditions (such as height and distance from the sea), *B. oleae* can produce one to four generations in the Mediterranean region (Marchi et al., 2016; Kounatidis et al., 2008). Temperature is a major environmental factor affecting the phenology of *B. oleae* and its relationship with *O. europea* (Ragaglini et al., 2005). Tsitsipis (1980) and Fletcher (1989) examined the development of olive fly larvae under laboratory conditions and constant temperatures and found that the lower temperature threshold of olive fly larvae ranged from 10 to 12.5C and the upper-temperature threshold from 30 to 32C. The lower and upper thresholds for egg development are 7.5 to 10 C and 30 to 32 C respectively (Tsitsipis, 1977). Egg, larval, and pupal development can occur in as little as 1, 8, and 9 days, respectively, at ideal temperatures (92, 124), while larval and pupal development may happen between 8 and 37 and 9 to 49 days, respectively (Tsitsipis, 1980). Even though immature development could be completed at 30°C, mortality was significant, with pupal death at about 48%. Its total thermal constant is around 415 degree days (Tsitsipis and Abatzis, 1980). The olive fruit fly seems to be the most adapted to develop in most places during the autumn season, when the olive fruit, is in its ideal state for larval growth (Tzanakakis, 2006). Because of its lack of ovarian maturation in late spring and early to mid-summer, when days are long, Tzanakakis (2003) considered the olive fly a "short-day species" in southern Europe and the Middle East. The end of its vernal-estival hibernation comes when the fruits of the new crop are ready in late summer and the days begin to grow shorter. On the other hand, since the physiological processes of the adults stop around 35 C, high summer temperatures cause the mortality of *B.*

oleae and reduce its pest activity (Gutierrez et al., 2009). Wang et al. (2009) also demonstrated that eggs deposited inside an olive fruit had a mortality rate of over 50% when exposed to 35°C, and no eggs survived when exposed to 38°C for 10 days. The olive fly thrives in coastal areas with cold summers and warm winters, and the likelihood of an infestation decreases if the olive is grown farther from the coast (Volpi et al., 2020).

As previously mentioned, the temperature has independent impacts on the development, reproduction, survival, and dormancy of the olive fly, although its dynamics are closely related to olive fruiting. Therefore, a number of predictive models are based on this climate factor to try to understand the impacts of climate change on the distributions of *Olea europaea* (host) and *Bactrocera oleae* (pest). Hence, (Gutierrez et al., 2009) used a physiologically based demographic model (PBDM) to examine the distribution and abundance of both species in the Italian ecological zones under observed weather and three climate change scenarios. With a 1°C increase in temperature, PBDM predicted greater damage to *B. oleae,* particularly in northern Italy where the climate was expected to be beneficial to both species. In contrast, increases of +2°C and +3°C appeared to provide unfavorable conditions for olive fly survival and reproduction, particularly in the southern part of the peninsula. The authors conclude that in the northern and high-altitude regions of Italy, cold winter temperatures limit olive and fly development, although this is expected to change as previously unfavorable locations become suitable due to global warming. Ponti et al. (2014) used the same model to assess the ecological and economic impacts of an anticipated 1.8°C climate warming on olives and their obligate pest, the olive fly across the Mediterranean Basin. The analysis predicts that in Europe, fly populations are expected to increase inland and at higher elevations as conditions improve for the fly and the plant. In warmer parts of the basin, fly populations are expected to decrease as temperatures approach or exceed upper thermal limits. According to the authors, there will be winners and losers in economic terms depending on how this global warming level affects olive yields and olive fly infestations throughout the Mediterranean basin. Reduced profitability will increase the possibility of abandonment of small farms in marginal areas, which is detrimental to the protection of soils and biodiversity as well as to the reduction of fire risks.

To illustrate the risk of olive fly damage with global warming, Ponti et al. (2009) simulated the percentage of fruit attacked using the PBDM model in the Mediterranean island of Sardinia under three global warming scenarios. They found that due to the detrimental impacts of rising summer temperatures on fly survival and reproduction, many areas where the risk of fly damage is currently high are anticipated to have decreased risk as a result of climate change. Hence, the simulated proportion of attacked fruit using observed weather conditions ranged from 38 to 87% and was lower in areas unfavorable to the olive fly and higher in areas with high fly density. They showed that warming scenarios of +1, +2, and +3 °C result in 40-85%, 43-82%, and 35-80% of attacked fruit, respectively.

Global warming can have various effects on the natural enemies of pest species. Previous research had established the presence of many natural enemies of the olive fly which belongs to the family Braconidae in the subfamily Opiinae. The braconid hymenopteran *Psyttalia concolor* (Szépligeti) has been the focus of most studies (Garantonakis et al. 2017). *P. concolor* releases have been used for nearly 50 years in Europe (Daane and Johnson, 2010), and although this parasitoid has become established in some southern locations (Miranda et al., 2008), it has failed to produce significant numbers and adequate control. This is due, in part, to the inability of *P. concolor* to acclimate to the climate of some olive-growing regions. Indeed, herbivore adaptation to various environmental changes, such as climatic variations can impact parasitoid-host interactions (Hawkins, 1994; Yang and Rudolf, 2010). In fact, climate adaptation plays an important role in determining whether an introduced parasitoid can establish and spread in the non-native location where its host resides (Hoelmer and Kirk, 2005). According to Abd El-Salam et al. (2019), the equilibrium between the olive fly larva and its parasitoid, *P. concolor*, was disturbed by unfavorable temperature and humidity variations in two Egyptian localities. Unlike relative humidity, the study found that temperature played an important role in predicting *P. concolor* survival.

Impacts of Climatic Factors on Olive Moths

The olive moth (*Prays oleae* (Bern)) is considered the most widespread pest of olive in the Mediterranean region, the Black Sea, the Middle East, and the Canary Islands (Kumral et al., 2005). Its life cycle includes three generations per year, each of them forming on a different olive tree organ. In the autumn, the first generation larvae mine olive leaves (phyllophagous). In the spring, the emerged females of the first generation lay their eggs on the flower buds and the resulting larvae feed on the buds and flowers (anthophagous). The larval stages of the third generation live and burrow in the stone of the olive (carpophagus) during summer. Between September and October, adults of the carpophagous generation emerge, and females lay their eggs on leaves to restart the cycle (Pelekassis, 1962). Anthophagous and carpophagous are the two generations that cause the most damage, reducing fruit set and accelerating olive drop. Nevertheless, the damage caused varies each year and appears to be related to weather variables (Rosales et al., 2006). Damage caused by this insect can result in a 49-63% reduction in olive production which is equivalent to 8-11 kg per tree of the current cultivars (Ramos et al., 1998; Patanita and Mexia, 2004). On the other hand, the population dynamics of *P. oleae* are controlled by complex system-level behavior that is influenced by environmental factors, host plant characteristics, and the presence of natural antagonists (Ramos et al., 1978; Tzanakakis, 2003). In fact, according to certain environmental factors, their development rate can be accelerated or retarded by non-linear functions depending on temperature (Gonzalez et al., 2015). In addition, there is a considerable intra- and inter-annual correlation between the temperature and phenology of olive trees and the olive moth (Armendáriz et al., 2007). A study conducted in the olive groves of southwestern Morocco revealed that adult population densities are high in coastal areas with moderate temperatures and high humidity. Whereas, emergence was delayed and densities were lower in inland areas, which have a warmer, less humid climate (Mansour et al., 2017). Kaplan and Alaserhat (2020), in a recent study conducted in 30 olive orchards in two Turkish localities, concluded that with the warming weather, the olive moth became active from the first or second week of April until late October or early November. Their study also showed that the damage rate of *P. oleae* ranged from 12-21% in leaf generations, 2-5% in flower generations, and 0.8-5% in fruit generations depending on the years and orchards.

Impacts of Climatic Factors on Olive Psyllid *Euphyllura olivina*

The olive psyllid, *Euphyllura olivina* (Costa) (Hemiptera: Liviidae), is a stinging and sucking insect that causes significant damage to olive orchards in the Mediterranean area during the pupal and adult phases. Its life cycle includes two (spring and autumn) or three generations (two in spring and one in autumn) (Chemseddine, 1988; Tajnari, 1992; Prophetou and Tzanakakis, 1977). Moreover, its biology is intimately linked to its host plant and to environmental factors, notably temperature. Both of these elements influence the explosion or decline of its populations (Ksantini, 2003). The psyllid overwinters as an adult in the inflorescence axils (Hilal et al., 1997). The first significant oviposition phase corresponds to the first spring generation, and female activity resumes at the same time as the host plant resumes growth (Chermiti, 1989). This first generation is followed by a second generation whose adults begin their summer resting period in June due to high temperatures and slower vegetative development of the tree. A third autumnal generation usually forms when the temperature drops below 27 °C. It can also result from rainfall that allows the olive trees to resume vegetative growth at the end of the growing season (Arambourg, 1984). On the other hand, several studies have shown that high temperatures have a lethal effect on *E. olivina* and that when the temperature exceeds or approaches 30°C, the average life span of *E. olivina* decreases (Chermiti, 1989; Tajnari, 2001; Ksantini et al.,2002; Ksantini, 2003). Furthermore, Ksantini (1986) and Chermiti (1989) observed that high temperatures combined with low relative humidities result in high egg and larval mortality and even cessation of oviposition in female *E. olivina*. Hence, when temperatures reach 32°C, oviposition is completely inhibited, whereas temperatures ≥35°C cause resorption of mature eggs and growing oocytes in 100% of mature females, resulting in adults entering reproductive diapause (Ksantini, 2003).

E. olivina affects the normal growth of the olive organ on which it feeds by sucking sap, both in the larval and imaginal stages, rendering flowers sterile (Jardak et al., 1985) and causing inflorescence and fruit drop (Chermiti, 1983). The nymphs secrete a cottony mass and honeydew that favors the installation of ectoparasitic fungi such as *Capnodium oleaginum*, which reduce the photosynthesis of the trees and their production (Arambourg and Chermiti, 1986). In some areas of the Mediterranean basin, it can cause up to 60% yield loss in the spring when the olive trees are flowering. According to IOC (2007), economic tolerance is approximately 2.5 to 3 nymphs per 100 flower clusters, which is equivalent to a cluster infestation rate of 50-60%. A recent study conducted by Guessab et al. (2021) in two olive farms in the Algerian Wilaya of Mascara revealed that the infestation rate was higher in the irrigated farm than in the poorly irrigated farm and that the activity of the insect was influenced by and is related to climatic factors. This study also showed that southern and western orientations seem to be more infested and that the evolution of this psyllid was influenced by seasonal weather conditions and the host plant, with spring being the most favorable season for their proliferation.

In addition, *Euphyllura olivina*, continues to expand its geographic range, posing risks in areas where it was not previously present. Indeed, in a recent study, Hougardy et al., (2020) reported that since *E. olivina* was first reported in 2007 in California, it has slowly spread in the state and could become an economic pest if it reaches the olive-growing areas of California, where the climate is very similar to that of southeastern Spain, where the psyllid is well established. Furthermore, in Central Europe, potted olive trees are popular, and olive trees infected by the psyllid have been found in the Czech Republic, Germany, and Switzerland. Although the Netherlands has also received the olive psyllid, there is no evidence that the species has become established there (Den Bieman, 2019). More recently, after its accidental introduction into South Korea in 2020 following the importation of olive seedlings, the presence of the insect was reported for the first time in the Eastern Palearctic region, but its establishment in South Korea has not been confirmed (Cho et al., 2021).

Impacts of Climatic Factors on the Olive Black Scale

The black scale *Saissetia oleae* (Olivier) (Hemiptera: Coccidae) is a cosmopolitan and polyphagous pest of more than 60 plant species in the Mediterranean region, including citrus and olives (Passos de Carvalho et al., 2003). Depending on the crop, the nutrition status of the tree, the agricultural practices, and especially the climate, this species can produce one generation per year, one and a partial second generation, or two complete generations in the Mediterranean countries (Blumberg et al., 1975;

Passos de Carvalho et al., 2003). The second generation is usually observed in coastal areas, which have more humidity and moderate temperatures than continental areas. The climatic conditions in these areas combined with the better tree conditions in irrigated olive groves allow for faster insect growth in summer and early autumn, allowing some members of the main generation to mature and begin ovulating in the autumn before the cold weather arrives (Tzanakakis, 2003).

The main damage caused by *S. oleae* is due to its feeding activity, the considerable amount of honeydew that the scale excretes and the subsequent development of black mold fungi. This limits the photosynthetic capacity of the plant, reduces the vigor of the tree and, in the case of severe infestations, it causes the death of branches (Bodenheimer, 1951, Ben-Dov and Hodgson, 1997). This damage varies periodically due to fluctuating populations of *Saissetia oleae*, which frequently exhibit wide variations in abundance both within and between years due to biotic and abiotic factors such as natural enemy activity, broad-spectrum insecticide use, and harsh weather conditions such as extreme temperatures (Panis 1977; Mendel et al. 1984). High mortality is caused by low winter temperatures and low humidity combined with high summer temperatures. First instar scale mortality can be as high as 80% at temperatures of 30°C and 30% relative humidity (De Freitas, 1972; Pucci et al. 1982).

INTEGRATED PESTS MANAGEMENT IN OLIVE GROVES UNDER CLIMATE CHANGE

The need to double global food production to meet a rapidly growing population will be the greatest challenge facing humanity in the 21st century. This will require using less land, water, and soil nutrients, and facing more droughts due to climate change (Deka et al., 2008). Therefore, these changes will have significant effects on the diversity and abundance of arthropods and on the magnitude of losses caused by insect pests, which will influence agricultural production, plant protection, and food security. Integrated pest management (IPM), a widely used and accepted practice, faces significant and largely unanticipated challenges related to climate change. IPM has been described as a comprehensive approach to pest and disease management using all techniques while reducing the use of chemical agents to control pest populations below economic thresholds (Wang et al., 2022). As the behavior, life cycles, and biology of many insect herbivore species are affected by climate change, this could impact a variety of IPM techniques, including host plant resistance, biological control, and chemical control (Petzoldt and Seamann, 2012). Thus, mitigating the effects of agricultural pests on crops in a changing climate requires changes in agricultural practices and adaptive management strategies

(Skendžic et al., 2021). Olive growers in the Mediterranean basin, which is a climate change “hot spot,” may face additional difficulties in managing pests and therefore need to adapt their pest management techniques to limit damage and yield losses.

Development of Climate And Pest-Resistant Varieties

One of the most environmentally beneficial strategies for managing crop pests is host plant resistance, which allows plants to minimize pest damage through a variety of processes, including anti-xenosis, antibiosis, and tolerance (Painter, 1968; Dhaliwal and Dilawari, 1993). Over the years, olive growers have chosen the varieties best suited to each site and climate. Some olive clones and cultivars have proven to be more resistant to drought and/or high temperatures. On the other hand, environmental factors such as temperature, sunlight, soil moisture, air pollution, etc. have a significant impact on how the host plant expresses resistance. Interactions between insect pests and their host plants can change following changes in certain climatic conditions (Sharma et al., 2010). The weakening of plant defenses under stress will make them more vulnerable to insect damage (Rhoades, 1985). Hence, the selection of olive varieties resistant to climatic conditions and insect pests could help olive growers in the Mediterranean region to better adapt to the potential negative effects of climate change.

Adaptative Cultural Practices

Changing agricultural practices and implementing adaptive management techniques could mitigate the effect of agricultural pests on crops (Thomson et al., 2010). Adaptation can encompass changes to agricultural systems in response to current or projected climate change or its consequences, in order to mitigate damages or take advantage of beneficial opportunities. For instance, the potential negative effects of climate change can be reduced by modifying planting patterns and increasing the geographic and temporal diversity of species and cultivars. This strategy can be adopted at the scale of the olive grove to improve the variety of the edge habitat to stimulate the population of natural enemies (Marino and Landis, 1996). The widespread use of no-till or low-till planting and the use of mulch to retain water in various crops are the results of the evolution of agricultural techniques (Thomson et al., 2010). Mulches can stimulate parasitoid and predator populations in a field. For example, after using straw and compost as mulch to prevent water loss, the number of carabids, parasitic Hymenoptera and spiders, predatory and parasitic Diptera, and predatory Hemiptera increased in a vineyard (Thomson and Hoffmann, 2007).

Promotion of Natural Enemy Abundance and Activity

Temperature extremes and fluctuations impact the biology of insect pests and their natural enemies in agroecosystems, including generation length, sex ratio, life span, fecundity, activity, distribution, and survival (Chidawanyika et al. 2012). The timing of daytime activity of various insect species will also change due to changes in temperature (Young, 1982), and changes in interspecific interactions may also affect the ability of natural enemies to control pests (Hill and Dymock, 1989). Thus, the relationship between insect pests and their natural enemies may not be synchronized, disrupting the balance of biological management (Thomson et al., 2010). Furthermore, if insects move into areas outside the range of their natural enemies, the effectiveness of the natural enemies in managing the pests will decrease, although a new community of enemies may then provide some degree of control (Thomson et al., 2010). In addition to being impacted by host plants and associated herbivores, the abundance and activity of natural enemies will fluctuate due to adaptive management practices used by growers to cope with climate change. These adaptation practices could make pests and natural enemies incompatible both geographically and temporally, reducing their effectiveness for biocontrol (Thomson et al., 2010). A greater richness and variety of plant species mimicking the natural ecosystem, both geographically and temporally, will help provide adequate habitat and maintain a large number of natural enemy species of crop pests (Altieri, 1999). To predict the responses of ecological groups, long-term monitoring of the demography of herbivores and their associated natural enemies in response to climate would be necessary, as well as more geographically accurate models of local climate change (Stireman et al., 2005).

Synthetic Insecticides and Biopesticides

The mechanism of action of an insecticide, the target species, the mode of action, and the amount of insecticide ingested or contacted have been shown to influence the temperature-efficacy relationship (Johnson, 1990). Insecticide toxicity, and therefore efficacy, is significantly influenced by temperature, either positively or negatively. For instance, at higher temperatures, the insect growth regulator (IGR) diflubenzuron produced rapid mortality and was most effective at 35°C (Amarasekare and Edelson, 2004). In contrast, the entomopathogenic fungus *Beauveria bassiana* (Balsamo), shows reduced biological activity at temperatures above 25°C (Amarasekare and Edelson, 2004). There is a huge need to develop new molecules with a new mode of action and new biopesticides that will be effective in the context of future global warming while taking into account their toxicity to human health and the environment.

CONCLUSION

The Mediterranean basin is particularly vulnerable to climate change, where the rates recorded for most climate indicators are higher than global trends (Giorgi, 2006). Despite being one of the species that has adapted the best to the Mediterranean climate, the co-adaptation between the olive tree and its pests may change with the considerable warming and drying trends in future climate projections. In fact, climate change will result in significant changes in arthropod diversity, the geographic distribution of insect pests, pest-plant interactions, natural enemy activity and abundance, and the effectiveness of crop protection systems for pest control. To maintain yields in the face of increasing pest pressure, olive growers may need to make further adjustments and adaptations, such as implementing various management techniques, including crop diversification, the use of climate and pest-resistant varieties, and the use of selective insecticides and biocontrol agents. Nevertheless, agricultural research can help determine how olive trees, insect pests, and their natural enemies respond to abiotic stress under climate change to plan and design adaptation and mitigation strategies for future pest management programs. Although the future of climate change is difficult to predict and there is much uncertainty about it, recent research has begun to reveal a consistent pattern in how crops, pathogens, insect pests, and their interactions adapt to climate change.

REFERENCES

Abou-Saaid, O., Yaacoubi, A. E., Moukhli, A., Bakkali, A. E., Oulbi, S., & Delalande, M. (2022). Statistical Approach to Assess Chill and Heat Requirements of Olive Tree Based on Flowering Date and Temperatures Data: Towards Selection of Adapted Cultivars to Global Warming. *Agronomy (Basel)*, *12*(12), 2975. doi:10.3390/agronomy12122975

Ainsworth, E. A., & Long, S. P. (2005). What have we learned from 15 years of free-air CO_2 enrichment (FACE)? A meta-analytic review of the responses of photosynthesis, canopy properties and plant production to rising CO_2. *The New Phytologist*, *165*(2), 351–372. doi:10.1111/j.1469-8137.2004.01224.x PMID:15720649

Altieri, M. A. (1999). The ecological role of biodiversity in agroecosystems. In *Invertebrate biodiversity as bioindicators of sustainable landscapes* (pp. 19–31). Elsevier. doi:10.1016/B978-0-444-50019-9.50005-4

Amarasekare, K. G., & Edelson, J. V. (2004). Effect of temperature on efficacy of insecticides to differential grasshopper (Orthoptera: Acrididae). *Journal of Economic Entomology, 97*(5), 1595–1602. doi:10.1603/0022-0493-97.5.1595 PMID:15568348

Arambourg, Y., & Chermiti, B. (1986). Euphyllura olivina Costa-Psyllidae. Traité d'entomologie oléicole. Espagne: Conseil oléicole international.

Arampatzis, G., Hatzigiannakis, E., Pisinaras, V., Kourgialas, N., Psarras, G., Kinigopoulou, V., Panagopoulos, A., & Koubouris, G. (2018). Soil water content and olive tree yield responses to soil management, irrigation, and precipitation in a hilly Mediterranean area. *Journal of Water and Climate Change, 9*(4), 672–678. doi:10.2166/wcc.2018.224

Armendáriz, A., De La Iglesia, L., Santiago, Y., Campillo, G., Alberte, C., Barroso, L. M., ... Sanz, A. P. (2007). Ciclo del prays del olivo (" Prays oleae" Bern.) en Arribes del Duero. *Boletin de Sanidad Vegetal, Plagas, 33*(4), 443–456.

Badawi, A., Awadallah, A. M., & Foda, S. M. (1976). On the biology of the olive moth Palpita unionalis (Hubner) (Lepidoptera: Pyralidae). *Zeitschrift für Angewandte Entomologie, 80*, 103–110. doi:10.1111/j.1439-0418.1976.tb03306.x

Balachowsky, A. S. (1972). *Lepidopteres* (Vol. 2). Entomologie appliquée a l'agriculture. Masson.

Bale, J., Masters, G. J., Hodkins, I. D., Awmack, C., Bezemer, T. M., Brown, V. K., Buterfield, J., Buse, A., Coulson, J. C., Farrar, J., Good, J. E. G., Harrigton, R., Hartley, S., Jones, T. H., Lindroth, R. L., Press, M. C., Symrnioudis, I., Watt, A. D., & Whittaker, J. B. (2002). Herbivory in global climate change research: Direct effects of rising temperature on insect herbivores. *Global Change Biology, 8*(1), 1–16. doi:10.1046/j.1365-2486.2002.00451.x

Bale, J. S., & Hayward, S. A. L. (2010). Insect overwintering in a changing climate. *The Journal of Experimental Biology, 213*(6), 980–994. doi:10.1242/jeb.037911 PMID:20190123

Batta, Y. (2019). New findings on infestation and phenology of Dasineura oleae Angelini (Diptera, Cecidomyiidae): An emerging pest on olive trees in the Palestinian Territories. *Journal of Plant Diseases and Protection, 126*(1), 55–66. doi:10.100741348-018-0196-y

Biel, C., De Herralde, F., Savé, R., & Evans, R. Y. (2008). Effects of CO2 atmospheric fertilization on greenhouse production of olive trees (Olea europaea L. 'Arbequina'). *European Journal of Horticultural Science, 73*(5), 227.

Bjorkman, C., Berggren, A., & Bylund, H. (2011). Causes behind insect folivory patterns in latitudinal gradients. *Journal of Ecology, 99*, 367–369. PMID:21479109

Blumberg, D., Swirski, E., & Greenberg, S. (1975). Evidence for bivoltine populations of the Mediterranean black scale Saissetia oleae (Olivier) on citrus in Israel. *Israel Journal of Entomology, 10*, 19–24.

Bonnemaison, L. (1976). *Enemigos animals de las plantas cultivadas y forestales*. II OIKOS-TAU, SA Endiciones.

Brkić Bubola, K., Krapac, M., & Sladonja, B. (2014). Effects of olive scale (Parlatoria oleae (Colvée)) attack on yield, quality and fatty acid profile of virgin olive oil. *Croatian Journal of Food Science and Technology, 6*(1), 1–6.

Bueno, A. M., & Jones, O. (2002). Alternative methods for controlling the olive fly, Bactrocera oleae, involving semiochemicals. *IOBC/WPRS Bulletin, 25*(9), 147–156.

Burgiel, S. W., & Muir, A. A. (2010). *Invasive species, climate change and ecosystem-based adaptation: addressing multiple drivers of global change*. Academic Press.

Chamseddine, M. (1988). *Les arthropodes frondicoles de l'oliveraie du Haouz (Maroc): évolution spatio-temporelle des peuplements et bioécologie des espèces dominantes* [Doctoral dissertation]. Thèse Doc. Es-sc. Univ. Cadi Ayyad.

Chermiti, B. (1983). *Contribution à l'étude bioécologique du Psylle de l'olivier Euphylluraolivina Costa (Hom; Psyllidae) et de son endoparasite Psyllaephagus euphyllurae Silv. (Hym; Encyrtidae)* [Thèse Doctorat Ingénieur]. Université d'Aix-Marseille, France.

Chermiti, B. (1989). *Dynamique des populations du psylle de l'olivier E. olivina, en conditions méditerranéennes* [Thèse Doctorat Es-Sciences]. Université d'Aix-Marseille, France.

Chidawanyika, F., Mudavanhu, P., & Nyamukondiwa, C. (2012). Biologically based methods for pest management in agriculture under changing climates: Challenges and future directions. *Insects, 3*(4), 1171–1189. doi:10.3390/insects3041171 PMID:26466733

Cho, G., Lee, J. H., & Choi, D. S. (2021). First interception of olive psyllid Euphyllura olivina (Costa, 1839)(Hemiptera: Psylloidea) in the East Palaearctic region. *Phytoparasitica, 49*(5), 865–868. doi:10.100712600-021-00924-w

Cleland, E. E., Chuine, I., Menzel, A., Mooney, H. A., & Schwartz, M. D. (2007). Shifting plant phenology in response to global change. *Trends in Ecology & Evolution, 22*(7), 357–365. doi:10.1016/j.tree.2007.04.003 PMID:17478009

Cornara, D., Saponari, M., Zeilinger, A. R., de Stradis, A., Boscia, D., Loconsole, G., Bosco, D., Martelli, G. P., Almeida, R. P. P., & Porcelli, F. (2017). Spittlebugs as vectors of Xylella fastidiosa in olive orchards in Italy. *Journal of Pest Science, 90*(2), 521–530. doi:10.100710340-016-0793-0 PMID:28275326

Coviella, C. E., & Trumble, J. T. (1999). Effects of elevated atmospheric carbon dioxide on insect-plant interactions. *Conservation Biology, 13*(4), 700–712. doi:10.1046/j.1523-1739.1999.98267.x

Daane, K. M., & Johnson, M. W. (2010). Olive fruit fly: Managing an ancient pest in modern times. *Annual Review of Entomology, 55*(1), 151–169. doi:10.1146/annurev.ento.54.110807.090553 PMID:19961328

Daane, K. M., Rice, R. E., Zalom, F. G., Barnett, W. W., & Johnson, M. W. (2005). Arthropod pests of olive. In Olive production manual (2nd ed.). University of California Agriculture and Natural Resources.

Daku, L., Norton, G. W., Pfeiffer, D. G., Luther, G. C., Pitts, C. W., Taylor, D. B., . . . Uka, R. (2000). Farmers' knowledge and attitudes towards pesticide use and olive pest management practices in Vlora, Albania: A Baseline Survey. IPM CRSP, Office of International Research and Development, Virginia Tech.

de Carvalho, J. P. (2003). *A cochonilha-negra: Saissetia oleae (olivier, 1791):(homoptera-coccidae)*. Academic Press.

De Freitas, A. (1972). A cochonilha-negra (Saissetia oleae (Oliv.)) em Oliveira. Bio-ecologiae enfluencia dos tratamentos antidacicos. *Agric. Lusit., 33*, 349–390.

Deka, S., Byjesh, K., Kumar, U., & Choudhary, R. (2009). Climate change and impacts on crop pests-a critique. In *Workshop Proceedings: Impact of Climate Change on Agriculture* (pp. 147-149). Academic Press.

Den Bieman, C. F. M., Malenovský, I., Burckhardt, D., & Heijerman, T. (2019). First checklist of the Dutch jumping plant lice since 93 years (Hemiptera: Psylloidea). *Nederlandse Faunistische Mededelingen, 53*, 55–118.

Deutsch, C. A., Tewksbury, J. J., Tigchelaar, M., Battisti, D. S., Merrill, S. C., Huey, R. B., & Naylor, R. L. (2018). Increase in crop losses to insect pests in a warming climate. *Science, 361*(6405), 916–919. doi:10.1126cience.aat3466 PMID:30166490

Dhaliwal, G. S., & Dilawari, V. K. (1993). *Advances in Host Resistance to Insects*. Kalyani Publishers.

Drake, B. G., González-Meler, M. A., & Long, S. P. (1997). *More efficient plants: A consequence of rising atmospheric CO2*. Academic Press.

El-Salam, A., Ezzat, A. M., Salem, S. A. W., Abdel-Rahman, R. S., El-Behery, H. H., & Magd Elden, M. A. (2019). Effects of climatic changes on olive fly, Bactrocera oleae (Rossi) population dynamic with respect to the efficacy of its larval parasitoid in Egyptian olive trees. *Bulletin of the National Research Center, 43*(1), 1–9. doi:10.118642269-019-0220-9

Environment Protection Agency (EPA). (1989). The Potential Effects of Global Climate Change on the United States: National Studies (vol. 2). Review of the Report to Congress, US Environmental Protection Agency.

FAO. Food and Agriculture Organization. Plant Pests and Diseases in the Context of Climate Change and Climate Variability, Food Security and Biodiversity Risks. (n.d.). Available online: https://www.fao.org/3/nb088/nb088.pdf

FAO/RNE. (2010). *Olive GAP Manual. Good agriculture practices for the Near East and North Africa Countries*. FAO, Food and Agriculture Organization of the United Nations Regional Office for the Near East.

FAOSTAT. (2019). *Food and Agriculture Organization of the United Nations Statistical Dataset*. FAO..

Fletcher, B. S. (1989). Temperature-development rate relationships of the immature stages and adults of tephritid fruit flies. In World Crop Pests: Fruit Flies—Their Biology, Natural Enemies and Control. Elsevier.

Fraga, H., Pinto, J. G., & Santos, J. A. (2020). Olive tree irrigation as a climate change adaptation measure in Alentejo, Portugal. *Agricultural Water Management, 237*, 106193. doi:10.1016/j.agwat.2020.106193

Galán, C., García-Mozo, H., Vázquez, L., Ruiz, L., De La Guardia, C. D., & Trigo, M. M. (2005). Heat requirement for the onset of the Olea europaea L. pollen season in several sites in Andalusia and the effect of the expected future climate change. *International Journal of Biometeorology, 49*(3), 184–188. doi:10.100700484-004-0223-5 PMID:15645246

Garantonakis, N., Varikou, K., & Birouraki, A. (2017). Parasitism of Psytallia concolor (Hymenoptera: Braconidae) on Bactrocera oleae (Diptera: Tephritidae) infesting different olive varieties. *Phytoparasitica, 45*(4), 461–469. doi:10.100712600-017-0597-7

Gill, R. J. (1997). *The scale insects of California. Part 3: the amored scales (Homoptera: Coccoidea: Diaspididae)*. California Department of Food and Agriculture Technical Series in Agricultural Biosystematics and Plant Pathology.

Giorgi, F. (2006). Climate change hot-spots. *Geophysical Research Letters*, *33*(8), L08707. doi:10.1029/2006GL025734

Gonzalez, D., Cabral, J. A., Torres, L., & Santos, M. (2015). A cohort-based modelling approach for managing olive moth Prays oleae (Bernard, 1788) populations in olive orchards. *Ecological Modelling*, *296*, 46–56. doi:10.1016/j.ecolmodel.2014.10.012

Gonzalez, R., & Campos, M. (1994).*A preliminary study of the effect of attacks by Phloeotribusscarabaeoides (BERN.)(Coleoptera: Scolytidae) on the productivity*. Academic Press.

Greven, M., Neal, S., Green, S., Dichio, B., & Clothier, B. (2009). The effects of drought on the water use, fruit development and oil yield from young olive trees. *Agricultural Water Management*, *96*(11), 1525–1531. doi:10.1016/j.agwat.2009.06.002

Guessab, A., Elouissi, M., Lazreg, F., & Elouissi, A. (2021). *Population dynamics, seasonal fluctuations and spatial distribution of the olive psyllid Euphyllura olivina Costa (Homoptera, Psyllidae) in Algeria*. Academic Press.

Gutierrez, A. P., Mills, N. J., & Ponti, L. (2010). Limits to the potential distribution of light brown apple moth in Arizona–California based on climate suitability and host plant availability. *Biological Invasions*, *12*(9), 3319–3331. doi:10.100710530-010-9725-8

Gutierrez, A. P., Ponti, L., & Cossu, Q. A. (2009). Effects of climate warming on olive and olive fly (Bactrocera oleae (Gmelin)) in California and Italy. *Climatic Change*, *95*(1), 195–217. doi:10.100710584-008-9528-4

Gutierrez, A. P., Ponti, L., d'Oultremont, T., & Ellis, C. K. (2008). Climate change effects on poikilotherm tritrophic interactions. *Climatic Change*, *87*(1), 167–192. doi:10.100710584-007-9379-4

Hahn, D. A., & Denlinger, D. L. (2007). Meeting the energetic demands of insect diapause: Nutrient storage and utilization. *Journal of Insect Physiology*, *53*(8), 760–773. doi:10.1016/j.jinsphys.2007.03.018 PMID:17532002

Haniotakis, E. G. (2005). Olive pest control: Present status and prospects. Integrated Protection of Olive Crops. *IOBC/WPRS Bulletin*, *28*, 1–9.

Hare, J. D. (1992). Herbivore-Natural Enemy. *Plant resistance to herbivores and pathogens: ecology, evolution, and genetics*, 278.

Harrington, R., Fleming, R. A., & Woiwod, I. P. (2001). Climate change impacts on insect management and conservation in temperate regions: Can they be predicted? *Agricultural and Forest Entomology*, *3*(4), 233–240. doi:10.1046/j.1461-9555.2001.00120.x

Hawkins, B. A. (1994). *Pattern and process in host-parasitoid interactions*. Cambridge University Press. doi:10.1017/CBO9780511721885

Hilal, A., Tajnari, H., Sekkat, A., & Chemseddine, M. (1997). Mise en évidence d'une diapause de reproduction de psylle de l'olivier Euphylluraolivina Costa (Homoptera, Psyllidae). *Al Awamia*, *98*, 23–33.

Hill, M. G., & Dymock, J. J. (1989). *Impact of climate change: Agricultural/ horticultural systems. In DSIR Entomology Division Submission to the New Zealand Climate Change Programme*. Department of Scientific and Industrial Research.

Hoelmer, K. A., & Kirk, A. A. (2005). Selecting arthropod biological control agents against arthropod pests: Can the science be improved to decrease the risk of releasing ineffective agents? *Biological Control*, *34*(3), 255–264. doi:10.1016/j.biocontrol.2005.05.001

Holgado, M. G. (2007). Bioecología de Hylesinusoleiperda F." taladrillodelolivo"(Coleoptera-Scolytidae) en Mendoza (Argentina). *Revista de la Facultad de CienciasAgrarias*, *39*(2), 51–54.

Hougardy, E., Wang, X., Hogg, B. N., Johnson, M. W., Daane, K. M., & Pickett, C. H. (2020). Current distribution of the olive psyllid, Euphyllura olivina, in California and initial evaluation of the Mediterranean parasitoid Psyllaephagus euphyllurae as a biological control candidate. *Insects*, *11*(3), 146. doi:10.3390/insects11030146 PMID:32110919

Hughes, C. L., Hill, J. K., & Dytham, C. (2003). Evolutionary trade-offs between reproduction and dispersal in populations at expanding range boundaries. *Proceedings of the Royal Society of London. Series B, Biological Sciences*, *270*(suppl_2), S147–S150. PMID:14667365

Iglesias, A., & Garrote, L. (2015). Adaptation strategies for agricultural water management under climate change in Europe. *Agricultural Water Management*, *155*, 113–124. doi:10.1016/j.agwat.2015.03.014

International Olive Council. (2007). *Production techniques in olive growing*. IOC.

Jardak, T., Moalla, M., Khalfallah, H., & Smiri, H. (1985, November). Essais d'évaluation des dégâts causés par le psylle de l'olivier Euphyllura olivina (Homoptera: Psyllidae), données préliminaires sur le seuil de nuisibilité. In *Proc. CEC/FAO/ IOBC. Int. Joint Meeting, Pisa (Italy)* (pp. 270-284). Academic Press.

Jarraya, A. (1986). Traite d'entomologie oleicole. Conseil Oleicole International.

Johnson, D. L. (1990). Influence of temperature on toxicity of two pyrethroids to grasshoppers (Orthoptera: Acrididae). *Journal of Economic Entomology, 83*(2), 366–373. doi:10.1093/jee/83.2.366

Joy Burrack, H., & Zalom, F. G. (2008). Olive fruit fly (Diptera: Tephritidae) ovipositional preference and larval performance in several commercially important olive varieties in California. *Journal of Economic Entomology, 101*(3), 750–758. doi:10.1093/jee/101.3.750 PMID:18613575

Kannan, R., & James, D. A. (2009). Effects of climate change on global diversity: A review of key literature. *Tropical Ecology, 50*, 31–39.

Kaplan, M., & Alaserhat, İ. (2020). Determination of distribution, population change, infestation and damage situation of Olive Moth, Prays oleae (Bernard)(Lepidoptera: Praydidae) causing damage in olive orchards. *Erwerbs-Obstbau, 62*(3), 301–307. doi:10.100710341-020-00493-y

Kassout, J., Ater, M., Ivorra, S., Barbara, H., Limier, B., Ros, J., Girard, V., Paradis, L., & Terral, J.-F. (2021). Resisting aridification: Adaptation of sap conduction performance in Moroccan wild olive subspecies distributed over an aridity gradient. *Frontiers in Plant Science, 12*, 1288. doi:10.3389/fpls.2021.663721 PMID:34276726

Kassout, J., Terral, J. F., El Ouahrani, A., Houssni, M., & Ivorra, S. (2022). Species Distribution Based-Modelling Under Climate Change: The Case of Two Native Wild Olea europaea Subspecies in Morocco, O. e. subsp. europaea var. sylvestris and O. e. subsp. maroccana. In *Climate Change in the Mediterranean and Middle Eastern Region* (pp. 21–43). Springer. doi:10.1007/978-3-030-78566-6_2

Katsoyannos, P. (1992). Olive pests and their control in the Near East. *FAO Plant Production and Protection Paper*, 115.

Kocmánková, E., Trnka, M., Eitzinger, J., Dubrovský, M., Štěpánek, P., Semerádová, D., Balek, J., Skalák, P., Farda, A., Juroch, J., & Žalud, Z. (2011). Estimating the impact of climate change on the occurrence of selected pests at a high spatial resolution: A novel approach. *The Journal of Agricultural Science, 149*(2), 185–195. doi:10.1017/S0021859610001140

Kocmánková, E., Trnka, M., Juroch, J., Dubrovský, M., Semerádová, D., Možný, M., & Žalud, Z. (2009). Impact of climate change on the occurrence and activity of harmful organisms. *Plant Protection Science, 45*, S48.

Kounatidis, I., Papadopoulos, N. T., Mavragani-Tsipidou, P., Cohen, Y., Tertivanidis, K., Nomikou, M., & Nestel, D. (2008). Effect of elevation on spatio-temporal patterns of olive fly (Bactrocera oleae) populations in northern Greece. *Journal of Applied Entomology, 132*(9-10), 722–733. doi:10.1111/j.1439-0418.2008.01349.x

Ksantini, M. (1986). Contribution à l'étude de l'influence des températures constantes sur le développement de l'œuf du psylle de l'olivier Euphyllura olivina Costa (Homoptera: Psyllidae). FAO.

Ksantini, M. (2003). *Contribution à l'étude de la dynamique des populations du psylle de l'olivier Euphylluraolivina Costa (Homoptera, Aphalaridae) et de sa nuisibilité dans la région de Sfax* [Thèse de Doctorat en Sciences biologiques]. Fac. Sc. Sfax.

Ksantini, M., Jardak, T., & BouAïn, A. (2002). Temperature Effect on the Biology of Euphylluraolivina Costa. *Acta Horticulturae*, (586), 827–829. doi:10.17660/ActaHortic.2002.586.179

Kumral, N. A., Kovanci, B., & Akbudak, B. (2005). Pheromone trap catches of the olive moth, Prays oleae (Bern.) (Lep., Plutellidae) in relation to olive phenology and degree-day models. *Journal of Applied Entomology, 129*(7), 375–381. doi:10.1111/j.1439-0418.2005.00985.x

Lehmann, P., Ammunét, T., Barton, M., Battisti, A., Eigenbrode, S. D., Jepsen, J. U., Kalinkat, G., Neuvonen, S., Niemelä, P., Terblanche, J. S., Økland, B., & Björkman, C. (2020). Complex responses of global insect pests to climate warming. *Frontiers in Ecology and the Environment, 18*(3), 141–150. doi:10.1002/fee.2160

Levitus, S., Antonov, J., Boyer, T., Baranova, O., Garcia, H., Locarnini, R., ... Zweng, M. (2017). NCEI ocean heat content, temperature anomalies, salinity anomalies, thermosteric sea level anomalies, halosteric sea level anomalies, and total steric sea level anomalies from 1955 to present calculated from in situ oceanographic subsurface profile data (NCEI Accession 0164586). *NOAA National Centers for Environmental Information, 10.*

Lewis, W. J., Van Lenteren, J. C., Phatak, S. C., & Tumlinson Iii, J. H. (1997). A total system approach to sustainable pest management. *Proceedings of the National Academy of Sciences of the United States of America, 94*(23), 12243–12248. doi:10.1073/pnas.94.23.12243 PMID:9356432

Lionello, P., & Scarascia, L. (2018). The relation between climate change in the Mediterranean region and global warming. *Regional Environmental Change, 18*(5), 1481–1493. doi:10.100710113-018-1290-1

Logan, J. A., Régnière, J., & Powell, J. A. (2003). Assessing the impacts of global warming on forest pest dynamics. *Frontiers in Ecology and the Environment, 1*(3), 130–137. doi:10.1890/1540-9295(2003)001[0130:ATIOGW]2.0.CO;2

Mafokoane, L. D., Zimmermann, H. G., & Hill, M. P. (2007). Development of Cactoblastis cactorum (Berg)(Lepidoptera: Pyralidae) on six north American Opuntia species. *African Entomology, 15*(2), 295–299. doi:10.4001/1021-3589-15.2.295

Mansour, A. A., Ouanaimi, F., Chemseddine, M., & Boumezzough, A. (2017). Study of the flight dynamics of Prays oleae (Lepidoptera: Yponomeutidae) using sexual trapping in olive orchards of Essaouira region, Morocco. *Journal of Entomology and Zoology Studies, 5*(2), 943–952.

Marchi, S., Guidotti, D., Ricciolini, M., & Petacchi, R. (2016). Towards understanding temporal and spatial dynamics of Bactrocera oleae (Rossi) infestations using decade-long agrometeorological time series. *International Journal of Biometeorology, 60*(11), 1681–1694. doi:10.100700484-016-1159-2 PMID:27044274

Marino, P. C., & Landis, D. A. (1996). Effect of landscape structure on parasitoid diversity and parasitism in agroecosystems. *Ecological Applications, 6*(1), 276–284. doi:10.2307/2269571

Mazomenos, B. E. (1989). Dacusoleae. In A. S. Robinson & G. Hooper (Eds.), *Word crop pests. Vol. 3B* (pp. 169–177). Elsevier Sc. Publishers.

Mendel, Z., Podoler, H., & Rosen, D. (1984). Population dynamics of the mediterranean black scale, Saissetia oleae (Olivier), on citrus in Israel. 4. The natural enemies. *Journal of the Entomological Society of Southern Africa, 47*(1), 1-21.

Miranda, M. A., Miquel, M., Terrassa, J., Melis, N., & Monerris, M. (2008). Parasitism of Bactrocera oleae (Diptera; Tephritidae) by Psyttalia concolor (Hymenoptera; Braconidae) in the Balearic Islands (Spain). *Journal of Applied Entomology, 132*(9-10), 798–805. doi:10.1111/j.1439-0418.2008.01358.x

Molina de la Rosa, J. L., Jiménez Herrera, B., Ruiz Coleto, F., García Zamorano, F., Cano Rodríguez, J., & Pérez García, J. (2004). *Técnicas de cultivo: Plagas y enfermedades del olivo. Viceconsejería, Servicio de Publicaciones y Divulgación.* J. de Andalucía.

Moussa, Z., Choueiri, E., Youssef, A., & El Riachy, M. (2017). First record of Olive Pyralid Moth, Euzopherapinguis (Haworth, 1811) on olive trees in Lebanon (Lepidoptera, Pyralidae). *Bulletin de la Société Entomologique de France, 122*(1), 57–60. doi:10.3406/bsef.2017.3174

Painter, R. H. (1968). Insect resistance in crop plants. *Insect Resistance in Crop Plants.*

Panis, A. (1977). Contribución al conocimiento de la biología de la "cochinilla negra de los agrios"(Saissetia oleae Olivier). *Boletín del Servicio contra Plagas, 3*(1), 157-173.

Parry, M.L., & Carter, T.R. (1989). An assessment of the effects of climatic change on agriculture. *Climate Change, 15*, 95-116.

Patanita, M., & Mexia, A. (2004). *Loss assessment due to Prays oleae Bern. andBactroceraoleaeGmelin in Moura's region Portugal.* http://pubol.ipbeja.pt/Artigos/Italia.htm

Pelekassis, C. D. (1962). *A contribution to the study or nomenclature, taxonomy, biology, ecology and the natural parasitisation of the olive kernel borer () Prays oleae (Bernard) Lesne.* A contribution to the study or nomenclature, taxonomy, biology, ecology and the natural parasitisation of the olive kernel borer () Prays oleae (Bernard) Lesne).

Pérez-López, D., Ribas, F., Moriana, A., Rapoport, H. F., & De Juan, A. (2008). Influence of temperature on the growth and development of olive (Olea europaea L.) trees. *The Journal of Horticultural Science & Biotechnology, 83*(2), 171–176. doi:10.1080/14620316.2008.11512366

Petridis, A., Therios, I., Samouris, G., Koundouras, S., & Giannakoula, A. (2012). Effect of water deficit on leaf phenolic composition, gas exchange, oxidative damage and antioxidant activity of four Greek olive (Olea europaea L.) cultivars. *Plant Physiology and Biochemistry, 60*, 1–11. doi:10.1016/j.plaphy.2012.07.014 PMID:22885895

Petzoldt, C., & Seaman, A. (2006). Climate change effects on insects and pathogens. *Climate change and agriculture: Promoting practical and profitable responses, 3*, 6-16.

Ponti, L., Cossu, Q. A., & Gutierrez, A. P. (2009). Climate warming effects on the Olea europaea–Bactrocera oleae system in Mediterranean islands: Sardinia as an example. *Global Change Biology, 15*(12), 2874–2884. doi:10.1111/j.1365-2486.2009.01938.x

Ponti, L., Gutierrez, A. P., Ruti, P. M., & Dell'Aquila, A. (2014). Fine-scale ecological and economic assessment of climate change on olive in the Mediterranean Basin reveals winners and losers. *Proceedings of the National Academy of Sciences of the United States of America, 111*(15), 5598–5603. doi:10.1073/pnas.1314437111 PMID:24706833

Porter, J., Parry, M., & Carter, T. (1991). The potential effects of climatic change on agricultural insect pests. *Agricultural and Forest Meteorology*, *57*(1-3), 221–240. doi:10.1016/0168-1923(91)90088-8

Prophetou, D. A., & Tzanakakis, M. E. (1977). Seasonal development and number of generations of Euphyllura olivina in Halkidiki (N. Greece). *Annals of the Entomological Society of America, 70*(5), 707–710. doi:10.1093/aesa/70.5.707

Pucci, C. (1982). *Incidenza dei fattori abiotici sulla mortalità della Saissetia oleae*. Oliv.

Quiroga, S., & Iglesias, A. (2009). A comparison of the climate risks of cereal, citrus, grapevine and olive production in Spain. *Agricultural Systems, 101*(1-2), 91–100. doi:10.1016/j.agsy.2009.03.006

Ragaglini, G., Tomassone, D., & Petacchi, R. (2007). Can spring-preventive adulticide treatments be assumed to improve Bactrocera oleae (Rossi) management? *IOBC/WPRS Bulletin*, *30*(9), 309.

Ramos, P., Campos, M., & Ramos, J. M. (1978). Bio-écologie de Prays oleae Bern (Lepidoptera, Plutellidae) dans la province de Grenade: Durée de dévelopment et variables climatiques. *Ann Zool Ecol, 10*(4), 589–601.

Ramos, P., Campos, M., & Ramos, J. M. (1998). Long-term study on the evaluation of yield and economic losses caused by Prays oleae Bern. in the olive crop of Granada (southern Spain). *Crop Protection (Guildford, Surrey), 17*(8), 645–647. doi:10.1016/S0261-2194(98)00065-9

Rapoport, H. F., Fabbri, A., & Sebastiani, L. (2016). Olive biology. *The olive tree genome*, 13-25.

Rehman, M. U., Rather, G. H., Gull, Y., Mir, M. R., Mir, M. M., Waida, U. I., & Hakeem, K. R. (2015). Effect of climate change on horticultural crops. In *Crop production and global environmental issues* (pp. 211–239). Springer. doi:10.1007/978-3-319-23162-4_9

Rhoades, D. F. (1985). Offensive-defensive interactions between herbivores and plants: Their relevance in herbivore population dynamics and ecological theory. *American Naturalist*, *125*(2), 205–238. doi:10.1086/284338

Rosales, R., Garrido, D., Ramos, P., & Ramos, J. M. (2006). Ethylene can reduce Prays oleae attack in olive trees. *Crop Protection (Guildford, Surrey)*, *25*(2), 140–143. doi:10.1016/j.cropro.2005.03.014

Servili, M., Esposto, S., Fabiani, R., Urbani, S., Taticchi, A., Mariucci, F., Selvaggini, R., & Montedoro, G. F. (2009). Phenolic compounds in olive oil: Antioxidant, health and organoleptic activities according to their chemical structure. *Inflammopharmacology*, *17*(2), 76–84. doi:10.100710787-008-8014-y PMID:19234678

Sharma, H. C., Srivastava, C. P., Durairaj, C., & Gowda, C. L. L. (2010). Pest management in grain legumes and climate change. In *Climate change and management of cool season grain legume crops* (pp. 115–139). Springer. doi:10.1007/978-90-481-3709-1_7

Shrestha, S. (2019). Effects of climate change in agricultural insect pest. *Acta Sci. Agric*, *3*(12), 74–80. doi:10.31080/ASAG.2019.03.0727

Skendžić, S., Zovko, M., Živković, I. P., Lešić, V., & Lemić, D. (2021). The impact of climate change on agricultural insect pests. *Insects*, *12*(5), 440. doi:10.3390/insects12050440 PMID:34066138

Spooner-Hart, R., Tesoriero, L., & Hall, B. H. (2007). *Field guide to olive pests, diseases and disorders in Australia*. Rural Industries Research and Development Corporation.

Staley, J. T., Hodgson, C. J., Mortimer, S. R., Morecroft, M. D., Masters, G. J., Brown, V. K., & Taylor, M. E. (2007). Effects of summer rainfall manipulations on the abundance and vertical distribution of herbivorous soil macro-invertebrates. *European Journal of Soil Biology*, *43*(3), 189–198. doi:10.1016/j.ejsobi.2007.02.010

Stireman, J. O. III, Dyer, L. A., Janzen, D. H., Singer, M. S., Lill, J. T., Marquis, R. J., Ricklefs, R. E., Gentry, G. L., Hallwachs, W., Coley, P. D., Barone, J. A., Greeney, H. F., Connahs, H., Barbosa, P., Morais, H. C., & Diniz, I. (2005). Climatic unpredictability and parasitism of caterpillars: Implications of global warming. *Proceedings of the National Academy of Sciences of the United States of America*, *102*(48), 17384–17387. doi:10.1073/pnas.0508839102 PMID:16293686

Tajnari, H. (1992). *Étude bio-écologique d'Euphyllura olivina Costa (Hom. Psyllidae) dans les régions du Haouz et d'Essaouira: mise en évidence d'un état de diapause ovarienne*. Meknès, Maroc: Thèse de troisième cycle, École nationale d'agriculture.

Tajnari, H. (2001). *Étude de la nuisibilité du psylle de l'olivier, Euphyllura olivina Costa*. Symposium sur la Protection Intégrée des Cultures dans la région Méditerranéenne, DPVCTRF, Rabat, Maroc.

Tauber, M. J., Tauber, C. A., & Masaki, S. (1986). *Seasonal adaptations of insects*. Oxford University Press on Demand.

Therios, I. N. (2009). *Olives* (Vol. 18). CABI.

Thomas, C. D., Cameron, A., Green, R. E., Bakkenes, M., Beaumont, L. J., Collingham, Y. C., Erasmus, B. F. N., de Siqueira, M. F., Grainger, A., Hannah, L., Hughes, L., Huntley, B., van Jaarsveld, A. S., Midgley, G. F., Miles, L., Ortega-Huerta, M. A., Townsend Peterson, A., Phillips, O. L., & Williams, S. E. (2004). Extinction risk from climate change. *Nature*, *427*(6970), 145–148. doi:10.1038/nature02121 PMID:14712274

Thomson, L. J., Macfadyen, S., & Hoffmann, A. A. (2010). Predicting the effects of climate change on natural enemies of agricultural pests. *Biological Control*, *52*(3), 296–306. doi:10.1016/j.biocontrol.2009.01.022

Tsitsipis, J. A. (1977). Effect of constant temperatures on eggs of olive fruit fly, Dacusoleae (Diptera: Tephritidae). *Ann. Zool. Ecol. Anim.*, *9*, 133–139.

Tsitsipis, J. A. (1980). Effect of constant temperatures on larval and pupal development of olive fruit flies reared on artificial diet. *Environmental Entomology*, *9*(6), 764–768. doi:10.1093/ee/9.6.764

Tsitsipis, J. A., & Abatzis, C. (1980). Relative humidity effects, at 20, on eggs of the olive fruit fly, Dacus oleae (Diptera: Tephritidae), reared on artificial diet. *Entomologia Experimentalis et Applicata*, *28*(1), 92–99. doi:10.1111/j.1570-7458.1980.tb02991.x

Tzanakakis, M. E. (2003). Seasonal development and dormancy of insects and mites feeding on olive: A review. *Netherlands Journal of Zoology*, *52*(2/4), 87–224. doi:10.1163/156854203764817670

Tzanakakis, M. E. (2006). *Insects and mites feeding on olive: Distribution, importance, habits, seasonal development, and dormancy*. Academic Press.

van Asch, B., Pereira-Castro, I., Rei, F. T., & da Costa, L. T. (2015). Marked genetic differentiation between western Iberian and Italic populations of the olive fly: Southern France as an intermediate area. *PLoS One*, *10*(5), e0126702. doi:10.1371/journal.pone.0126702 PMID:25951107

Visser, M. E., & Both, C. (2005). Shifts in phenology due to global climate change: The need for a yardstick. *Proceedings. Biological Sciences*, *272*(1581), 2561–2569. doi:10.1098/rspb.2005.3356 PMID:16321776

Volney, W. J. A., & Fleming, R. A. (2000). Climate change and impacts of boreal forest insects. *Agriculture, Ecosystems & Environment*, *82*(1-3), 283–294. doi:10.1016/S0167-8809(00)00232-2

Volpi, I., Guidotti, D., Mammini, M., Petacchi, R., & Marchi, S. (2020). Managing complex datasets to predict Bactrocera oleae infestation at the regional scale. *Computers and Electronics in Agriculture*, *179*, 105867. doi:10.1016/j.compag.2020.105867

Vono, G., Bonsignore, C. P., Gullo, G., & Marullo, R. (2020). Olive Production Threatened by a Resurgent Pest Liothrips oleae (Costa, 1857) (Thysanoptera: Phlaeothripidae) in Southern Italy. *Insects*, *11*(12), 887. doi:10.3390/insects11120887 PMID:33339275

Walther, G. R., Roques, A., Hulme, P. E., Sykes, M. T., Pyšek, P., Kühn, I., ... Settele, J. (2009). Alien species in a warmer world: Risks and opportunities. *Trends in Ecology & Evolution*, *24*(12), 686–693. doi:10.1016/j.tree.2009.06.008 PMID:19712994

Wang, B. X., Hof, A. R., & Ma, C. S. (2022). Impacts of climate change on crop production, pests and pathogens of wheat and rice. *Frontiers of Agricultural Science and Engineering*, *9*(1), 4–18. doi:10.15302/J-FASE-2021432

Wang, X. G., Johnson, M. W., Daane, K. M., & Nadel, H. (2009). High summer temperatures affect the survival and reproduction of olive fruit fly (Diptera: Tephritidae). *Environmental Entomology*, *38*(5), 1496–1504. doi:10.1603/022.038.0518 PMID:19825305

Woiwod, I. (1997). Detecting the effects of climate change on Lepidoptera. *Journal of Insect Conservation*, *1*(3), 149–158. doi:10.1023/A:1018451613970

Wolf, A., Kozlov, M. V., & Callaghan, T. V. (2008). Impact of non-outbreak insect damage on vegetation in northern Europe will be greater than expected during a changing climate. *Climatic Change*, *87*(1-2), 91–106. doi:10.100710584-007-9340-6

Yamamura, K., & Kiritani, K. (1998). A simple method to estimate the potential increase in the number of generations under global warming in temperate zones. *Applied Entomology and Zoology*, *33*(2), 289–298. doi:10.1303/aez.33.289

Yang, L. H., & Rudolf, V. H. W. (2010). Phenology, ontogeny and the effects of climate change on the timing of species interactions. *Ecology Letters*, *13*(1), 1–10. doi:10.1111/j.1461-0248.2009.01402.x PMID:19930396

Yihdego, Y., Salem, H. S., & Muhammed, H. H. (2019). Agricultural pest management policies during drought: Case studies in Australia and the state of Palestine. *Natural Hazards Review, 20*(1), 05018010. doi:10.1061/(ASCE)NH.1527-6996.0000312

Young, A. M. (1982). *Population Biology of Tropical Insects*. Plenum Press. doi:10.1007/978-1-4684-1113-3

Zhang, J., Xing, G. M., Liao, J. X., Hou, Z. D., Wang, G. X., & Wang, Y. F. (2003). Effects of different atmospheric CO2 concentrations and soil moistures on the populations of bird cherry-oat aphid (Rhopalosiphum padi) feeding on spring wheat. *European Journal of Entomology*, *100*(4), 521–530. doi:10.14411/eje.2003.080

Zittis, G., Almazroui, M., Alpert, P., Ciais, P., Cramer, W., Dahdal, Y., ... Lelieveld, J. (2022). Climate change and weather extremes in the Eastern Mediterranean and Middle East. *Reviews of Geophysics, 60*(3).

KEY TERMS AND DEFINITIONS

Alien Species: An individual or population, at any life stage, or a viable part of an organism that is non-indigenous to an area and that has entered by the human agency into the area.

Climate Change: A phenomenon that involves changes over long periods in environmental factors such as temperature, humidity, and precipitation.

Diapause: A period of physiologically controlled dormancy or inactivity in insects or other invertebrates. It allows insects to survive unfavorable conditions.

Economic Threshold: The point at which a pest has become numerous or damaging enough to require management action to prevent intolerable economic loss.

Global Warming: The observed increase in average temperature near the Earth's surface and the lowest layer of the atmosphere.

Host Range: Species capable, under natural conditions, of sustaining a specific pest or another organism.

Outbreak: A recently detected pest population, including an incursion, or a sudden significant increase of an established pest population in an area.

Parasitoid: An insect parasitic only in its immature stages, killing its host in the process of its development, and free living as an adult.

Poikilotherm: An organism whose internal temperature varies with the temperature of its surroundings.

Population Dynamics: Major abiotic and biotic factors that tend to increase or decrease the population size, age, and sex composition of a species.

Chapter 3
Impact of Climate Change on the Insect and Mite Pests of Moroccan Citrus

Alahyane Hassan
https://orcid.org/0000-0002-4105-8327
Cadi Ayyad University, Morocco

Ouknin Mouhamed
https://orcid.org/0000-0001-5251-3033
Moulay Ismail University, Morocco

Mounir Abdelaziz
Cadi Ayyad University, Morocco

Majidi Lhou
Moulay Ismail University, Morocco

ABSTRACT

Climate change has greatly affected citrus production and pests. They affect agriculture by expanding the geographical range of insect and mite pests, and by modifying the lifecycle inducing an increase in the number of generations and survival during the wintering period. These changes in climatic parameters altered interspecific and intraspecific interactions between plants and pests and increased the risk of invasive species and reduced efficacy of biological control, especially those used by natural enemies, causing serious risks as well as crop damages. To mitigate the negative effects of climate change in plant agriculture, management strategies must be taken to deal with the new status of agricultural pests.

DOI: 10.4018/978-1-6684-4824-3.ch003

INTRODUCTION

The Moroccan citrus sector plays a major socio-economic role in the national economy, with an area of around 105,000 ha and 2.28 million tons of annual production and 3 billion dirhams of income per year (MAPMDREF, 2020). They contributed to the creation of 21 million working days per year (MAPMDREF, 2020). However, the national production yield remains low (20 to 26 tons / ha) compared with those achieved by many countries (more than 30 tons / ha) (Alahyane et al., 2020). This weakness in production yield is due to pathogens and pests' attacks (Smaili et al., 2013) and probably the effects of climate change that negatively acting on the quality and quantity of fruit production. Several pest species attacked the Moroccan citrus trees such as the fruit fly, mealy bugs, the citrus leafminer, mites, aphids and snails (Mazih, 2011). Recently, the damage caused by these pests has become more and more important, especially with the variations of climatic conditions which create suitable conditions for their development, and the management strategies are inappropriate (Haokip et al., 2020).

Actually, more than 100 species of pests were recorded on Moroccan citrus plants causing serious damage of about 13.6% of crop losses (Alahyane et al., 2019), and they increase from year to year. To limit these damages, several strategies have been taken. It is predicted that climate change is likely to have a substantial effect on citrus pest management strategies, such as biopesticides or natural plant products, host plant resistance, natural enemies, and chemical pesticides efficacy (Sharma, 2016). The increased temperature directly affects several parameters of pest species such as reproduction, spread, survival, and population dynamics as well as the interactions between pests, the natural enemies and the environment (Skendžić et al., 2021). In fact, the big challenge in the Moroccan citrus is how to mitigate the effects of climate change. Therefore, these strategies must be suitable, in order to meet market requirements, attenuation of losses, environmental preservation, and production cost reduction.

In the present review, we discuss the major pests associated with the Moroccan citrus and the IPM tactics used to control the economic pests, and how to prevent the invasive species reaching economic levels. We also present the effects of climate change on citrus pest species.

CITRUS SECTOR AND THE NATIONAL ECONOMY

The Moroccan citrus are considered as strategic sectors playing an important socio-economic role and generating an important source of foreign currency of around 2 billion dirhams per year (MAPMDREF, 2020). Morocco is considered as one of the major producers in the world with an average annual production of 2.28 million tons. On the social level, this sector guarantees the employment of a large workforce providing more than 34 million working days in different production units (MAPMDREF, 2020). Thus, the citrus fruits occupied around 105,000 ha, approximately 12% of the area sown in arboriculture. The planted area is shared between five regions: the Souss Massa zone, Gharb, Moulouya, Tadla and Haouz, which alone hold more than 93% of the national planted area (MAPMDREF, 2020). The Moroccan citrus are very diversified and dominated by 3 groups, namely the Clementine with 23.8%, Morocco Late with 20.3%, Navels with 15.7%, Nador Cott with 6.7% and others with 33.5% (MAPMDREF, 2020).

INSECT AND MITE SPECIES AND THEIR ECONOMIC IMPORTANCE IN THE MOROCCAN CITRUS CROPS

In Morocco, about 30 insect and mite pests are associated with citrus orchards and some of them are considered key pests (Mazih, 2008; 2015). Among these species, we find the citrus aphids (*Aphis spiraecola*, *Toxoptera aurantii*, *Myzus persicae*, *Aphis gossypii*), Mediterranean fruit fly (*Ceratitis capitata*), whiteflies (*Parabemisia myricae*, *Aleurothrixus floccosus*, *Dialeurodes citri*, *Aleurodicus dispersus*), scale insects (*Aonidiella aurantii*, *Aspidiotus nerii*, *Chrysomphalus dictyospermi*, *Parlatoria zizyphi*, *Parlatoria pergandii*, *Lepidosaphes gloverii*, *Icerya purchasi*), thrips (*Frankliniella occidentalis*), *Limothrips cerealium*, citrus leafminer (*Phyllocnistis citrella*) and mites (*Eutetranychus orientalis*, *Panonychus citri*, *Tetranychus urticae*) (Mazih 2015; Alahyane et al., 2019).

Aphids

The main aphids present on Moroccan citrus are *Aphis spireacola*, *Taxoptera aurantii* (Figure 1), *A. gossypii*, *Myzus persicae*, *A. craccivora* and *A. fabae* (Smaili et al., 2013) and only the first three species are considered the most important in citrus orchards (Benziane et al., 2001; Smaili et al., 2001, 2008; Sekkat, 2008; Alahyane, 2020). Aphids induce important damages on the attacked trees: the growth of young shoots can be stunted, and the plant weakens. Also, flower abortion, leaf drops or drying of young shoots can be observed (Jaouad et al., 2020). As a result, we assist in decreasing shoot growth and tree productivity. The toxic action of injected saliva causes deformations of leaves and twigs. This ranges from simple deformation of the foliage to the formation of cankers or galls. Aphids can transmit and disseminate pathogenic viruses. In fact, *A. gossypii* is a vector of more than 44 viruses, especially tristeza virus which is considered as a serious threat to national citrus. In addition, the secretion of honeydew, rich in sugar, causes the development of fumagine on young leaves and reduces transpiration and photosynthesis (Sekkat, 2008).

Figure 1. Taxoptera aurantii on citrus leaves

Ceratitis capitata

The Mediterranean fruit fly, *Ceratitis capitata* (Figure 2), is one of the most damaging fruit pests in the world, due to its ability to tolerate various types of climates and its polyphagous specificity. It attacks a wide range of host species, approximately 250 plants. The majority of attacked plants are represented by fruit trees including citrus, apple, apricot, peach, plum, pear, prickly pear and some vegetable crops (Mazih, 2011). This species causes considerable economic losses for Mediterranean citrus growers and mainly those in the Maghreb.

The fruit fly damages on citrus are in the form of puncture points and galleries on fruits (Marchiori, 2021). These led to the rapid rotting and dropping of the fruits (Mazih, 2015). The puncture sites provide a pathway for fungi and bacteria responsible for fruit decay. The fruit fly damage is a major impediment to exports due to the devaluation of the commodity and quarantine measures imposed by some importing countries.

Figure 2. Mediterranean fruit fly (Ceratitis capitata)

In Morocco, the economic and social importance of the Mediterranean fruit fly lies to the great socio-economic impact to the citrus industry. The invasion of orchards by fruit fly would lead to a socio-economic crisis in infested regions, as it participates year after year in the degradation of fruit quality and consequently the decrease of production and the limitation of export of citrus products (Aboussaid et al., 2011).

Leafminer

The citrus leafminer, *Phyllocnistis citrella* is a micro lepidopteran belonging to the family Gracillariidae (Balakchowsky, 1966). It's considered one of the most important pests recorded on citrus in the Mediterranean regions. The appearance of *P. citrella* as the primary pest in Morocco dates from 1994. Adults having a wingspan of 3 mm, after overwintering, appear in young shoots in spring. The female of *P. citrella* preferentially looks for young leaves, rich in nutrients, to lay her eggs and neglects old leaves. The eggs of *P. citrella* are deposited one by one on the underside of the leaves. *P. citrella* has several annual generations and the length of lifecycle varies with temperature variations. Development from egg to adult can be completed in less than two weeks (Wilson, 1991; Ware, 1994; Nawaz et al., 2021).

In Morocco, *P. citrella* has five generations in the Gharb region, three in summer and two in winter (Smaili, 2001). This leafminer lives mainly on citrus trees. Analysis of the young leaves of three citrus varieties (Navel, Nour and Maroc Late), from summer and autumn shoots showed that the leaf area destroyed by a caterpillar varies between 20 and 26% depending on the variety and the shoot (Figure 3). The attack due to leafminer can reach almost 100% (Mazih and Chitoukli, 2001). Only three larva can destroy more than 40 to 50% of the leaf area, which can lead to leaf drops (Nath & Sikha, 2019). Also, the citrus leafminer is very sensitive to high heat, significant mortality has been recorded in the region of Marrakech during the summer of about 54% (Boussaade, 2003).

Figure 3. Damage of leafminer on citrus leaves

Mealybugs

Several species belonging to the Homoptera family are harmful to Moroccan citrus. Among these species, we found *Aonidiella aurantii, Lepidosaphe beckii*, *Parlatoria pergandii*, *Parlatoria ziziphi*, *Lepidosaphe gloverii*, and *Planococcus citri* (Mazih, 2015). The diaspid mealybugs represent the most damaging species in Morocco. They are known by the shape of the pygidium of the females provided with comb and by the presence of protective shield encompassing the larval and nymphal exuviae (Irvin et al., 2021).

The mealybugs induce significant damage to the citrus plants. *A. aurantii* is the most damaging pest of all varieties (Loussert, 1987). This pest causes the destruction of chlorophyll under the effect of toxins injected by the stylets on the leaves (Delabie et al., 2008) (Figure 4). All infested fruits are deformed and marked by pest bites. In addition, the females induce severe desiccation to infested branches. Whereas, in heavy attacks the whole tree is weakened and can be destroyed in two years.

Figure 4. Damage of Aonidiella aurantii on Citrus leaves

Mites

Several pest mites are associated with Moroccan citrus: *Tetranychus urticae* (Koch) (Figure 5) or yellow mite is a polyphagous species, attacking more than 15 crops including citrus (Appert, 1982; Alahyane et al., 2020). *Panonychus citri* is a major species of Moroccan citrus found in all regions known by a citrus plantation.

Figure 5. Adult of Tetranychus urticae (Acari: Tetranychidae)

The mite stages feed on plant leaves and cause the destruction of palisade cells and those of the spongy parenchyma of the leaf (Mazih, 2011). These alterations result in a reduction in the number of chloroplasts and therefore a reduction in photosynthetic activity. This is attributed to the damage of the protective surface of the leaf and to alterations in the functioning of the stomata.

Recently, new pests have been detected in Moroccan orchards, including *Eutetranychus orientalis*, which has become a major citrus pest and adds to the existing spider mites on Moroccan citrus (Mazih, 2015).

Thrips

Thrips are highly polyphagous insects. On citrus, many species of thrips have been identified, but few are considered the key pests (Lewis, 1997). *Pezothrips kellyanus* is one of the economically important pests in citrus in worldwide (Blank & Gill, 1997; Mound & Jackman 1998) including the Mediterranean regions (Conti et al., 2001; Navarro et al., 2011). In fact, *P. kellyanus* caused damage to citrus fruits, followed by a qualitative depreciation reducing the commercial value of the fruits (Navarro et al., 2008).

In Morocco, *P. kellyanus* caused serious damages to citrus fruits and actually is considered as a key pest (Abbassi, 2011). Indeed, in orchards the presence of *P. kellyanus* larvae under the calyx of fruits, devalues their marketable quality. In addition, other species of thrips have also been identified on Moroccan citrus such as *Frankiniella occidentalis* and *Scirothrips sp*.

Other pests

In Moroccan citrus, dozens of pest species are considered as secondary pests. The status of some pests has changed, as they induce visible and sometimes economically important damages, such as snails, whiteflies (*Parabemisia myricae*, *Aleurothrixus floccosus*, *Dialeurodes citri*) and tarsonemes (*Polyphagotarsonemus latus*) (Mazih, 2015).

IMPACT OF CLIMATE CHANGE ON MITE AND INSECT PESTS

Temperature Increase

Temperature directly affects insects and mites biology, the development rates of their life stages are highly dependent on temperature variations. In fact, when the temperature increases the lifecycle becomes fast and vice versa. These findings are consistent with several experiences (Cammell & Knight, 1991; Fleming & Volney, 1995).

Each degree increase in temperature shortens the life cycle of insects and mites. The faster the life cycle, the more the pest population increases during the hottest time of the year. Therefore, species with a broad niche defined by the climate regime will respond more predictably to climate change than those with a limited niche, which will respond less predictably (Sutherst, 2000; Bale et al., 2002).

Increased temperature, affect crop pest populations in several ways. They increase their geographic range, overwintering and population growth rate. They also, changes the number of generations and development season, and alteration of interspecific interactions. When the climatic conditions change, we also observed the appearance of invasive species and introduction of alternate hosts and overwintering hosts.

Diversity of Insect and Mite Pests

The insect and mite diversity in their habitats are good indicators of environmental change and indicate the status of the ecosystem health (Gregory et al., 2009). The relative abundance of these species may be affected by climate change and those unable to adapt to these changes may disappear over time (Thomas et al., 2004).

In the current scenario, many species are actually threatened due to degradation of natural vegetation for various anthropogenic development activities (Costanza et al., 1987; Storch et al., 2022; Sidhu and Mehta, 2008). Therefore, it is important to increase the functional diversity in agro-ecosystems vulnerable to climate change to improve system resilience and reduce the magnitude of pest losses (Newton et al., 2009). However, changes in cropping patterns resulting from climate change will dramatically affect the balance between insect and mite pests and their natural enemies. The main effects of climate change on pests and natural enemy communities are a deterioration in the relative abundance of decomposers and predators, and an increase in herbivory, which may have negative consequences for overall ecosystem structure and services. Responses of insect and mite pests and their natural enemies are dependent on both temperature and precipitation. Whereas, the negative effects on the overall ecosystem are likely to increase under anticipated climate change (Zvereva & Kozlov, 2010).

Large-scale changes in precipitation due to climate change will have a significant effect on the abundance and diversity of insect and mite pests and natural enemies. Under extreme drought conditions due to climate change, it is likely that multi-trophic diversity will decrease, and ecosystem composition will change. Habitat fragmentation results in the subdivision of habitats into smaller units, increasing their insularity and resulting in a loss of total habitat area. Such fragmentation alters the microenvironment at the edges of the fragmented area, and the effects of the edges include microclimate, changes in light, temperature, and humidity, and each of these effects can have a significant impact on the vitality and composition of species within the fragment. Due to climate change, current extinction rates are 100 to 1000 times greater than what has occurred in the past, with nearly 45 to 275 species going extinct every day (Sharma, 2010).

Geographic Range Extension

The geographic distribution of each animal and plant is determined by the climatic conditions including their survival, growth, and reproduction. The variation of temperature affects the abundance and distribution of insects and mites. An increase in temperature of about 1 °C would lead to an increase of distribution areas of 200 km northward (in altitude). Areas that are actually not favorable due to low temperature could be appropriate with increasing temperature. Maximum temperature, rather than minimum temperature, plays an important role in determining the distribution of insect and mite species. Thus, increases in temperature will result in variations in the ability of species to overwinter at higher altitudes resulting in migrations of pests from south to north. Many pest species have geographic ranges that are not directly limited by vegetation, but mainly by temperature.

Several studies reported that with increasing temperature, pest insects and mites will increase their distribution area, resulting in an increase in the number of host plants (Sharma et al., 2002; 2010; Skendžić et al., 2021; Pérez-Hedo et al., 2021). This can lead to increased abundance of insects and mites (Bale et al., 2002; Diffenbaugh et al., 2008) and sudden outbreaks of insect and mite pests can wipe out some crop species entirely (Kannan & James, 2009). Meanwhile, warming in temperate regions can lead to a decrease in the relative abundance of temperature-sensitive pest populations (Petzoldt & Seaman, 2010: Sharma, 2010, 2016). The majority of Polar Regions are unaffected by insect epidemics due to low temperatures and frequent frosts (Volney & Fleming, 2000; Schneider et al., 2022). In the future, predicted global warming (Carroll et al., 2004) and increased incidence of drought (Logan et al., 2003; Ding and Wen, 2022) are expected to result in more frequent insect epidemics in temperate regions as well.

Insect and Mite Phenology Changes

Climate change also induced changes in species phenology. This impact of climate change is considered the most easily monitored case and the most documented in this regard for a wide range of organisms (Root et al., 2003; Buckley et al., 2022). With increasing temperatures, insects are expected to pass through their larval stages more quickly and become adults earlier. Therefore, expected responses in insects and mites could include an advance in the timing of larval and adult emergence and an increase in the duration of the flight period (Menéndez, 2007). Members of the order Lepidoptera are the best examples of these phenological changes. Changes in butterfly phenology have been reported in the United Kingdom, where 26 of 35 species have advanced their first appearance (Roy & Sparks, 2000). Early adult emergence and early arrival of migratory species have also been reported for aphids in the UK (Harrington et al., 2007). Gordo & Sanz (2005) studied the impacts of climate on four Mediterranean insect species (butterfly, bee, fly and beetle), and indicated that all species changes their date of first appearance over the last 50 years, which was correlated with increasing spring temperature.

Changes in insect and mite phenology can be studied by long-term experiments with varying sowing dates to observe the appearance of pests on crops. Similarly, the timing of the arrival of insect species can also be recorded using light traps, suction traps or pheromone traps. Analysis of long-term phenology data would reveal changes in pest onset dates under climate change (Pathak et al., 2012). Suction traps have been used to monitor aphids at the Rothamsted Insect Survey since 1964. Analysis of suction trap data revealed that spring flights of the potato aphid, *Myzus persicae* (Sulzer) began two weeks earlier for every 1°C increase in the combined January and February mean temperature. Similarly, long-term data from several insect recording programs in Europe and North America have provided evidence that species become active, migrate, or breed earlier in the year as a direct result of increased temperatures leading to increased growth rates or earlier emergence from winter inactivity (Roy & Sparks, 2000).

Increased Overwintering Survival

Insects possess a limited capacity for homeostasis in the face of external temperature changes. As a result, they have evolved a range of strategies such as behavioral avoidance through migration and physiological adaptations such as diapause to maintain life in thermally stressful environments (Bale & Hayward, 2010). Diapause refers to a period of suspension of developmental activities, the manifestation of which is determined by environmental factors such as humidity, temperature, etc. Diapause, as an adaptive trait, plays a vital role in the seasonal regulation of insect life cycles, allowing them to better survive many environmental adversities. Two main types of diapause exist in insects: aestivation and hibernation, which maintain life under extremely high and low temperature conditions, respectively (Chapman, 1998). Studies have shown that global warming occurs more particularly in winter than in summer and is more significant at high latitudes (IPCC, 2007; IMD, 2010).

Looking at the climate profile of the Mediterranean region over the past 100 years, the warming has been more pronounced during the winter season, and it is the minimum, not the maximum temperature that has increased significantly (IMD, 2010). This makes winter diapause insects susceptible to the most significant changes in their thermal environment (Bale & Hayward, 2010). Warming in winter may delay the onset of diapause and early summer may cause insects to end diapause more quickly. They allow them to resume active growth and development. This implies that temperature increases between 1°C and 5°C would increase insect survival due to low winter mortality, increased populations, early infestations, and crop damage by insect pests under the global warming scenario (Sharma et al., 2002; 2010).

Increase in Number of Generations

Some crop pests develop on an intermittent basis according to temperature. They grow faster during periods of proper temperature. Rising temperatures accelerate the development of these types of insects, which may result in more production cycles per year (Awmack et al., 1997; Mozaffari, 2022). With a temperature increase of 2°C, insects are estimated to have 1-5 additional life cycles per season (Tariq, 2020). Warming may reduce the frequency of severe cold events, potentially extending the overwintering range of insect pests (Patterson et al., 1999). Thus, the general increase in temperature within a given favorable range may accelerate the development, reproduction, and survival rates of tropical and subtropical insects. They allow them to complete a greater number of generations per year (Yamamura & Kiritani, 1998; Petzoldt & Seaman, 2010).

Pest Population Dynamics

Seasonal effects of climatic conditions lead directly to alterations in the dispersal and development of insect and mite species. Changes in surrounding temperatures certainly lead to changes in insect and mite development and survival rates and then inducing variations in population size, density, and genetic composition (Bale et al., 2002; Jactel et al., 2019). This can result in disturbances in the ecological balance due to unpredictable changes in the pest populations such as their existing and potential natural enemies (Rao et al., 2009). Changes in climatic factors have resulted in an increase in the frequency and intensity of pest infestations. These situations of increased damage and frequency of pest infestation on crops have negatively affected the performance of farmers who are already in difficulty by increasing the cost of plant protection and reducing farmers' incomes.

Crop-Pest Interactions

The ability of an insect and mite to complete their lifecycle development depends on their ability to adapt to environmental conditions. Climate change has been found to exert both bottom-up and top-down effects on tri-trophic interactions between crops, insects, mites, and their natural enemies, by many physiological changes related in particular to host adaptation and nutritional status (Coviella & Trumble, 1999; Skendžić et al., 2021). The large outbreaks observed in expansion areas on new hosts may be explained either by the high susceptibility of the hosts or by the inability of natural enemies to locate the borer larvae on an unusual host or environment (Battisti et al., 2006). Temperature and photoperiod deeply affect critical events such as stem elongation, flowering, and fruiting in the plant life cycle (Cleland et al., 2007). Rising temperature due to global warming can accelerate the lifecycles of some plants (Parmesan & Yohe, 2003; Willis et al., 2008), who could significantly affect the feeding and reproduction patterns of associated insect pests such as aphids, jassids, mealy bugs, etc. Such increases can significantly exacerbate negative ecological and economic consequences (MEA, 2005).

Pest Management Practices

Actually, the control strategies implemented to control citrus pests tend to combine many methods such as chemical and non-chemical control, with the aim of achieving integrated pest management.

Chemical Control

The chemical control of citrus pests is essentially based on the spreading of synthetic pesticides based on several active ingredients such as Bifenazate, Spirodiclofen, Etoxazole and hexythiazox (Mazih, 2015; Alahyane, 2020). Application of the products is recommended when the infestation threshold reaches 30% of the leaves bearing more than 3 mobile forms (Benziane et al., 2003). The number of treatments can reach two or more per year depending on the type of pests. The overuse of pesticide has led to the emergence of many cases of pesticides resistance by pests which have been reported in the Arthropod Pesticide Resistance Database (APRD, 2018). The use of these chemicals has also generated a number of disadvantages, such as adverse effects on natural enemies (Urbaneja et al., 2008), on human health and the environment (Cavalcanti et al., 2010; Kumral et al., 2010).

Cultural Control

Cultural control consists of using specific cultural techniques or practices, with the aim of helping to prevent pest attacks and making the environment less favorable to their development. Among the techniques used crop rotation, fertilization and irrigation management techniques, use of resistant varieties, pruning, soil cover, etc.

Proper management of irrigation can lead to a reduction in the damage caused by pests. In the case of citrus trees subjected to salt stress, the plants accumulate certain compounds such as metabolites which act as a defense mechanism (Arbona et al., 2003; Partel et al., 2019) and these compounds can affect pest's development. Aucejo-Romero et al. (2004) reported that a low or medium concentration of NaCl in irrigation water could increase reproduction of pests on clementine trees. Ansaloni et al. (2008) also reported that cultural practices (in particular an adequate irrigation program) and good control of pests that attack new leaves (miners and aphids), can promote the presence of tender shoots in summer and autumn, and both prevent spider mites from migrating to the fruit and damaging it. In perennial crops such as citrus, soil cover plays an important role in regulating mite populations, as mites find refuge in weeds and can migrate to crops and vice versa (Bassanezi et al., 2020).

Biological Control by Natural Enemies

The natural enemies used in the biological control of pests are taxonomically and functionally very diverse. They include groups of parasitoids and predators (Hajek, 2004; Hull et al., 2010; Bouvet et al., 2019; Staton et al., 2021). The objective of interactions between natural enemies and their hosts is to keep the hosts as quickly as possible below levels of economic damage, and before the appearance of these. However, an effective natural enemy must meet certain characteristics such as: 1) host specificity; 2) synchrony with the pest; 3) high growth rate; 4) the ability to survive for long periods with less food; and 5) a good ability to search for prey (Hajek, 2004; Sahraoui, 2012). Before the implementation of a biological control method against citrus pests, it is important to assess whether the relationships between pests and chosen natural enemy are synchrony (Holt et al., 2000).

Biological Control by Botanical Insecticides

Plant products as biopesticides have long been considered a very useful tool in the biological control of mites. They contain several active substances with insecticides and acaricides effects and can be used in pest control (Ghalbane et al. 2022). Recently, several natural products of botanical origin have been developed and used in the management strategies of crop pests. For example, some of these products are based on plant extracts (Khan et al. 2011), and others are based on plant essential oils (Kumar and Archana, 2015; Ouknin et al., 2021; Alahyane et al., 2022). Botanical extracts and essential oils are increasingly used as alternative or complementary approaches to insecticide and acaricide treatments in integrated pest management strategies. They show a broad spectrum of activity and a great effectiveness to control several pest species.

CONCLUSION

Climate change has serious effects on the distribution, diversity, reproduction, development, growth, and phenology of citrus pest species (insects and mites). It also has effects on plant defense activity, biopesticides, synthetic chemicals and invasive species. Without local and global collaborative efforts, pest control, such as biological control, host plant resistance, and chemical insecticides, could become ineffective. Studying the effectiveness of natural enemies, insect-plant interactions, host plant resistance, biopesticides and insecticides under changing climates will be imperative for designing appropriate pest management strategies.

RESEARCH FUNDING

This research received no specific grant from any funding agency in the public, commercial, or not-for-profit sectors.

REFERENCES

Abbassi, M. (2011). Acarien tétranyque menaces permanent sur les agrumes: Cas de l'acarien rouge oriental *Eutetranychus orientalis*. *Agriculture du Maghre, 50*.

Aboussaid, H., El-Aouame, L., El-Messoussi, S., & Oufdou, K. (2010). Biological Activity of *Bacillus Thuringiensis* (Berliner) Strains on Larvae and Adults of *Ceratitis Capitata* (Wiedemann) (Diptera: Tephritidae). *Journal of Environmental Protection, 1*(4), 337–345. doi:10.4236/jep.2010.14040

Alahyane, H. (2020). *Dynamique des populations de l'acarien rouge oriental, Eutetranychus orientalis (Klein) (Acari: Tetranychidae) ravageur des agrumes, et lutte biologique dans la région de Marrakech-Safi* [Thèse]. Cadi Ayyad University.

Alahyane, H., El Alaoui, A., Abousaid, H., Aimrane, A., Atibi, Y., Oufdou, K., & El Messoussi, S. (2019). Biological activity of some native *Bacillus thuringiensis* Berliner strains against *Eutetranychus orientalis* (Klein) (Acari: Tetranychidae). *Applied Ecology and Environmental Research, 17*(2), 1967–1977. doi:10.15666/aeer/1702_19671977

Alahyane, H., Ouknin, M., Aboussaid, H., El Messoussi, S., Costa, J., & Majidi, L. (2022). Biological activities of essential oils from Moroccan plants against the honey bee ectoparasitic mite, *Varroa destructor. International Journal of Acarology, 48*(1), 50–56. doi:10.1080/01647954.2021.2015436

Ansaloni, T., Pascual-Ruiz, S., Hurtado, M. A., & Jacas, J. A. (2008). Can summer and fall vegetative growth regulate the incidence of *Tetranychus urticae* Koch on Clementine fruit? *Crop Protection (Guildford, Surrey), 27*(3-5), 459–464. doi:10.1016/j.cropro.2007.07.016

Appert, J. (1982). Les acariens. In *Les ravageurs des cultures vivrières et maraîchères sous les tropiques* (p. 413). Edit.G.P. Maisonneuve et Larose.

APRD. (2018). *Arthropod Pesticide Resistance Database (APRD)*. Michigan State University. https://www.pesticideresistance.org/

Arbona, V., Flors, V., Jacas, J., García-Agustín, P., & Gómez-Cadenas, A. (2003). Enzymatic and Nonenzymatic Antioxidant Responses of Carrizo citrange, a Salt-Sensitive Citrus Rootstock, to Different Levels of Salinity. *Plant & Cell Physiology, 44*(4), 388–394. doi:10.1093/pcp/pcg059 PMID:12721379

Aucejo-Romero, S., Gómez-Cadenas, A., & Jacas-Miret, J. A. (2004). Effects of NaCl-stressed citrus plants on life-history parameters of *Tetranychus urticae* (Acari: Tetranychidae). *Experimental & Applied Acarology, 33*(1–2), 55–67. doi:10.1023/B:APPA.0000030026.77800.0c PMID:15285138

Awmack, C. S., Woodcock, C. M., & Harrington, R. (1997). Climate change may increase vulnerability of aphids to natural enemies. *Ecological Entomology, 22*(3), 366–368. doi:10.1046/j.1365-2311.1997.00069.x

Balakchowsky, A. S. (1966). *Entomologie appliquée à l'agriculture lépidoptères* (Vol. 2). Ed Masson et Cie.

Bale, J., Masters, G. J., Hodkins, I. D., Awmack, C., Bezemer, T. M., Brown, V. K., Buterfield, J., Buse, A., Coulson, J. C., Farrar, J., Good, J. E. G., Harrigton, R., Hartley, S., Jones, T. H., Lindroth, R. L., Press, M. C., Symrnioudis, I., Watt, A. D., & Whittaker, J. B. (2002). Herbivory in global climate change research: Direct effects of rising temperature on insect herbivores. *Global Change Biology*, *8*(1), 1–16. doi:10.1046/j.1365-2486.2002.00451.x

Bale, J. S., & Hayward, S. A. L. (2010). Insect overwintering in a changing climate. *The Journal of Experimental Biology*, *213*(6), 980–994. doi:10.1242/jeb.037911 PMID:20190123

Bassanezi, R. B., Lopes, S. A., de Miranda, M. P., Wulff, N. A., Volpe, H. X. L., & Ayres, A. J. (2020). Overview of citrus huanglongbing spread and management strategies in Brazil. *Tropical Plant Pathology*, *45*(3), 251–264. doi:10.100740858-020-00343-y

Battisti, A., Stastny, M., Buffo, E., & Larsson, S. (2006). A rapid altitudinal range expansion in the pine processionary moth produced by the 2003 climatic anomaly. *Global Change Biology*, *12*(4), 662–671. doi:10.1111/j.1365-2486.2006.01124.x

Benziane, T., Abbassi, M., & Jahaj, A. (2001). Comparaison de deux méthodes de lutte intégrée contre les ravageurs en vergers d'agrumes dans la région du gharb. Symp.Prot.Int.Cul.Rég.Mé., 257–269.

Benziane, T., Abbassi, T., & Bihi, M. (2003). Evaluation de deux méthodes de lutte intégrée contre les ravageurs en vergers d'agrumes. *Journal of Applied Entomology*, *127*(1), 51–63. doi:10.1046/j.1439-0418.2003.00713.x

Blank, R. H., & Gill, G. S. C. (1997). Citrus thrips in New Zealand. *Proceedings of International Society of Citriculture. South Africa*, *1*, 515–518.

Boussaade, L. (2003). *Etude comparative des principaux ravageurs des agrumes et leurs ennemis naturels en vergers bio, sous protection intégrée et conventionnelle* [Thesis]. Mediterranean Agronomic Institute Bari (IAMB).

Bouvet, J. P. R., Urbaneja, A., Pérez-Hedo, M., & Monzó, C. (2019). Contribution of predation to the biological control of a key herbivorous pest in citrus agroecosystems. *Journal of Animal Ecology*, *88*(6), 915–926. doi:10.1111/1365-2656.12982 PMID:30895609

Buckley, L. B. (2022). Temperature-sensitive development shapes insect phenological responses to climate change. *Current Opinion in Insect Science*, *52*, 100897. doi:10.1016/j.cois.2022.100897 PMID:35257968

Cammell, M. E., & Knight, J. D. (1991). Effects of climate change on the population dynamics of Crop pests. *Advances in Ecological Research*, *22*, 117–162. doi:10.1016/S0065-2504(08)60135-X

Carroll, A. L., Taylor, S. W., Regniere, J., & Safranyik, D. L. (2004). Effects of climate change on range expansion by the mountain pine beetle in British Columbia. In *Mountain Pine Beetle Symposium: Challenges and Solutions*. Natural Resources Canada, Canadian Forest Service, Pacific Forestry Centre.

Cavalcanti, S. C., Niculau, E. S., Blank, A. F., Câmara, C. A., Araújo, I. N., & Alves, P. B. (2010). Composition and acaricidal activity of *Lippia sidoides* essential oil against two-spotted spider mite (*Tetranychus urticae* Koch). *Bioresource Technology*, *101*(2), 829–832. doi:10.1016/j.biortech.2009.08.053 PMID:19758799

Chapman, R. F. (1998). *The Insects-Structure and Function* (4th ed.). Cambridge University Press. doi:10.1017/CBO9780511818202

Cleland, E. E., Chuine, I., Menzel, A., Mooney, H. A., & Schwartz, M. D. (2007). Shifting plant phenology in response to global change. *Trends in Ecology & Evolution*, *22*(7), 357–365. doi:10.1016/j.tree.2007.04.003 PMID:17478009

Conti, F., Tumminelli, R., Amico, C., Fisicaro, R., Frittitta, C., Perrotta, G., & Marullo, R. (2001). Monitoring of *Pezothrips kellyanus* on citrus in eastern Sicily. In *Proceedings of the 7th International Symposium on Thysanoptera*. University of Reggio Calabria.

Costanza, R., d'Arge, R., de Groot, R., Farber, S., Grasso, M., Hannon, B., Limburg, K., Naeem, S., O'Neill, R. V., Paruelo, J., Raskin, R. G., Sutton, P., & van den Belt, M. (1987). The Value of the World's Ecosystem Services and Natural Capital. *Nature*, *387*(6630), 253–260. doi:10.1038/387253a0

Coviella, C. E., & Trumble, J. T. (1999). Effects of elevated atmospheric carbon dioxide on insect-plant interactions. *Conservation Biology*, *13*(4), 700–712. doi:10.1046/j.1523-1739.1999.98267.x

Delabie, J. H. C., Serrao, J. E., Mariano, C. S. F., & Matile-Ferrero, D. (2008). Comportements de communication de la cochenille neotropicale *Neochavesia caldasiae* (Balachowsky 1957) (*Pseudococcidae*: Rhizoecinae) avec sa fourmi symbiote *Acropyga fuhrmanni* (Forel 1914) (Formicidae: Formicinae). *Annales de la Société Entomologique de France*, *44*(4), 471–475. doi:10.1080/00379271.2008.10697582

Diffenbaugh, N. S., Krupke, C. H., White, M. A., & Alexander, C. E. (2008). Global warming presents new challenges for maize pest management. *Environmental Research Letters*, *3*(4), 1–9. doi:10.1088/1748-9326/3/4/044007

Fleming, R. A., & Volney, W. J. (1995). Effects of climate change on insect defoliator population processes in Canada's boreal forests: Some plausible scenarios. *Water, Air, and Soil Pollution*, *82*(1-2), 445–454. doi:10.1007/BF01182854

Ghalbane, I., Alahyane, H., Aboussaid, H., Chouikh, N. E., Costa, J., Romane, A., & El Messoussi, S. (2022). Chemical composition and insecticidal properties of Moroccan *Lavandula dentata* and *Lavandula stoechas* essential oils against Mediterranean Fruit Fly, *Ceratitis capitata*. *Neotropical Entomology*, *51*(4), 628–636. doi:10.100713744-022-00975-6 PMID:35737225

Gordo, O., & Sanz, J. J. (2005). Phenology and climate change: A long-term study in a Mediterranean locality. *Oecologia*, *146*(3), 484–495. doi:10.100700442-005-0240-z PMID:16170564

Gregory, P. J., Johnson, S. N., Newton, A. C., & Ingram, J. S. I. (2009). Integrating pests and pathogens into the climate change/food security debate. *Journal of Experimental Botany*, *60*(10), 2827–2838. doi:10.1093/jxb/erp080 PMID:19380424

Hajek, A. E. (2004). Introduction to biological control. In A. E. Hajek Book (Ed.), *Natural Enemies: An Introduction to Biological Control* (pp. 19–38). Cambridge University Press. doi:10.1017/CBO9780511811838.003

Haokip, S. W., Shankar, K., & Lalrinngheta, J. (2020). Climate change and its impact on fruit crops. *Journal of Pharmacognosy and Phytochemistry*, *9*(1), 435–438.

Harrington, R., Clark, S. J., Weltham, S. J., Virrier, P. J., Denhol, C. H., Hullé, M., Maurice, D., Roun-sevell, M. D., & Cocu, N. (2007). Environmental change and the phenology of European aphids. *Global Change Biology*, *13*(8), 1556–1565. doi:10.1111/j.1365-2486.2007.01394.x

Holt, J. G., Krieg, R. N., Sneath, A. P. H., Staley, J. T., & Williams, S. T. (2000). In W. R. Hensyl (Ed.), *Bergey's Manual of Determinative Bacteriology* (9th ed., pp. 559–562). William and Willkins.

Hull, M., D'Acier, A. C., Bankhead-Dronnet, S., & Harrington, R. (2010). Aphids in the face of global changes. *Comptes Rendus Biologies*, *333*(6-7), 497–503. doi:10.1016/j.crvi.2010.03.005 PMID:20541161

IMD. (2010). *Annual Climate Summary. India Meteorological Department*. Government of India, Ministry of Earth Sciences.

IPCC. (2007). Climate Change-Impacts, Adaptation and Vulnerability. *Cambridge University Press*.

Irvin, N. A., Pierce, C., & Hoddle, M. S. (2021). Evaluating the potential of flowering plants for enhancing predatory hoverflies (Syrphidae) for biological control of *Diaphorina citri* (Liviidae) in California. *Biological Control, 157*, 104574. doi:10.1016/j.biocontrol.2021.104574

Jactel, H., Koricheva, J., & Castagneyrol, B. (2019). Responses of forest insect pests to climate change: Not so simple. *Current Opinion in Insect Science, 35*, 103–108. doi:10.1016/j.cois.2019.07.010 PMID:31454625

Jaouad, M., Moinina, A., Ezrari, S., & Lahlali, R. (2020). Key pests and diseases of citrus trees with emphasis on root rot diseases: An overview. *Moroccan Journal of Agricultural Sciences, 1*(3).

Kannan, R., & James, D. A. (2009). Effects of climate change on global diversity: A review of key literature. *Tropical Ecology, 50*, 31–39.

Khan, Z. H., Khan, A. A., Ahmad, S. B., Dar, S., & Khan, Z. H. (2011). *Biopesticides, its prospects and limitations: An Overview Shahnawaz a biodiversity and management of aphid in temperate horticultural ecosystem of Kashmir View project Mass Production and Supply of bio-agent for the management of horticultural Pests View project*. Division of Entomology, Sher-eKashmir University of Agricultural Sciences and Technology of Kashmir, Shalimar Campus.

Kumar, S., & Archana, S. (2015). Biopesticides: Present status and the future prospects. *Journal of Fertilizer Pesticide, 6*(2), 12. doi:10.4172/2471-2728.1000e129

Kumral, N. A., Çobanoglu, S., & Yalçın, C. (2010). Acaricidal, repellent and oviposition deterrent activities of *Datura stramonium* L. against adult *Tetranychus urticae* (Koch). *Journal of Pest Science, 83*(2), 173–180. doi:10.100710340-009-0284-7

Lewis, T. (1997). Flight and dispersal. In T. Lewis (Ed.), *Thrips as crop pests* (pp. 175–196). CAB International. doi:10.1079/9780851991788.0000

Logan, J. A., Regniere, J., & Powell, J. A. (2003). Assessing the impacts of global warming on forest pest dynamics. *Frontiers in Ecology and the Environment, 1*(3), 130–137. doi:10.1890/1540-9295(2003)001[0130:ATIOGW]2.0.CO;2

Loussert, R. (1987). Les agrumes, Vol 2, production, technique agricole méditerranéenne. Academic Press.

MAPMDREF. (2020). *Statistique du ministère de l'agriculture et de la pêche maritimes, du développement rural et des eaux et forêts*. MAPMDREF.

Marchiori, C. H. (2021). *Study of the ecology of the Tephritidae family (Insecta: Diptera)*. Academic Press.

Mazih, A. (2008). Current situation of Citrus pests and the control methods in use in Morocco. *IOBC/WPRS Bulletin, 38*, 10–16.

Mazih, A. (2011). Citrus IPM in Morocco: Current status. In Proceedings of Meeting of Working Group "Integrated Control in Citrus Fruit Crops" (Agadir, Morocco). *Bulletin, IOBC/WPRS, 62*, 266.

Mazih, A. (2015). Status of citrus IPM in the southern Mediterranean basin Morocco, North Africa. In *ISHS Acta Horticulturae 1065: XII International Citrus Congress*. International Society of Citriculture.

Mazih, A., & Chitoukli, M. S. (2001). Studies on annual population dynamics and damage of citrus leafminer (*Phyllocnistis citrella*) in sous area (Morocco). *Proceedings du symposium sur protection intégrée des cultures dans la région méditerranéenne*, 159-171.

MEA (Millennium Ecosystem Assessment). (2005). *Ecosystems and Human Well-being: Synthesis*. Island Press.

Menéndez, R. (2007). How are insects responding to global warming? *Tijdschrift voor Entomologie*, *150*, 355–365.

Mound, L. A., & Jackman, D. J. (1998). Thrips in the economy and ecology of Australia. *Proceedings of the Sixth Australian Applied Entomological Research Conference*, 472–478.

Mozaffari, G. A. (2022). Climate Change and Its Consequences in Agriculture. *The Nature, Causes, Effects and Mitigation of Climate Change on the Environment*, 83.

Nath, R., & Sikha, D. (2019). Insect pests of citrus and their management. *International Journal of Plant Production*, *12*(2), 188–196. doi:10.15740/HAS/IJPP/12.2/188-196

Navarro, C. C., Aguilar, A., & Marí, F. G. (2008). *Pezothrips Kellyanus, Thrips causante* de daños en frutos de cítricos. *Levante agrícola: Revista internacional de cítricos, troisième trimestre, 299*, 298–303.

Navarro, C. C., Aguilar, A., & Marí, F. G. (2011). Population trend and fruit damage of *Pezothrips kellyanus* in Citrus orchards in Valencia (Spain). Integrated Control in Citrus Fruit Crops. *IOBC/WPRS Bulletin*, *38*, 204–220.

Nawaz, R., Abbasi, N. A., Hafiz, I. A., Khan, M. F., & Khalid, A. (2021). Environmental variables influence the developmental stages of the citrus leafminer, infestation level and mined leaves physiological response of Kinnow mandarin. *Scientific Reports*, *11*(1), 1–20. doi:10.103841598-021-87160-8 PMID:33833311

Newton, A. C., Begg, G., & Swanston, J. S. (2009). Deployment of diversity for enhanced crop function. *Annals of Applied Biology*, *154*(3), 309–322. doi:10.1111/j.1744-7348.2008.00303.x

Ouknin, M., Alahyane, H., Aboussaid, H., Oufdou, K., Messoussi, S. E., Costa, J., & Majidi, L. (2021). Acaricidal properties of essential oils from Moroccan Thyme against Oriental Red mite, *Eutetranychus orientalis* (Klein) (Acari: Tetranychidae). *Journal of Essential Oil-Bearing Plants*, *24*(2), 329–341. doi:10.1080/0972060X.2021.1927854

Parmesan, C., & Yohe, G. (2003). A globally coherent fingerprint of climate change impacts across natural systems. *Nature*, *421*(6918), 37–42. doi:10.1038/nature01286 PMID:12511946

Partel, V., Nunes, L., Stansly, P., & Ampatzidis, Y. (2019). Automated vision-based system for monitoring Asian citrus psyllid in orchards utilizing artificial intelligence. *Computers and Electronics in Agriculture*, *162*, 328–336. doi:10.1016/j.compag.2019.04.022

Pathak, H., Aggarwal, P. K., & Singh, S. D. (2012). *Climate Change Impact, Adaptation and Mitigation in agriculture: Methodology for Assessment and Application Division of Environmental Sciences*. Indian Agricultural Research Institute New Delhi.

Patterson, D. T., Westbrook, J. K., Joyce, R. J. V., Lingren, P. D., & Rogasik, J. (1999). Weeds, insects and diseases. *Climatic Change*, *43*(4), 711–727. doi:10.1023/A:1005549400875

Pérez-Hedo, M., Riahi, C., & Urbaneja, A. (2021). Use of zoophytophagous mirid bugs in horticultural crops: Current challenges and future perspectives. *Pest Management Science*, *77*(1), 33–42. doi:10.1002/ps.6043 PMID:32776672

Petzoldt, C., & Seamann, A. (2010). Climate Change Effects on Insects and Pathogens. *Climate change and Agriculture: Promoting Practical and Profitable responses*, 6-16. Accessed online at http://www.climateandfarming.org/pdfs/ FactSheets/III.2Insects.Pathogens. pdf

Rao, G. G. S. N., Rao, A. V. M. S., & Rao, V. U. M. (2009). Trends in rainfall and temperature in rainfed India in the previous century. In P. K. Aggarwal (Ed.), *Global climate change and Indian Agriculture case studies from ICAR network project* (pp. 71–73). ICAR Publication.

Root, T. L., Price, J. T., Hall, K. R., Schneider, S. H., Rosenzweig, C., & Pounds, J. A. (2003). Fingerprints of global warming on wild animals and plants. *Nature, 421*(6918), 57–60. doi:10.1038/nature01333 PMID:12511952

Roy, D. B., & Sparks, T. H. (2000). Phenology of British butterflies and climate change. *Global Change Biology*, *6*(4), 407–416. doi:10.1046/j.1365-2486.2000.00322.x

Sahraoui, H. (2012). *Influence des pratiques agroécologiques et de la protection phytosanitaire sur les communautés d'acariens Phytoseiidae (Acari : mesostigmata) dans les vergers agrumicoles tunisiens. Thésis of doctorat*. Institut National Agronomique.

Schneider, L., Rebetez, M., & Rasmann, S. (2022). The effect of climate change on invasive crop pests across biomes. *Current Opinion in Insect Science*, *50*, 100895. doi:10.1016/j.cois.2022.100895 PMID:35240333

Sekkat, A. (2008). *Les pucerons des agrumes au Maroc*. Symposium Méditerranéen sur la Protection Phytosanitaire des agrumes, Rabat, Maroc.

Sharma, H. C. (2002). Host plant resistance to insects: principles and practices. In Resources Management in Plant Protection, Vol. 1. Plant Protection Association of India. Rajendranagar.

Sharma, H. C. (2010). *Effect of climate change on IPM in grain legumes*. In Fifth International Food Legumes Research Conference (IFLRC V), and the Seventh European Conference on Grain Legumes (AEP VII), Anatalaya, Turkey.

Sharma, H. C. (2016). *Climate Change vis-à-vis Pest Management*. In Conference on National Priorities in Plant Health Management (Organised by Plant Protection Association of India, NBPGR, Hyderabad), Tirupati.

Sidhu, A. K., & Mehta, H. S. (2008). Role of butterflies in the natural ecosystem with special reference to high altitude (Pangi Valley, Himachal Pradesh). In *Proceedings of International Conference on Climate Change, Biodiversity and Food Security in the South Asian Region*. Punjab State Council for Science and Technology, Chandigarh and United Nations Educational, Scientific and Cultural Organization, New Delhi.

Skendžić, S., Zovko, M., Živković, I. P., Lešić, V., & Lemić, D. (2021). The impact of climate change on agricultural insect pests. *Insects*, *12*(5), 440. doi:10.3390/insects12050440 PMID:34066138

Smaili, C., Affelah, M., & Aarab, A. (2001). *Biologie, écologie et lutte biologique de la mineuse des agrumes Phyllocnistis citrella Lep. (Gracillariidae) sur clémentinier dans la région du Gharb.1ère colloque de la région du Gharb Kénitra*. Academic Press.

Smaili, M. C., Abbassi, M., Boutaleb, J. A., & Blenzar, A. (2013). Richesse spécifique des ennemis naturels associés aux vergers d'agrumes au Maroc: Intérêt et implication pour la lutte biologique. *Bulletin OEPP. EPPO Bulletin. European and Mediterranean Plant Protection Organisation*, *43*(1), 155–166. doi:10.1111/epp.12013

Smaili, M. C., & Wadjinny, J. (2008). Evaluation de la kaolinite comme insecticide naturels contre la cératite des agrumes *Ceraitis capitata* (Dip, Tephritidae) dans la région du Nord du Gharb. *Symposium méditerranéen sur la protection phytosanitaire des agrumes*.

Staton, T., Walters, R. J., Smith, J., Breeze, T. D., & Girling, R. D. (2021). Evaluating a trait-based approach to compare natural enemy and pest communities in agroforestry vs. arable systems. *Ecological Applications*, *31*(4), e02294. doi:10.1002/eap.2294 PMID:33427350

Storch, D., Šímová, I., Smyčka, J., Bohdalková, E., Toszogyova, A., & Okie, J. G. (2022). Biodiversity dynamics in the Anthropocene: How human activities change equilibria of species richness. *Ecography*, *44*(4), 1–19. doi:10.1111/ecog.05778

Sutherst, R. (2000). Climate change and invasive species: a conceptual framework. In H. A. Mooney & R. J. Hobbs (Eds.), *Invasive Species in a Changing World* (pp. 211–240). Island Press.

Tariq, A. M. (2020). Global warming and climate change: Impact on biodiversity, pest management and food security. *Plant Archives*, *20*(2), 110–115.

Thomas, C. D., Cameron, A., Green, R. E., Bakkenes, M., Beaumont, L. J., Collingham, Y. C., Erasmus, B. F. N., Fereira de Sigueira, M., & Grainger, A. (2004). Extinction risk from climate change. *Nature*, *427*(6970), 145–148. doi:10.1038/nature02121 PMID:14712274

Urbaneja, A., Pascual-Ruiz, S., Pina, T., Abad-Moyano, R., Vanaclocha, P., Montón, H., Dembilio, O., Castañera, P., & Jacas, J. A. (2008). Efficacy of five selected acaricides against *Tetranychus urticae* (Acari: Tetranychidae) and their side effects on relevant natural enemies occurring in citrus orchards. *Pest Management Science, 64*(8), 834–842. doi:10.1002/ps.1572 PMID:18383196

Volney, W. J. A., & Fleming, R. A. (2000). Climate change and impacts of boreal forest insects. *Agriculture, Ecosystems & Environment*, *82*(1-3), 283–294. doi:10.1016/S0167-8809(00)00232-2

Ware, A. (1994). The biology and control of citrus leafminer. *Citrus Journal*, *4*(4), 26–28.

Willis, C. G., Ruhfel, B., Primack, R. B., Miller Rushing, A. J., & Davis, C. C. (2008). Phylogenetic patterns of species loss in Thoreau's woods are driven by climate change. *Proceedings of the National Academy of Sciences of the United States of America, 105*(44), 17029–17033. doi:10.1073/pnas.0806446105 PMID:18955707

Wilson, C. G. (1991). Notes on *Phyllocnistis citrella* Stainton (Lepidoptera: Phyllocnistidae) attacking four citrus varieties in Darwin. *Journal of the Australian Entomological Society*, *30*(1), 77–78. doi:10.1111/j.1440-6055.1991.tb02198.x

Yamamura, K., & Kiritani, K. (1998). A simple method to estimate the potential increase in the number of generations under global warming in temperate zones. *Journal Applied Entomology and Zoology*, *33*(2), 289–298. doi:10.1303/aez.33.289

Zvereva, E. L., & Kozlov, M. V. (2010). Responses of terrestrial arthropods to air pollution: A metaanalysis. *Environmental Science and Pollution Research International*, *17*(2), 297–311. doi:10.100711356-009-0138-0 PMID:19319587

Chapter 4
The Impact of Climate Change on Insect Pests Damaging Crops:
How Insect Pests Damage Crops

Nassima Amiri
Faculty of Sciences Semlalia, Cadi Ayyad University, Morocco

Mohammed Yacoubi Khebiza
Cady Ayyad University, Morocco

Mohammed Messouli
Cady Ayyad University, Morocco

ABSTRACT

Climate change affects the physiology of insects by accelerating their metabolism, growth, and reproduction, which increases their density, and ultimately leads to the consumption and infestation of crops damaged, this action, contributes to economic losses, especially in countries where agriculture plays an important role in their economic sector. This chapter aims to identify how the climatic parameters, in particular the rise in temperature that promotes the development of its insect pests, also causes the action of expanding their distribution, and further reduces biological control, especially natural enemies. And also this chapter shows how insects interact with crops to cause agricultural infestations and damage. The control of agricultural pests can only be established through in-depth studies on the impact of climate change on the interaction between agricultural pests and crops and the contribution of strategies that incorporate the environment and physical resources, and social and financial benefits for farmers. Moreover, the application

DOI: 10.4018/978-1-6684-4824-3.ch004

of adaptation practices, such as improved pest management, monitoring of climate and pest populations, and development of integrated models for predicting climate change, agriculture, and pests, are also considered relevant and effective solutions for the pest control.

BACKGROUND

Agriculture is the primary source of food and plays an important role in the economies of countries. In 2018, the representation of the global gross domestic product (GDP) of agriculture is 4% as an economic sector, and 25% in some developing countries among the least developed (Banque Mondiale, Agriculture) and in Morocco, agriculture occupies 15% (GDP) (MAPM, 2015), which reflects its great economic importance.

Agronomic crop productivity is primarily determined by local and regional climatic conditions (Amiri et al., 2021), especially in countries that adopt traditional practices and grow on rainfed land, these climatic conditions can also indirectly influence agricultural insect pests to develop and damage crops.

This chapter shows how agricultural insect pests interact with climate parameters, especially temperature, to develop and damage crops. This chapter also deals with solutions as many measures and actions to apply and adopt to minimize the damage caused by these pests.

This chapter shows how agricultural insect pests interact with climate parameters, especially temperature, to survive and reproduce and damage crops. This chapter also deals with solutions as many measures and actions to apply and adopt to minimize the damage caused by these pests.

INTRODUCTION

Climate change affects several systems such as water resources, at global and regional levels (Ayt Ougougdal et al., 2020) and also productive sectors such as agriculture (Karmaoui, 2019), which is one of the sectors most vulnerable to climate change (Eddoughri et al., 2022), as there is a strong interrelation between agriculture and climate change drivers, which include both meteorological variables (temperature, precipitation) and greenhouse gases (Amiri et al., 2021). Since the climate is a key determinant of agricultural productivity, any change will influence crop growth and yield. Agricultural production may already have been affected by climate change, with significant yield losses estimated in the historical past (Sultan et al., 2019). Average yield reduction is 8% in Africa due to climate change. Climate change is one of the major risks facing developing countries in Africa for which agriculture is a predominant part of the economy (Amiri et al., 2021).

Agro-ecosystems support an enormous diversity and biomass of insect species because agricultural landscapes can provide large insect habitats if managed accordingly (Busse et al., 2021). But agriculture also suffers from the invasion of pests that survive and reproduce at the expense of plant species and this causes serious damage reducing the production of the main crops (Metcalf, 1996; Pimental, 1976; Strickland, 1969) and losses from these insects can occur in the field and during storage (Oerke, 2006).

Since the mid-XXe century, several researchers have estimated global insect losses of 13.8% in 1967 for cop production generally (Strickland, 1969), and 13.0% in 1985 (Pimentel, 1986), and for 2006 global losses were estimated to be between 7.9% and 15.1% for wheat, rice, corn, potatoes, soybeans, and cotton, even with effective controls (Oerke, 2006). Agricultural damage is caused by several factors related to environmental conditions, crop species, farmers' socioeconomic conditions, and the level of technology used (Oliveira et al., 2014).

Several studies revealed that the growth of insect pests in agriculture remains an often-forgotten consequence of global warming (Tonnang et al., 2022; Ma & Ma, 2022; Jasrotia et al., 2021; Schneider et al., 2022; Stengaard Hansen et al., 2012; Outhwaite et al., 2022).

The temperature has a very important role in regulating the physiology and metabolism of insects, and since temperature increases with climate change, it accelerates the metabolic activity of insects and increases their consumption (Parmesan & Yohe, 2003; Tonnang et al., 2022).

Their food, reproduction, and shelter are dependent on agriculture, causing damage to crops already affected by climatic conditions (Le Gall & Behmer, 2014; Dyer et al., 2013). And always looking for favorable conditions for their growth, insect pests expand their geographical distribution (Meynard et al., 2013).

Climatic conditions may also promote the development of insect pests by causing their predators to decrease, and the trophic interaction may be disturbed (Welch & Harwood, 2014).

This Chapter aims to identify how the climatic parameters, in particular, the increase of temperature, favors the invasion of the pests by their development, reproduction, and widening of their geographical distribution, and also tries to understand the interaction between these insects, climate change and crops to propose solutions that can be adopted to reduce the problems of invasions such as pest control, the use of environmentally friendly plant protection products and the contribution of predictive models to control and manage situations properly.

Questions, Controversies and Problems About the Interaction of Insects with Climate Change to Transform to Agricultural Pests

For many countries, especially developed ones, including Morocco, agriculture is a very important economic sector it ranks first (Amiri et al., 2021). So economic development and food security depend on the farming sector. However, climate change is already directly and strongly affecting this agricultural sector and has serious consequences for agricultural production, and again indirectly, crops are damaged by the invasion of insect pests interacting with climate change factors. Given that temperature is the most important environmental factor affecting insect population dynamics, it is expected that global warming promotes their dynamics, development, and expansion as well as changes in their interaction with their predators.

This chapter clarifies how the variation of climatic parameters (significantly increasing temperature) favours the invasion of insects and also explains the interaction of these pests with the variation of climatic parameters to damage crops; finally, this chapter suggests the various solutions be applied to fight against these parasites.

Associations Between Climate Change and Insects

1. Impact of Climate Change on Insect Pests

Change in insect biodiversity is linked to land use change and is also strongly related to climate change (Newbold et al., 2015; Spooner et al., 2018). Decline and renewal of species composition without decline are aspects of this impact on biodiversity; this was shown by several research (Lister & Garcia, 2018; Soroye et al., 2020; Outhwaite et al., 2020; Crossley et al., 2020; Janzen & Hallwachs, 2021) working on this theme. As the climate and land-use changes are expected to increase further their influence over the coming decades in many parts of the world (IPBES secretariat, 2019), their combined impacts on insect population trends are interesting (Oliver & Morecroft, 2014). But the response of insects through climate change may be detrimental by a shorter generation time, acceleration of metabolism to have higher fertility and survival, resulting in increased range expansion and epidemics that will result in an invasion with severe crop damage.

Climate change involves changing precipitation patterns, soil moisture storage, evaporation and runoff (Skendžić et al., 2021). Furthermore, according to the 6th IPCC report, there will be dramatic increases in temperature (IPCC, 2022), the earth will reach a warming limit of 1.5°C and a decrease in precipitation, and for the 5th IPCC report, atmospheric CO2 have increased from 280 to 400 ppm over the last century and could reach 550 ppm by 2050 (IPCC, 2014). Also, the frequency and intensity of extreme droughts have increased (Diffenbaugh et al., 2017) and are expected to increase in the future (Lehner et al., 2017).

The impact of climate change on insect pests is directly on the reproduction, development, survival and dispersal of pests while indirectly affecting the relationships between pests, their environment and other insect species such as natural enemies, competitors, vectors and mutualists (Prakash et al., 2014).

The body temperature of the insects depends on the temperature of the environment, they are poikilothermic organisms, so it is probably that the temperature is the most important environmental factor affecting the behaviour, insect distribution, development and reproduction (Kocmánková et al., 2010). As a consequence, and in a significative way, the dynamics of insect populations may be affected by the main drivers of climate change (increase in atmospheric CO2, increase in temperature and decrease in soil moisture) (Frand et al., 2012). These engines create new ecological niches that allow insects to spread to new geographic areas and move from one region to another, which poses a significant threat to food security for all countries and regions of the world (FAO, 2008).

2. Temperature

a. The pest fruit fly (*Bactrocera tau* (Walker))

The high temperatures due to climate change are among the main factors that affect the growth and reproduction of insects. The species *Bactrocera tau* (Walker) (Diptera: Tephritidae), is the fruit fly that attacks over 80 different fruit and vegetable

Figure 1. Dorsal view of Bactrocera tau
Source: (CABI, 2021)

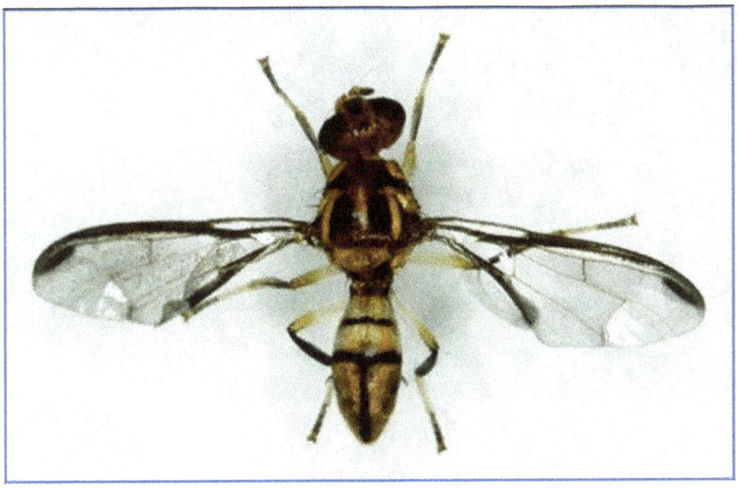

cultivars, for example, pumpkin, luffa, cucumber, melon, eggplant, tomato, mango, orange, papaya, etc., (Jaleel et al., 2018; Huang et al., 2020). Figure 1 shows a dorsal view of this insect. The insect is a major pest of cucurbit plants, deteriorating vegetable and fruit yields with estimated losses of 21-34% and 21-32% of *Siraitia grosvenorii* and *Cucurbita moschata* respectively in Taiwan and about 5.6% of *Luffa acutangula* infestations in Thailand (Huang et al., 2020). The larvae cause most of the damage by feeding inside the host plants.

Today, the pest is of global importance, as it is a serious and economically significant pest that has spread and established itself in many parts of the world (Liu et al., 2022).

b. The fall armyworm (*Spodoptera frugiperda*)

Temperature plays again an essential role in the activity of *Spodoptera frugiperda* since it does not enter diapause, this insect *Spodoptera frugiperda* is of tropical origin that has recently invaded Africa, the Far East, and Australia and is called The fall armyworm (*Spodoptera frugiperda*) (Du Plessis et al., 2020). Figure 2 shows a dorsal view of the caterpillar of this insect.

Figure 2. Dorsal view of Spodoptera frugiperda
Source: (Fall Armyworm, Spodoptera Frugiperda (J.E. Smith))

This caterpillar has a very vague expansion in the world, at the beginning, this species appeared for the first time in Africa (Goergen et al., 2016). Then it started to invade South and Southeast Asia (Early et al., 2018), and in 2018 and 2019, this pest spread to these countries: India, Thailand, China, Republic of Korea (Sharanabasappa et al., 2018; FAO, 2018; FAO, 2019) and recently also in Australia (Allyson Horn, 2020).

c. The Mediterranean citrus fly (*Ceratitis capitata)*

Native to tropical regions of Africa, *Ceratitis capitata* is also found in southern Europe, the Middle East, Central America, the Caribbean, Australia, and parts of Oceania, and is conspicuously absent from colder regions of the world (Malavasi et al. 2000). Dorsal view of the fly *Ceratitis capitate* was shown in Figure 3.

Figure 3. Dorsal view of Ceratitis capitata
Source: (Ecosostenibile, 2021)

Insect growth and development are affected mainly by temperature, which directly affects physiological processes and enzymatic activity (Trudgill et al., 2005). Insects develop more rapidly under fluctuating temperatures when maximum and minimum temperatures are within their optimum range of development (Hagstrum

& Hagstrum, 1970), however, this development may be beneficial to an invasion by insect pests that destroy crops.

Among the large and major agricultural insect pests in the world is the Mediterranean fruit fly *Ceratitis capitata* (Liquido et al., 1990; Papadopoulos et al., 2001; Weldon et al., 2018), which attacks more than 300 host plants, including citrus, tomatoes, figs, and others (Morales et al., 2004; Meats & Smallridge, 2007).

The deterioration of the fruits is caused by the adults who lay their eggs by facilitating also the transmission of bacteria and fungi and the trophic activity of the larvae (Cayol et al., 1994).

3. Expansion of the Geographical Distribution of Insects

According to studies (Ezcurra et al., 1978), insect pest repair has been determined by the following factors: natural biogeography, crop distribution, agricultural practices, climate, trade, and cultural patterns. With climate change, there will be several changes in the environmental factors affecting the repair of insects that always seek favorable conditions for their development, but in the case of insect pests, their displacement will have negative effects on crops by damaging them (Meynard et al., 2013) and farmers will face new and serious pest problems.

Bebber et al. (2013) showed that through climate change, pests are moving poleward by an average of 2.7 ± 0.8 km yr−1 since 1960. Taking the example of the corn borer (*Ostrinia nubilalis Hubner*), this insect has moved more than 1000 km northwards (Porter et al., 1991). In the south of Europe, however, the population of this insect will decrease due to climate change. Parmesan et al. (1999) demonstrated by 35 non-migratory European butterflies, their poleward shifts due to regional warming, 63% have ranges that have moved northward by 35 to 240 km during this century, and only 3% have moved southward (Parmesan et al., 1999).

4. Insect Predators

The natural enemies of agricultural pests can be influenced by climate change, through the alteration of the ability of natural enemies in response to changes in the quality and size of herbivores induced by the effects of temperature and CO_2 on plants (Thomson et al., 2010). Also if there is a displacement of agricultural pests to areas outside their natural enemies, the effectiveness of the control of the latter will decrease, although new enemies may control (Thomson et al., 2010).

Insect pests and their predators respond differently to changes in temperature such as aphids being insect pests and their enemies being parasitic wasps and ladybugs (Harrington et al., 1999; Hullé et al., 2010).

Spahn et al. (2022) demonstrated that higher temperatures reduce parasitoid survival, but not host survival, such as that of Spahn and colleagues who conducted a study on the common agricultural pest *Plutella xylostella* and its larval parasitoid *Diadegma* insular, at higher temperatures, both host and parasitoid developed more rapidly. Host survival was not affected by temperature, but higher temperatures resulted in increased mortality of parasitoids in the pupal stage (Spahn & Lill, 2022).

ASPECTS OF AGRICULTURAL INSECT PEST INTERACTIONS WITH CLIMATE CHANGE

Climate change strongly affects agriculture because plant metabolic processes are controlled by meteorological variables, including temperature, solar radiation, carbon dioxide concentration, and water availability (Chen et al., 2020; Li et al., 2020; Ali et al., 2017; Appiah et al., 2018), particularly heat, which has a significant negative impact on crop yields (Birthal et al., 2021) and several studies (Dell et al., 2012; Lobell et al., 2012; Birthal et al., 2015a; Birthal et al., 2014; Taraz, 2018; Birthal et al., 2015b; Zampieri et al., 2017; Taraz, 2018; Birthal & Hazrana, 2019; Zaveri & B. Lobell, 2019; Word Bank, 2013) showed that excessive temperatures cause more damage to agricultural productivity than insufficient precipitation.

Climatic factors limit food production in various countries and climate change affects the yield of different crops differently (Song et al., 2022), as crop yields are critically dependent on climatic conditions (Saxena et al., 2023).

Agricultural intensification and changes in host plants have been conducive to transforming some herbivorous insects into crop pests (Bernal & Medina, 2018), causing losses and damage that reduce the productivity of any crop species (Oliveira et al., 2014), also aware that other insects play an irreplaceable role in ecosystem services such as pollination, soil decomposition of organic matter or seed dispersal, which are also crucial for agricultural production (Isaacs et al., 2009; Bardenhagen et al., 2020; Dainese et al., 2019; Cardoso et al., 2020). But of all taxonomic groups, insect pests are the most destructive and costly (Diagne et al., 2021; Paini et al., 2016), and the most affected countries have the most significant agricultural production areas in the world (Paini et al., 2016).

Crop pests are strongly influenced by the diverse and complex effects of climate change worldwide (Schneider et al., 2022), warmer temperatures and increased precipitation and CO_2 levels have altered plant communities, phenotypes, and insect interactions (Macfadyen & Bohan, 2010; Piperno, 2006; Piperno et al. 2007; Cotton et al., 2016; Chen et al., 2015; Chen, 2016; Whitehead et al., 2017; Turcotte et al., 2017). The evolution of these parameters is likely to favor or limit the pest species, depending on their ecological context (Schneider et al., 2022). The effect of climate change depends on the specific variable on the interactions between insect pests and crops of the specific variable.

Several studies on the relationship between temperature and population dynamics have shown that temperature influences fecundity, population size, range, and generation time, which tends to accelerate the consumption, development, and movement of insects (Bale et al., 2002). By distinguishing between aerial and soil insects, it is assumed that aerial insects are more responsive to temperature increases than those that spend most of their life cycle in the soil because the soil reduces the impact of temperature on its insects since it is a thermally insulating medium (Bale et al., 2002).

Taking the example of the Mediterranean fly *Ceratitis capitata*, research has been carried out to show the relationship of temperature with population dynamics since temperature is the most crucial factor that increases the developmental rates of most insects (Shoukry & Hafez, 1979; Tassan et al., 1983; Crovetti et al. 1986; Vargas et al., 1984). Additionally, more recent studies have modeled the development time of *Ceratitis capitata* pupae based on thermal summation (Duyck & Quilici, 2002) and determined that at the following constant temperatures: 15, 20, 25, 30, and 35°C (±1°C), 50% of the individuals reach the development time.

With climate change, biological invasions on a global scale have become very frequent (G. Ma & Ma, 2022) with substantial economic costs and serious ecological and social consequences (Diagne et al., 2021; Paini et al., 2016). Pest range shifts by impacts of environmental factors related to climate change are assessed by microclimate use, thermoregulation, life history variation, and evolutionary responses (Bush et al., 2016; Faye et al., 2014; Barton et al., 2019). These processes can be clarified by that Large-scale climate change alters small-scale microclimates during their pest life cycle, microhabitats provide insects with thermal refuges by reducing climate change damping (Ma & Ma, 2022; Suggitt et al., 2018; Ma et al., 2021; Faye et al., 2017). And in hot areas, some pests can develop a great ability to survive at extremely high temperatures (Ma et al., 2014; Mesas et al., 2021; Muñoz-Valencia et al., 2016).

Climate change also influences the natural enemies of agricultural pests. The link between the appearance and abundance of these enemies at times when crops are under pest pressure, and the tolerance of environmental extremes determines the ability of natural enemies to follow their hosts (Thomson et al., 2010). Increased atmospheric temperature and CO_2, and changes in precipitation will alter plant phenology, influencing the growth and abundance of herbivores and indirectly affecting the abundance of prey and hosts of natural enemies (Tonnang et al., 2022). At higher CO_2 conditions, the consumption of cultivated plants may decrease the performance of invertebrate feeders (Zvereva & Kozlov, 2006). A decrease in prey size will not necessarily lead to a reduction in the success of natural enemies, in some cases the number of prey consumed by predators could be increased and lead to better pest control (Coll & Hughes, 2008). The effectiveness of natural enemies may be reduced because the enemies must feed on a lower quality host and in situations where the hosts are more difficult to locate, and also the effectiveness of the enemies may increase if the host is available and easy to manipulate (Thomson et al., 2010). Natural enemies can withstand high temperatures, some studies (Wanderley et al., 2007; Amarendra K. Pandey & Tripathi, 2008) have shown that these natural enemies can often develop over a wide thermal range, and Hance et all (2007) have shown that some enemies can also develop at short-term thermal extremes. Because rising temperatures affect trophic levels by altering the interaction between species (Thomson et al., 2010; Kollberg et al., 2015; Gilman et al., 2010; Abarca & Spahn, 2021). The effects of climate change on natural enemies that are mediated by the effects of CO2, temperature, and moisture on plants can be complex (Tonnang et al., 2022).

Modeling the impact of climate change on crops has been frequently developed by several studies (Asseng et al., 2009; Asseng et al., 2014; Asseng et al., 2015). However, very few of these models take into account the impacts on insect pests, despite their great damage in a direct way through pest infestations and indirectly through pesticide use (Gregory et al., 2009; Oerke, 2006). Some models for quantifying damage caused by insect pests range from empirical functions to mechanistic simulation models (Donatelli et al., 2017; Tonnang et al., 2017).

Pest modeling is further developed in the form of software applications such as the creation of population models by evaluating the life cycle of insects by introducing phenological parameters (development, survival, reproduction) and estimating the size of the pest population independently of the actual growth of the crop (H. E. Z. Tonnang et al., 2015). Other studies on the population dynamics have been carried out by models such as the FRUFLY model for the complete life cycle of the fruit fly (Diptera: Tephritidae) described by El Messoussi et al (2003) based on biotic and abiotic factors and adjusting a limiting factor such as temperature, humidity, parasitism and predation (Messoussi et al., 2007). It is important to draw the attention of researchers to develop models on the impact of climate change on interactions between insect pests and crops specifically and on the importance of including the effects of pests in the design of models of cropping systems to design sustainable pest management strategies to achieve good yields.

SOLUTIONS AND RECOMMENDATIONS

Strategies must be adopted to combat the invasion of agricultural pests. In many of these cases, crop control measures can be implemented, especially with the most modern tools such as phytosanitary products for the types of insects that target certain crops and the introduction of new crop species resistant to pest infestations, to understand the behavior of pests, especially in a changing environment to properly control and manage the constraints of its dynamics. These actions may be drawn up in various fields and aspects:

- Scientific research:

Good visibility for effective pest management requires encouraging scientific research to study the invasion of insect pests by expanding topics in all scientific fields, for example:

- Genetic engineering to have plant species resistant to pest infestations;
- In the field of chemistry, we must produce pesticides that respect the environment and are targeted to the types of pests;
- Developing predictive models that integrate climate change, insect pests, and agriculture, specifying plant and pest species to adequately predict insect interaction with changing climatic conditions to manage pest control interventions;
- Contribution of socioeconomic studies by carrying out surveys with farmers especially those who own small farms to understand the social and physical constraints.
- Grants and Investments:

Farmers must be helped and supported financially, mainly by financial aid for small farms and by public and private investment to improve conditions for farmers to protect their crops.

FUTURE RESEARCH DIRECTIONS

Research should focus on studies of the interaction of insect pests with the variability of climate parameters, to cause severe economic damage by infesting and consuming crops, especially in developing countries that are most vulnerable to climate change, and scientific research must be directed towards:

- Pest management by considering the interaction of insect pests with variations in climate parameters;
- Develop climate models combined with environmental requirements of pests (by pest type and target crop type);

- Use of possible new formulations of effective, powerful insecticides that meet environmental protection standards;
- Establish socio-economic studies to understand the constraints of farmers to manage financial support for pest management.

CONCLUSION

The variability of climate parameters, in particular the increase in temperature caused by climate change, leads to an infestation of pests directly affecting their production, survival, spread, and population dynamics, damaging crops causing significant economic losses, and posing a challenge to human food security. Therefore, the control of insect pests is a necessity by applying pre- and post-harvest power measures adapted to the types of insects and strengthening the physical, social, and financial resources of farmers.

In addition, there is a need for climate and insect pest monitoring strategies and the use of predictive modeling tools and antiparasitic control. Techniques such as trapping, push-pull strategy, and sexual confusion can also be used. It is also necessary to encourage scientific research in its various fields, to have good visibility of the interaction of the pests with the climatic conditions to cause economic damage to control and manage pest control.

FUNDING

This research received no specific grant from any funding agency in the public, commercial, or not-for-profit sectors.

ACKNOWLEDGMENT

This research was supported by the Laboratory of Hydrobiology Ecology and Ecotoxicology and Global Changes, Faculty of Sciences Semlalia, Cadi Ayyad University.

REFERENCES

Abarca, M., & Spahn, R. (2021). Direct and indirect effects of altered temperature regimes and phenological mismatches on insect populations. *Current Opinion in Insect Science*, *47*, 67–74. doi:10.1016/j.cois.2021.04.008 PMID:33989831

Agriculture et alimentation. (n.d.). World Bank. Retrieved 28 November 2022, from https://www.banquemondiale.org/fr/topic/agriculture/overview

Ali, S., Liu, Y., Ishaq, M., Shah, T., Abdullah, Ilyas, A., & Din, I. (2017). Climate Change and Its Impact on the Yield of Major Food Crops: Evidence from Pakistan. *Foods*, *6*(6), 6. Advance online publication. doi:10.3390/foods6060039 PMID:28538704

Horn, A. (2020, February 12). A very hungry caterpillar that decimated crops around the world has arrived in Australia. *ABC News*. https://www.abc.net.au/news/2020-02-12/worm-moth-fall-armywo rm-detected-found-australia-torres-strait/11957838

Amarendra, K. (2008). Effect of temperature on the development, fecundity, progeny sex ratio and life-table of Campoletis chlorideae, an endolarval parasitoid of the pod borer, Helicoverpa armigera. *BioControl*, *53*(3), 461–471. doi:10.100710526-007-9083-3

Amiri, N., Lahlali, R., Amiri, S. E. L., Jarroudi, M., Khebiza, M. Y., & Messouli, M. (2021). Development of an Integrated Model to Assess the Impact of Agricultural Practices and Land Use on Agricultural Production in Morocco under Climate Stress over the Next Twenty Years. *Sustainability*, *13*(21), 21. Advance online publication. doi:10.3390u132111943

Appiah, K., Du, J., & Poku, J. (2018). Causal relationship between agricultural production and carbon dioxide emissions in selected emerging economies. *Environmental Science and Pollution Research International*, *25*(25), 24764–24777. doi:10.100711356-018-2523-z PMID:29926327

Asseng, S., Cao, W., Zhang, W., & Ludwig, F. (2009). Crop Physiology, Modelling and Climate Change: Impact and Adaptation Strategies. In V. Sadras & D. Calderini (Eds.), *Crop Physiology* (pp. 511–543). Academic Press. doi:10.1016/B978-0-12-374431-9.00020-7

Asseng, S., Zhu, Y., Basso, B., Wilson, T., & Cammarano, D. (2014). Simulation Modeling: Applications in Cropping Systems. In N. K. Van Alfen (Ed.), *Encyclopedia of Agriculture and Food Systems* (pp. 102–112). Academic Press. doi:10.1016/B978-0-444-52512-3.00233-3

Asseng, S., Zhu, Y., Wang, E., & Zhang, W. (2015). Crop modeling for climate change impact and adaptation. In V. O. Sadras & D. F. Calderini (Eds.), *Crop Physiology* (2nd ed., pp. 505–546). Academic Press. doi:10.1016/B978-0-12-417104-6.00020-0

Ayt Ougougdal, H., Yacoubi Khebiza, M., Messouli, M., & Lachir, A. (2020). Assessment of Future Water Demand and Supply under IPCC Climate Change and Socio-Economic Scenarios, Using a Combination of Models in Ourika Watershed, High Atlas, Morocco. *Water (Basel)*, *12*(6), 6. Advance online publication. doi:10.3390/w12061751

Bale, J. S., Masters, G. J., Hodkinson, I. D., Awmack, C., Bezemer, T. M., Brown, V. K., Butterfield, J., Buse, A., Coulson, J. C., Farrar, J., Good, J. E. G., Harrington, R., Hartley, S., Jones, T. H., Lindroth, R. L., Press, M. C., Symrnioudis, I., Watt, A. D., & Whittaker, J. B. (2002). Herbivory in global climate change research: Direct effects of rising temperature on insect herbivores. *Global Change Biology*, *8*(1), 1–16. doi:10.1046/j.1365-2486.2002.00451.x

Bardenhagen, C. J., Howard, P. H., & Gray, S. A. (2020). Farmer Mental Models of Biological Pest Control: Associations With Adoption of Conservation Practices in Blueberry and Cherry Orchards. *Frontiers in Sustainable Food Systems*, *4*, 54. doi:10.3389/fsufs.2020.00054

Barton, M. G., Clusella-Trullas, S., & Terblanche, J. S. (2019). Spatial scale, topography and thermoregulatory behaviour interact when modelling species' thermal niches. *Ecography*, *42*(2), 376–389. doi:10.1111/ecog.03655

Bebber, D. P., Ramotowski, M. A. T., & Gurr, S. J. (2013). Crop pests and pathogens move polewards in a warming world. *Nature Climate Change*, *3*(11), 11. Advance online publication. doi:10.1038/nclimate1990

Bernal, J. S., & Medina, R. F. (2018). Agriculture sows pests: How crop domestication, host shifts, and agricultural intensification can create insect pests from herbivores. *Current Opinion in Insect Science*, *26*, 76–81. doi:10.1016/j.cois.2018.01.008 PMID:29764664

Birthal, P. S., & Hazrana, J. (2019). Crop diversification and resilience of agriculture to climatic shocks: Evidence from India. *Agricultural Systems*, *173*, 345–354. doi:10.1016/j.agsy.2019.03.005

Birthal, P. S., Hazrana, J., Negi, D. S., & Bhan, S. C. (2021). Climate change and land-use in Indian agriculture. *Land Use Policy*, *109*, 105652. doi:10.1016/j.landusepol.2021.105652

Birthal, P. S., Negi, D. S., Khan, M., & Agarwal, S. (2015a). Is Indian agriculture becoming resilient to droughts? Evidence from rice production systems. *Food Policy*, *56*, 1–12. doi:10.1016/j.foodpol.2015.07.005

BirthalP. S.NegiD. S.KumarS.AggarwalS.SureshA.KhanT. (2014). How Sensitive Is Indian Agriculture to Climate Change? https://papers.ssrn.com/abstract=3867263

Birthal, P. S., Roy, D., & Negi, D. S. (2015b). Assessing the Impact of Crop Diversification on Farm Poverty in India. *World Development*, *72*, 70–92. doi:10.1016/j.worlddev.2015.02.015

Bush, A., Mokany, K., Catullo, R., Hoffmann, A., Kellermann, V., Sgrò, C., McEvey, S., & Ferrier, S. (2016). Incorporating evolutionary adaptation in species distribution modelling reduces projected vulnerability to climate change. *Ecology Letters*, *19*, 1468–1478.

Busse, M., Zoll, F., Siebert, R., Bartels, A., Bokelmann, A., & Scharschmidt, P. (2021). How farmers think about insects: Perceptions of biodiversity, biodiversity loss and attitudes towards insect-friendly farming practices. *Biodiversity and Conservation*, *30*(11), 3045–3066. doi:10.100710531-021-02235-2

Cardoso, P., Barton, P. S., Birkhofer, K., Chichorro, F., Deacon, C., Fartmann, T., Fukushima, C. S., Gaigher, R., Habel, J. C., Hallmann, C. A., Hill, M. J., Hochkirch, A., Kwak, M. L., Mammola, S., Ari Noriega, J., Orfinger, A. B., Pedraza, F., Pryke, J. S., Roque, F. O., ... Samways, M. J. (2020). Scientists' warning to humanity on insect extinctions. *Biological Conservation*, *242*, 108426. doi:10.1016/j.biocon.2020.108426

CABI. (2021). Bactrocera tau. CABI Compendium. *CABI Compendium*, *8741*. Advance online publication. doi:10.1079/cabicompendium.8741

Cayol, J. P., Causse, R., Louis, C., & Barthes, J. (1994). Medfly Ceratitis capitata Wiedemann (Dipt., Trypetidae) as a rot vector in laboratory conditions. *Journal of Applied Entomology*, *117*(1–5), 338–343. doi:10.1111/j.1439-0418.1994.tb00744.x

Chen, X., Wang, W., Chen, J., Zhu, X., Shen, M., Gan, L., & Cao, X. (2020). Does any phenological event defined by remote sensing deserve particular attention? An examination of spring phenology of winter wheat in Northern China. *Ecological Indicators*, *116*, 106456. doi:10.1016/j.ecolind.2020.106456

Chen, Y. H. (2016). Crop domestication, global human-mediated migration, and the unresolved role of geography in pest control. *Elementa*, *4*, 000106. doi:10.12952/journal.elementa.000106

Chen, Y. H., Gols, R., & Benrey, B. (2015). Crop Domestication and Its Impact on Naturally Selected Trophic Interactions. *Annual Review of Entomology*, *60*(1), 35–58. doi:10.1146/annurev-ento-010814-020601 PMID:25341108

Coll, M., & Hughes, L. (2008). Effects of elevated CO2 on an insect omnivore: A test for nutritional effects mediated by host plants and prey. *Agriculture, Ecosystems & Environment*, *123*(4), 271–279. doi:10.1016/j.agee.2007.06.003

Cotton, J. M., Cerling, T. E., Hoppe, K. A., Mosier, T. M., & Still, C. J. (2016). Climate, CO2, and the history of North American grasses since the Last Glacial Maximum. *Science Advances*, *2*(3), e1501346. doi:10.1126ciadv.1501346 PMID:27051865

Crossley, M. S., Meier, A. R., Baldwin, E. M., Berry, L. L., Crenshaw, L. C., Hartman, G. L., Lagos-Kutz, D., Nichols, D. H., Patel, K., Varriano, S., Snyder, W. E., & Moran, M. D. (2020). No net insect abundance and diversity declines across US Long Term Ecological Research sites. *Nature Ecology & Evolution*, *4*(10), 10. Advance online publication. doi:10.103841559-020-1269-4 PMID:32778751

Crovetti, A., Conti, B., & Delrio, G. (1986). *Effect of abiotic factors on Ceratitis capitata (Wied.) (Diptera: Tephritidae). 2. Pupal development under constant temperatures*. Working Group on Fruit Flies of Economic Importance; CEC/IOBC ad-hoc meeting, Hamburg (Germany, F.R.), 23 Aug 1984. https://scholar.google.com/scholar_lookup?title=Effect+of+abiotic+factors+on+Ceratitis+capitata+%28Wied.%29+%28Diptera%3A+Tephritidae%29.+2.+Pupal+development+under+constant+temperatures&author=Crovetti%2C+A.+%28Pisa+Univ.+%28Italy%29.+Istituto+di+Entomologia+Agraria%29&publication_year=1986

Dainese, M., Martin, E. A., Aizen, M. A., Albrecht, M., Bartomeus, I., Bommarco, R., Carvalheiro, L. G., Chaplin-Kramer, R., Gagic, V., Garibaldi, L. A., Ghazoul, J., Grab, H., Jonsson, M., Karp, D. S., Kennedy, C. M., Kleijn, D., Kremen, C., Landis, D. A., Letourneau, D. K., ... Steffan-Dewenter, I. (2019). A global synthesis reveals biodiversity-mediated benefits for crop production. *Science Advances*, *5*(10), eaax0121. Advance online publication. doi:10.1126ciadv.aax0121 PMID:31663019

Dell, M., Jones, B. F., & Olken, B. A. (2012). Temperature Shocks and Economic Growth: Evidence from the Last Half Century. *American Economic Journal. Macroeconomics*, *4*(3), 66–95. doi:10.1257/mac.4.3.66

Diagne, C., Leroy, B., Vaissière, A.-C., Gozlan, R. E., Roiz, D., Jarić, I., Salles, J.-M., Bradshaw, C. J. A., & Courchamp, F. (2021). High and rising economic costs of biological invasions worldwide. *Nature*, *592*(7855), 7855. Advance online publication. doi:10.103841586-021-03405-6 PMID:33790468

Diffenbaugh, N. S., Singh, D., Mankin, J. S., Horton, D. E., Swain, D. L., Touma, D., Charland, A., Liu, Y., Haugen, M., Tsiang, M., & Rajaratnam, B. (2017). Quantifying the influence of global warming on unprecedented extreme climate events. *Proceedings of the National Academy of Sciences of the United States of America, 114*(19), 4881–4886. doi:10.1073/pnas.1618082114 PMID:28439005

Donatelli, M., Magarey, R. D., Bregaglio, S., Willocquet, L., Whish, J. P. M., & Savary, S. (2017). Modelling the impacts of pests and diseases on agricultural systems. *Agricultural Systems, 155*, 213–224. doi:10.1016/j.agsy.2017.01.019 PMID:28701814

Du Plessis, H., Schlemmer, M.-L., & Van den Berg, J. (2020). The Effect of Temperature on the Development of Spodoptera frugiperda (Lepidoptera: Noctuidae). *Insects, 11*(4), 4. Advance online publication. doi:10.3390/insects11040228 PMID:32272548

Duyck, P. F., & Quilici, S. (2002). Survival and development of different life stages of three Ceratitis spp. (Diptera: Tephritidae) reared at five constant temperatures. *Bulletin of Entomological Research, 92*(6), 461–469. doi:10.1079/BER2002188 PMID:17598297

Dyer, L. A., Richards, L. A., Short, S. A., & Dodson, C. D. (2013). Effects of CO2 and Temperature on Tritrophic Interactions. *PLoS One, 8*(4), e62528. doi:10.1371/journal.pone.0062528 PMID:23638105

Early, R., González-Moreno, P., Murphy, S. T., & Day, R. (2018). *Forecasting the global extent of invasion of the cereal pest Spodoptera frugiperda, the fall armyworm* (p. 391847). bioRxiv. doi:10.1101/391847

Ecosostenibile. (2021, April 7). Ceratitis capitata: Sistematica, Habitat, Ciclo biologico, Lotta. Un Mondo Ecosostenibile. https://antropocene.it/2021/04/07/ceratitis-capitata/

Eddoughri, F., Lkammarte, F. Z., El Jarroudi, M., Lahlali, R., Karmaoui, A., Yacoubi Khebiza, M., & Messouli, M. (2022). Analysis of the Vulnerability of Agriculture to Climate and Anthropogenic Impacts in the Beni Mellal-Khénifra Region, Morocco. *Sustainability, 14*(20), 20. Advance online publication. doi:10.3390u142013166

El Messoussi, S., Lahrouni, A., & Afif, M. (2003). Simulation of the population dynamics of insect pests about biotic and abiotic factors. *Proceedings of the Journée Scientifique Sur la Lutte Biologique*.

El Messoussi, S. E., Hafid, H., Lahrouni, A., & Afif, M. (2007). Simulation of Temperature Effect on the Population Dynamic of the Mediterranean Fruit Fly Ceratitis capitata (Diptera; Tephritidae). *Journal of Agronomy*, *6*(2), 374–377. doi:10.3923/ja.2007.374.377

Ezcurra, E., Rapoport, E. H., & Marino, C. R. (1978). The Geographical Distribution of Insect Pests. *Journal of Biogeography*, *5*(2), 149–157. doi:10.2307/3038169

Fall armyworm, Spodoptera frugiperda (J.E. Smith). (n.d.). Retrieved 23 October 2022, from https://entnemdept.ufl.edu/creatures/field/fall_armyworm.htm

Fand, B. B., Kamble, A. L., & Kumar, M. (2012). Will climate change pose serious threat to crop pest management: A critical review. *International Journal of Scientific Research*, *2*, 1–14.

FAO. (2008). *Climate Related Transboundary Pests and Diseases*. https://www.fao.org/3/a-ai785e.pdf

Faye, E., Herrera, M., Bellomo, L., Silvain, J.-F., & Dangles, O. (2014). Strong Discrepancies between Local Temperature Mapping and Interpolated Climatic Grids in Tropical Mountainous Agricultural Landscapes. *PLoS One*, *9*(8), e105541. https://doi.org/10.1371/journal.pone.0105541

Faye, E., Rebaudo, F., Carpio, C., Herrera, M., & Dangles, O. (2017). Does heterogeneity in crop canopy microclimates matter for pests? Evidence from aerial high-resolution thermography. *Agriculture, Ecosystems & Environment*, *246*, 124–133. https://doi.org/10.1016/j.agee.2017.05.027

First Detection of Fall Armyworm in China. (n.d.). International Plant Protection Convention. Retrieved 23 October 2022, from https://www.ippc.int/fr/news/first-detection-of-fall-armyworm-in-china/

Gilman, S. E., Urban, M. C., Tewksbury, J., Gilchrist, G. W., & Holt, R. D. (2010). A framework for community interactions under climate change. *Trends in Ecology & Evolution*, *25*(6), 325–331. https://doi.org/10.1016/j.tree.2010.03.002

Goergen, G., Kumar, P. L., Sankung, S. B., Togola, A., & Tamò, M. (2016). First Report of Outbreaks of the Fall Armyworm Spodoptera frugiperda (J E Smith) (Lepidoptera, Noctuidae), a New Alien Invasive Pest in West and Central Africa. *PLoS One*, *11*(10), e0165632. https://doi.org/10.1371/journal.pone.0165632

Gregory, P. J., Johnson, S. N., Newton, A. C., & Ingram, J. S. I. (2009). Integrating pests and pathogens into the climate change/food security debate. *Journal of Experimental Botany*, *60*(10), 2827–2838. https://doi.org/10.1093/jxb/erp080

Hagstrum, D. W., & Hagstrum, W. R. (1970). A Simple Device for Producing Fluctuating Temperatures, with an Evaluation of the Ecological Significance of Fluctuating Temperatures1. *Annals of the Entomological Society of America*, *63*(5), 1385–1389. https://doi.org/10.1093/aesa/63.5.1385

Hance, T., van Baaren, J., Vernon, P., & Boivin, G. (2007). Impact of Extreme Temperature on Parasitoids in a Climate Change Perspective. *Annual Review of Entomology*, *52*, 107–126. https://doi.org/10.1146/annurev.ento.52.110405.091333

Harrington, R., Woiwod, I., & Sparks, T. (1999). Climate change and trophic interactions. *Trends in Ecology & Evolution*, *14*(4), 146–150. https://doi.org/10.1016/S0169-5347(99)01604-3

Huang, Y., Gu, X., Peng, X., Tao, M., Chen, G., & Zhang, X. (2020). Effect of short-term high-temperatures on the growth, development and reproduction in the fruit fly, Bactrocera tau (Diptera: Tephritidae). *Scientific Reports*, *10*, 6418. https://doi.org/10.1038/s41598-020-63502-w

Hullé, M., Cœur d'Acier, A., Bankhead-Dronnet, S., & Harrington, R. (2010). Aphids in the face of global changes. *Comptes Rendus Biologies*, *333*(6), 497–503. https://doi.org/10.1016/j.crvi.2010.03.005

IPBES Secretariat. (2019). *Summary for Policymakers of the Global Assessment Report on Biodiversity and Ecosystem Services of the Intergovernmental Science-Policy Platform on Biodiversity and Ecosystem Services*. Author.

IPCC. (2014). *AR5 Synthesis Report: Climate Change 2014*. Retrieved 26 December 2022, from https://www.ipcc.ch/report/ar5/syr/

IPCC. (2022). *AR6 Synthesis Report: Climate Change 2022*. Retrieved 28 November 2022, from https://www.ipcc.ch/report/sixth-assessment-report-cycle/

Isaacs, R., Tuell, J., Fiedler, A., Gardiner, M., & Landis, D. (2009). Maximizing arthropod-mediated ecosystem services in agricultural landscapes: The role of native plants. *Frontiers in Ecology and the Environment*, *7*(4), 196–203. https://doi.org/10.1890/080035

Jaleel, W., Lu, L., & He, Y. (2018). Biology, taxonomy, and IPM strategies of Bactrocera tau Walker and complex species (Diptera; Tephritidae) in Asia: A comprehensive review. *Environmental Science and Pollution Research International, 25*(20), 19346–19361. https://doi.org/10.1007/s11356-018-2306-6

Janzen, D. H., & Hallwachs, W. (2021). To us insectometers, it is clear that insect decline in our Costa Rican tropics is real, so let's be kind to the survivors. *Proceedings of the National Academy of Sciences of the United States of America, 118*(2), e2002546117. https://doi.org/10.1073/pnas.2002546117

Jasrotia, P., Yadav, J., Lal Kashyap, P., Kumar Bhardwaj, A., Kumar, S., & Singh, G. P. (2021). Impact of climate change on insect pests of rice–wheat cropping system: Recent trends and mitigation strategies. In S. Sareen, P. Sharma, C. Singh, P. Jasrotia, G. Pratap Singh, & A. K. Sarial (Eds.), Improving Cereal Productivity Through Climate Smart Practices (pp. 225–239). Woodhead Publishing. https://doi.org/10.1016/B978-0-12-821316-2.00013-3

Karmaoui, A. (2019). Drought and desertification in Moroccan Pre-Sahara, Draa valleys: Exploring from the perspective of young people. *Geoenvironmental Disasters, 6*(1), 2. doi:10.1186/s40677-019-0118-8

Kocmánková, E., Trnka, M., Juroch, J., Dubrovský, M., Semerádová, D., Možný, M., & Žalud, Z. (2010). Impact of climate change on the occurrence and activity of harmful organisms. *Plant Protection Science, 45*(Special Issue), S48–S52. https://doi.org/10.17221/2835-PPS

Kollberg, I., Bylund, H., Jonsson, T., Schmidt, A., Gershenzon, J., & Björkman, C. (2015). Temperature affects insect outbreak risk through tritrophic interactions mediated by plant secondary compounds. *Ecosphere, 6*(6), art102. https://doi.org/10.1890/ES15-000021.1

Latest Pest Reports. (n.d.). International Plant Protection Convention. Retrieved 23 October 2022, from https://www.ippc.int/en/countries/thailand/pestreports/2018/12/first-detection-of-fall-army-worm-on-the-border-of-thaila nd/

Lehner, F., Coats, S., Stocker, T. F., Pendergrass, A. G., Sanderson, B. M., Raible, C. C., & Smerdon, J. E. (2017). Projected drought risk in 1.5°C and 2°C warmer climates. *Geophysical Research Letters, 44*(14), 7419–7428. https://doi.org/10.1002/2017GL074117

Le Gall, M., & Behmer, S. T. (2014). Effects of Protein and Carbohydrate on an Insect Herbivore: The Vista from a Fitness Landscape. *Integrative and Comparative Biology, 54*(5), 942–954. https://doi.org/10.1093/icb/icu102

Lister, B. C., & Garcia, A. (2018). Climate-driven declines in arthropod abundance restructure a rainforest food web. *Proceedings of the National Academy of Sciences of the United States of America*, *115*(44), E10397–E10406. https://doi.org/10.1073/pnas.1722477115

Liu, H., Wang, X., Chen, Z., & Lu, Y. (2022). Characterization of Cold and Heat Tolerance of Bactrocera tau (Walker). *Insects*, *13*(4), 4. https://doi.org/10.3390/insects13040329

Li, Y., Hou, R., & Tao, F. (2020). Interactive effects of different warming levels and tillage managements on winter wheat growth, physiological processes, grain yield and quality in the North China Plain. *Agriculture, Ecosystems & Environment*, *295*, 106923. https://doi.org/10.1016/j.agee.2020.106923

Liquido, N. J., Cunningham, R. T., & Nakagawa, S. (1990). Host Plants of Mediterranean Fruit Fly (Diptera: Tephritidae) on the Island of Hawaii (1949-1985 Survey). *Journal of Economic Entomology*, *83*(5), 1863–1878. https://doi.org/10.1093/jee/83.5.1863

Lobell, D. B., Sibley, A., & Ivan Ortiz-Monasterio, J. (2012). Extreme heat effects on wheat senescence in India. *Nature Climate Change*, *2*(3), 3. https://doi.org/10.1038/nclimate1356

Ma, C.-S., Ma, G., & Pincebourde, S. (2021). Survive a Warming Climate: Insect Responses to Extreme High Temperatures. *Annual Review of Entomology*, *66*(1), 163–184. https://doi.org/10.1146/annurev-ento-041520-074454

Ma, F.-Z., Lü, Z.-C., Wang, R., & Wan, F.-H. (2014). Heritability and Evolutionary Potential in Thermal Tolerance Traits in the Invasive Mediterranean Cryptic Species of Bemisia tabaci (Hemiptera: Aleyrodidae). *PLoS One*, *9*(7), e103279. https://doi.org/10.1371/journal.pone.0103279

Ma, G., & Ma, C.-S. (2022). Potential distribution of invasive crop pests under climate change: Incorporating mitigation responses of insects into prediction models. *Current Opinion in Insect Science*, *49*, 15–21. https://doi.org/10.1016/j.cois.2021.10.006

Malavasi, A., Zucchi, R. A., & Sugayama, R. L. (2000). Biogeografia. In A. Malavasi & R. A. Zucchi (Eds.), *Moscas-das-frutas de importênciaeconômica no Brasil - conhecimento básico e aplicado* (pp. 41–48). Ribeião Preto. https://agris.fao.org/agris-search/search.do?recordID=US2013 00072841

MAPM. (2015). *Investor's guide in the agricultral sector in Morocco.pdf.* Retrieved 22 December 2022, from https://www.agriculture.gov.ma/sites/default/files/investors_guide_in_the_agricultural_sector_in_morocco.pdf

Macfadyen, S., & Bohan, D. A. (2010). Crop domestication and the disruption of species interactions. *Basic and Applied Ecology, 11*(2), 116–125. https://doi.org/10.1016/j.baae.2009.11.008

Meats, A., & Smallridge, C. J. (2007). Short- and long-range dispersal of medfly, Ceratitis capitata (Dipt., Tephritidae), and its invasive potential. *Journal of Applied Entomology, 131*(8), 518–523. https://doi.org/10.1111/j.1439-0418.2007.01168.x

Mesas, A., Jaramillo, A., & Castañeda, L. E. (2021). Experimental evolution on heat tolerance and thermal performance curves under contrasting thermal selection in Drosophila subobscura. *Journal of Evolutionary Biology, 34*(5), 767–778. https://doi.org/10.1111/jeb.13777

Metcalf, R. L. (1996). Applied Entomology in the Twenty-First Century. *American Entomologist (Lanham, Md.), 42*(4), 216–227. https://doi.org/10.1093/ae/42.4.216

Meynard, C. N., Migeon, A., & Navajas, M. (2013). Uncertainties in Predicting Species Distributions under Climate Change: A Case Study Using Tetranychus evansi (Acari: Tetranychidae), a Widespread Agricultural Pest. *PLoS One, 8*(6), e66445. https://doi.org/10.1371/journal.pone.0066445

Morales, P., Cermeli, M., Godoy, F., & Salas, B. (2004). A list of Mediterranean fruit fly Ceratitis capitata Wiedemann (Diptera: Tephritidae) host plants based on the records of INIA-CENIAP Museum of Insects of Agricultural Interest. *Entomotrópica, 19*(1), 51–54.

Muñoz-Valencia, V., González-Obando, R., & Díaz, F. (2016). Rapid responses to a strong experimental selection for heat hardening in the invasive whitefly Bemisia tabaci MEAM 1. *Entomologia Experimentalis et Applicata, 160*(2), 147–155. https://doi.org/10.1111/eea.12466

Newbold, T., Hudson, L. N., Hill, S. L. L., Contu, S., Lysenko, I., Senior, R. A., Börger, L., Bennett, D. J., Choimes, A., Collen, B., Day, J., De Palma, A., Díaz, S., Echeverria-Londoño, S., Edgar, M. J., Feldman, A., Garon, M., Harrison, M. L. K., Alhusseini, T., … Purvis, A. (2015). Global effects of land use on local terrestrial biodiversity. *Nature, 520*(7545). doi:10.1038/nature14324

Oerke, E.-C. (2006). Crop losses to pests. *The Journal of Agricultural Science, 144*(1), 31–43. https://doi.org/10.1017/S0021859605005708

Oliveira, C. M., Auad, A. M., Mendes, S. M., & Frizzas, M. R. (2014). Crop losses and the economic impact of insect pests on Brazilian agriculture. *Crop Protection (Guildford, Surrey)*, *56*, 50–54. https://doi.org/10.1016/j.cropro.2013.10.022

Oliver, T. H., & Morecroft, M. D. (2014). Interactions between climate change and land use change on biodiversity: Attribution problems, risks, and opportunities. *Wiley Interdisciplinary Reviews: Climate Change*, *5*(3), 317–335. https://doi.org/10.1002/wcc.271

Outhwaite, C. L., Gregory, R. D., Chandler, R. E., Collen, B., & Isaac, N. J. B. (2020). Complex long-term biodiversity change among invertebrates, bryophytes and lichens. *Nature Ecology & Evolution*, *4*(3), 3. https://doi.org/10.1038/s41559-020-1111-z

Outhwaite, C. L., McCann, P., & Newbold, T. (2022). Agriculture and climate change are reshaping insect biodiversity worldwide. *Nature*, *605*(7908), 7908. https://doi.org/10.1038/s41586-022-04644-x

Paini, D. R., Sheppard, A. W., Cook, D. C., De Barro, P. J., Worner, S. P., & Thomas, M. B. (2016). Global threat to agriculture from invasive species. *Proceedings of the National Academy of Sciences of the United States of America*, *113*(27), 7575–7579.

Papadopoulos, N. T., Katsoyannos, B. I., Carey, J. R., & Kouloussis, N. A. (2001). Seasonal and Annual Occurrence of the Mediterranean Fruit Fly (Diptera: Tephritidae) in Northern Greece. *Annals of the Entomological Society of America*, *94*(1), 41–50. https://doi.org/10.1603/0013-8746(2001)094[0041:SAAOOT]2.0.CO;2

Parmesan, C., Ryrholm, N., Stefanescu, C., Hill, J. K., Thomas, C. D., Descimon, H., Huntley, B., Kaila, L., Kullberg, J., Tammaru, T., Tennent, W. J., Thomas, J. A., & Warren, M. (1999). Poleward shifts in geographical ranges of butterfly species associated with regional warming. *Nature*, *399*(6736), 6736. https://doi.org/10.1038/21181

Parmesan, C., & Yohe, G. (2003). A globally coherent fingerprint of climate change impacts across natural systems. *Nature*, *421*(6918), 6918. https://doi.org/10.1038/nature01286

Pimental, D. (1976). World Food Crisis: Energy and Pests. *Bulletin of the Entomological Society of America*, *22*(1), 20–26. https://doi.org/10.1093/besa/22.1.20

Pimentel, D. (1986). *Agroecology and economics*. https://scholar.google.com/scholar_lookup?title=Agroecology+and+economics.&author=Pimentel+D.&publication_year=1986

Piperno, D. R. (2006). Quaternary Environmental History and Agricultural Impact on Vegetation in Central America. *Annals of the Missouri Botanical Garden*, *93*(2), 274–296.

Piperno, D. R., Moreno, J. E., Iriarte, J., Holst, I., Lachniet, M., Jones, J. G., Ranere, A. J., & Castanzo, R. (2007). Late Pleistocene and Holocene environmental history of the Iguala Valley, Central Balsas Watershed of Mexico. *Proceedings of the National Academy of Sciences of the United States of America*, *104*(29), 11874–11881. https://doi.org/10.1073/pnas.0703442104

Porter, J. H., Parry, M. L., & Carter, T. R. (1991). The potential effects of climatic change on agricultural insect pests. *Agricultural and Forest Meteorology*, *57*(1), 221–240. https://doi.org/10.1016/0168-1923(91)90088-8

Prakash, A., Rao, J., Mukherjee, A. K., Jeyaveeran, B., Pokhare, S., Adak, T., Munda, S., & R., S. (2014). *Climate Change: Impact on Crop Pests*. Academic Press.

Saxena, R., Joshi, A., Joshi, S., Borkotoky, S., Singh, K., Rai, P. K., Mueed, Z., & Sharma, R. (2023). The role of artificial intelligence strategies to mitigate abiotic stress and climate change in crop production. In A. Srivastav, A. Dubey, A. Kumar, S. Kumar Narang, & M. Ali Khan (Eds.), Visualization Techniques for Climate Change with Machine Learning and Artificial Intelligence (pp. 273–293). Elsevier. https://doi.org/10.1016/B978-0-323-99714-0.00006-6.

Schneider, L., Rebetez, M., & Rasmann, S. (2022). The effect of climate change on invasive crop pests across biomes. *Current Opinion in Insect Science*, *50*, 100895. https://doi.org/10.1016/j.cois.2022.100895

Sharanabasappa, K. C. M., Asokan, R., Swamy, H. M. M., Maruthi, M. S., Pavithra, H. B., Hegbe, K., Navi, S., Prabhu, S. T., & Goergen, G. E. (2018). *First report of the fall armyworm, Spodoptera frugiperda (J E Smith) (Lepidoptera: Noctuidae), an alien invasive pest on maize in India*. https://biblio1.iita.org/handle/20.500.12478/5181

Shoukry, A., & Hafez, M. (1979). Studies on the Biology of the Mediterranean Fruit Fly Ceratitis Capitata. *Entomologia Experimentalis et Applicata*, *26*(1), 33–39. https://doi.org/10.1111/j.1570-7458.1979.tb02894.x

Skendžić, S., Zovko, M., Živković, I. P., Lešić, V., & Lemić, D. (2021). The Impact of Climate Change on Agricultural Insect Pests. *Insects*, *12*(5), 5. https://doi.org/10.3390/insects12050440

Song, Y., Zhang, B., Wang, J., & Kwek, K. (2022). The impact of climate change on China's agricultural green total factor productivity. *Technological Forecasting and Social Change*, *185*, 122054. https://doi.org/10.1016/j.techfore.2022.122054

Soroye, P., Newbold, T., & Kerr, J. (2020). Climate change contributes to widespread declines among bumble bees across continents. *Science, 367*(6478), 685–688. https://doi.org/10.1126/science.aax8591

Spahn, R., & Lill, J. T. (2022). Higher temperatures reduce the efficacy of a key biocontrol parasitoid. *Biological Control, 176*, 105079. https://doi.org/10.1016/j.biocontrol.2022.105079

Spooner, F. E. B., Pearson, R. G., & Freeman, R. (2018). Rapid warming is associated with population decline among terrestrial birds and mammals globally. *Global Change Biology, 24*(10), 4521–4531. https://doi.org/10.1111/gcb.14361

Stengaard Hansen, L., Åkerlund, M., Grøntoft, T., Ryhl-Svendsen, M., Schmidt, A. L., Bergh, J.-E., & Vagn Jensen, K.-M. (2012). Future pest status of an insect pest in museums, Attagenus smirnovi: Distribution and food consumption in relation to climate change. *Journal of Cultural Heritage, 13*(1), 22–27. https://doi.org/10.1016/j.culher.2011.05.005

Strickland, A. H. (1969). Plant Protection and World Crop Production By H. H. Cramer Leverkusen: 'Bayer' Pflanzenschutz (1967), pp. 524. *Experimental Agriculture, 5*(1), 82–83. https://doi.org/10.1017/S0014479700010036

Suggitt, A. J., Wilson, R. J., Isaac, N. J. B., Beale, C. M., Auffret, A. G., August, T., Bennie, J. J., Crick, H. Q. P., Duffield, S., Fox, R., Hopkins, J. J., Macgregor, N. A., Morecroft, M. D., Walker, K. J., & Maclean, I. M. D. (2018). Extinction risk from climate change is reduced by microclimatic buffering. *Nature Climate Change, 8*(8), 8. https://doi.org/10.1038/s41558-018-0231-9

Sultan, B., Defrance, D., & Iizumi, T. (2019). Evidence of crop production losses in West Africa due to historical global warming in two crop models. *Scientific Reports, 9*(1), 1. https://doi.org/10.1038/s41598-019-49167-0

Taraz, V. (2018). Can farmers adapt to higher temperatures? Evidence from India. *World Development, 112*, 205–219. https://doi.org/10.1016/j.worlddev.2018.08.006

Thomson, L. J., Macfadyen, S., & Hoffmann, A. A. (2010). Predicting the effects of climate change on natural enemies of agricultural pests. *Biological Control, 52*(3), 296–306. https://doi.org/10.1016/j.biocontrol.2009.01.022

Tonnang, H. E., Sokame, B. M., Abdel-Rahman, E. M., & Dubois, T. (2022). Measuring and modelling crop yield losses due to invasive insect pests under climate change. *Current Opinion in Insect Science, 50*, 100873. https://doi.org/10.1016/j.cois.2022.100873

Tonnang, H. E. Z., Hervé, B. D. B., Biber-Freudenberger, L., Salifu, D., Subramanian, S., Ngowi, V. B., Guimapi, R. Y. A., Anani, B., Kakmeni, F. M. M., Affognon, H., Niassy, S., Landmann, T., Ndjomatchoua, F. T., Pedro, S. A., Johansson, T., Tanga, C. M., Nana, P., Fiaboe, K. M., Mohamed, S. F., … Borgemeister, C. (2017). Advances in crop insect modelling methods—Towards a whole system approach. *Ecological Modelling, 354*, 88–103. doi:10.1016/j.ecolmodel.2017.03.015

Tonnang, H. E. Z., Sporleder, M., Juarez, H., Carhuapoma, P., & Krosc, J. (2015). Insect Life Cycle Modelling (ILCYM) software—A generic platform for developing insect phenology models, population analysis and risk mapping. *Potato and Sweetpotato in Africa: Transforming the Value Chains for Food and Nutrition Security*, 350–361. doi:10.1079/9781780644202.0350

Trudgill, D., Honek, A., Li, D., & van Straalen, N. M. (2005). Thermal time—Concepts and utility. *Annals of Applied Biology, 146*, 1–14. https://doi.org/10.1111/j.1744-7348.2005.04088.x

Turcotte, M. M., Araki, H., Karp, D. S., Poveda, K., & Whitehead, S. R. (2017). The eco-evolutionary impacts of domestication and agricultural practices on wild species. *Philosophical Transactions of the Royal Society B: Biological Sciences, 372*(1712), 20160033. doi:10.1098/rstb.2016.0033

Turn Down the Heat: Climate Extremes, Regional Impacts, and the Case for Resilience. (n.d.). Retrieved 6 December 2022, from https://elibrary.worldbank.org/doi/abs/10.1596/978-1-4648-0055-9

Vargas, R. I., Miyashita, D., & Nishida, T. (1984). Life history and demographic parameters of three laboratory-reared tephritids (Diptera: Tephritidae). *Annals of the Entomological Society of America, 77*(6), 651–656.

Wanderley, P. A., Ramalho, F. S., Zanuncio, J. C., & Leite, G. L. D. (2007). Thermal requirements and development ofBracon vulgaris, a parasitoid of the cotton boll weevil. *Phytoparasitica, 35*(4), 336–345. https://doi.org/10.1007/BF02980696

Welch, K. D., & Harwood, J. D. (2014). Temporal dynamics of natural enemy–pest interactions in a changing environment. *Biological Control, 75*, 18–27. https://doi.org/10.1016/j.biocontrol.2014.01.004

Weldon, C. W., Nyamukondiwa, C., Karsten, M., Chown, S. L., & Terblanche, J. S. (2018). Geographic variation and plasticity in climate stress resistance among southern African populations of Ceratitis capitata (Wiedemann) (Diptera: Tephritidae). *Scientific Reports, 8*(1), 1. https://doi.org/10.1038/s41598-018-28259-3

Whitehead, S. R., Turcotte, M. M., & Poveda, K. (2017). Domestication impacts on plant–herbivore interactions: A meta-analysis. *Philosophical Transactions of the Royal Society B: Biological Sciences, 372*(1712), 20160034. doi:10.1098/rstb.2016.0034

Zampieri, M., Ceglar, A., Dentener, F., & Toreti, A. (2017). Wheat yield loss attributable to heat waves, drought and water excess at the global, national and subnational scales. *Environmental Research Letters*, *12*(6), 064008. https://doi.org/10.1088/1748-9326/aa723b

Zaveri, E., & Lobell, B. D. (2019). The role of irrigation in changing wheat yields and heat sensitivity in India. *Nature Communications, 10*(1), Article 1. doi:10.1038/s41467-019-12183-9

Zvereva, E. L., & Kozlov, M. V. (2006). Consequences of simultaneous elevation of carbon dioxide and temperature for plant–herbivore interactions: A metaanalysis. *Global Change Biology*, *12*(1), 27–41. https://doi.org/10.1111/j.1365-2486.2005.01086.x

ADDITIONAL READING

Harrington, R., Woiwod, I., & Sparks, T. (1999). Climate change and trophic interactions. *Trends in Ecology & Evolution*, *14*(4), 146–150. doi:10.1016/S0169-5347(99)01604-3 PMID:10322520

Messoussi, S. E., Hafid, H., Lahrouni, A., & Afif, M. (2007). Simulation of Temperature Effect on the Population Dynamic of the Mediterranean Fruit Fly Ceratitis capitata (Diptera; Tephritidae). *Journal of Agronomy*, *6*(2), 374–377. doi:10.3923/ja.2007.374.377

Meynard, C. N., Migeon, A., & Navajas, M. (2013). Uncertainties in Predicting Species Distributions under Climate Change: A Case Study Using Tetranychus evansi (Acari: Tetranychidae), a Widespread Agricultural Pest. *PLoS One*, *8*(6), e66445. doi:10.1371/journal.pone.0066445 PMID:23799105

Papadopoulos, N. T., Katsoyannos, B. I., Carey, J. R., & Kouloussis, N. A. (2001). Seasonal and Annual Occurrence of the Mediterranean Fruit Fly (Diptera: Tephritidae) in Northern Greece. *Annals of the Entomological Society of America*, *94*(1), 41–50. doi:10.1603/0013-8746(2001)094[0041:SAAOOT]2.0.CO;2

Thomson, L. J., Macfadyen, S., & Hoffmann, A. A. (2010). Predicting the effects of climate change on natural enemies of agricultural pests. *Biological Control*, *52*(3), 296–306. doi:10.1016/j.biocontrol.2009.01.022

Welch, K. D., & Harwood, J. D. (2014). Temporal dynamics of natural enemy–pest interactions in a changing environment. *Biological Control*, *75*, 18–27. doi:10.1016/j.biocontrol.2014.01.004

KEY TERMS AND DEFINITIONS

Climate Change: Refers to all the variations in climate characteristics in a given place over time: warming or cooling.

Damage to Crops: Loss of crop potential (reduced yield in quantity and/or quality).

Insect Predators: Insects that feed on and capture live prey. They help regulate the populations of other insects known as pests.

Insects Distribution: Insects geographical distribution.

Temperature: All atmospheric conditions, variable, subjectively translated into relative sensations of hot and cold, the same appreciation of which is provided by the thermometer.

Section 3

Insect, Human, and Environmental Health

Chapter 5
Cutaneous Leishmaniasis Due to *Leishmania tropica*: Investigating Climatic Factors Acting on *Phlebotomus sergenti* Density and Dispersion

Ahmed Karmaoui
https://orcid.org/0000-0003-3881-4029
Moulay Ismail University of Meknes, Morocco & Moroccan Center for Culture and Science, Morocco

Sebastian Cruft
Independent Researcher, UK

Issam Ifaadassan
Cadi Ayyad University, Morocco

Denis Sereno
https://orcid.org/0000-0002-0034-9291
Institut de Recherche pour le Développement, Université de Montpellier, France

Samir El Jaafari
Cluster of Competency on Health and Environment, Moulay Ismail University, Morocco

Hassan Chaachouay
Faculty of Sciences and Techniques of Errachidia, Moulay Ismail University, Morocco

Lhoussain Hajji
https://orcid.org/0000-0002-7785-8991
Bioactives, Health and Environmental Lab, University of Moulay Ismail, Morocco

ABSTRACT

Climate change is suspected to affect health by influencing the distribution and the activity of arthropod insects involved in vector-borne disease transmission, such as Phlebotomine Sandflies. In this chapter, the authors investigate the possible interplay between climatic parameters and the activity of Phlebotomus sergenti, the proven

DOI: 10.4018/978-1-6684-4824-3.ch005

vector of L. tropica causing cutaneous leishmaniasis affections. To this end, they collected data from the literature on the monthly vector density along with selected climatic parameters from three sites belonging to the North African (Morocco), Southern European (Spain), and Western Asian (Iran) areas.

INTRODUCTION

Globally, humans have always exposed to vector borne diseases. The leishmaniasis parasitic disease is among the most spread in topical and also in some subtropical regions. It can be found in Asia, North Africa, and in the Southern side of Europe and is gaining new foci. Unfortunately, the eradication is still far from being completed due to the increasing socioeconomic and climatic impacts. This disease is among the most dangerous health problem, particularly, in the countries of the global south. According to the World health organization, leishmaniasis is caused by protozoan, transmitted by the female of sandflies and can be divided into three types: visceral (the most dangerous), cutaneous (the most popular), and mucocutaneous (WHO, 2022). The leishmaniasis is linked to climatic, land use and land cover changes that influence the distribution of the reservoir hosts and vectors (Karmaoui et al., 2022). Globally, a large number of species of vectors were identified and *phlebotomus papatasi* and *phlebotomus sergenti* are the most frequent and serious. The current study was focused on *P. sergenti,* the principal vector of the *Leishmania tropica.* This vector was identified in many countries, in Iran (Afshar et al., 2017), Morocco (Ajaoud et al., 2015), Turkey (Belen et al., 2011), Portugal (Branco et al., 2013), Saudi Aribia (Doha & Samy, 2010), Italy (Lisi et al., 2014), and Spain (Alten et al., 2016). This insect is sensitive to climatic change including variation in precipitation, temperature, relative humidity, and wind speed. Starting from this fact, an analysis of the relationship between some climatic variables and the seasonal abundance of *P.* sergenti was performed. To explore the associations between this vector and climate, monthly maximum and minimum temperatures (°C), monthly humidity (%), and monthly wind speed (Km/h) were used and correlated in three representative sites of the three regions explored in this study. A comparison in three countries belonging to three different regions from Asia, Europe and Africa was conducted to show the seasonal activity of this vector.

MATERIAL AND METHOD

Study Area

To explore the Link between climate and the activity of *P. sergenti*, three sites were selected from three selected countries from three regions, Morocco from Africa, Spain from Southern Europe, and Iran from Asia (Figure 1 and Table 1).

Figure 1. The selected sites of P. sergenti activity from Spain (Madrid), Morocco (Sefrou), and Iran (Khorramabad)
Source: the authors

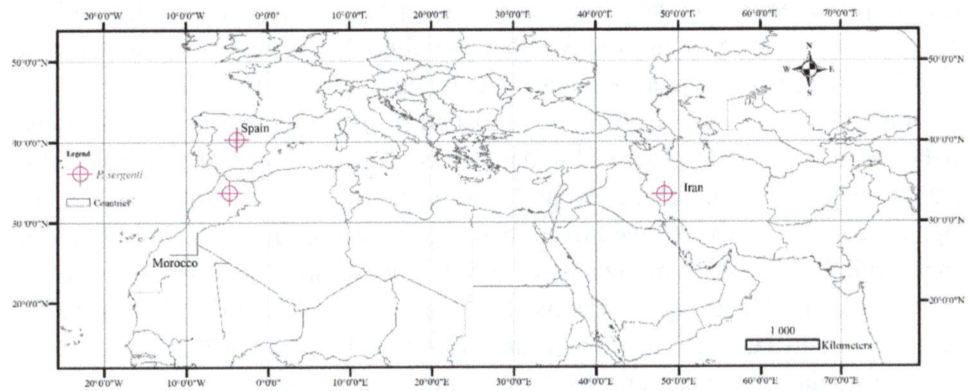

Table 1. The studied sites from three selected countries, Morocco from Africa, Spain from Southern Europe, and Iran from Asia

Country	Site	Latitude	Longitude	Altitude	Year	Reference
Morocco	Sefrou	33°39N	04°38W	809	2012	Talbi et al., (2015a)
Spain	Fuenlabrada, Madrid	40°17'53"N	3°47'31"W	635–691	2011-2012	Alten et al., (2016)
Iran	Khorramabad (Lorestan: KL) Human dwellings	~33°26' N	~48°20' E	~1100-1600	2010	Kayedi et al., (2016)

Source: Based on earlier studies

DATA AND METHOD

Data Collection

Two types of data was exploited, the climatic data, including monthly maximum and minimum temperatures (°C), monthly humidity (%), and monthly wind speed (Km/h), and the data on monthly activity of the sand fly vector. To explore the associations between *P. sergenti* and these climatic data, three representive sites of the three regions were explored in this study. Data on climatic parameters of the main weather stations of the three selected sites were obtained from the online Meteorological database available on Tutiempo Network, S.L.

The three sites include Sefrou in Morocco with a bimodal distribution using Sticky traps collection method in different biotopes, Khorramabad in Iran with a monomodal distribution using sticky traps as a collection method, and Fuenlabrada, Madrid in Spain with a monomodal activity using light and sticky traps.

Collected data on weather were downloaded from the meteorological database and the monthly data of the vector abundance was extracted from three studies (Table 1), and transferred to an excel sheet and then represented in one single figure. The geographic information retrieved from these studies was converted to the shapefile using ArcGIS 10.3.1. A normalization equation (For example, the normalized vector activity in January, see Eq. 1) was used to standardize the compiled data for the three selected sites using as scale that ranges from 0 to 1 (0 refers to the absence of density or no data available, and 1 refers to very high density)."

Table 2. Normalized values of the monthly data of the abundance of P. sergenti

Months	Spain (Fuenlabrada)	Morocco (Sefrou)	Iran (Khorramabad)
January	0.00	0.00	0.00
February	0.00	0.00	0.00
March	0.00	0.00	0.00
April	0.00	0.00	0.00
May	0.00	0.33	0.00
June	0.00	0.56	0.05
July	1.00	1.00	0.33
August	1.00	0.12	1.00
September	0.00	0.46	0.05
October	0.00	0.17	0.00
November	0.00	0.00	0.00
December	0.00	0.00	0.00

Table 3. Monthly data of maximal temperature in the three selected sites

Months	Spain (Madrid)	Morocco (Sefrou)	Iran (Khorramabad)
	Maximal temperature (°C)	Maximal temperature (°C)	Maximal temperature (°C)
January	11.3	21.2	13.0
February	13.4	13.7	13.0
March	16.5	22.4	19.2
April	19.4	18.0	22.1
May	25.6	28.7	23.1
June	30.8	31.7	35.8
July	33.3	34.9	36.6
August	33.9	36.4	38.2
September	28.8	30.1	36.0
October	22.5	23.8	29.3
November	14.6	19.3	20.7
December	11.1	19.1	16.1

Table 4. Monthly data of minimal temperature in the three selected sites

Months	Spain (Madrid)	Morocco (Sefrou)	Iran (Khorramabad)
	Minimal temperature (°C)	Minimal temperature (°C)	Minimal temperature (°C)
January	0.4	-2.0	-0.7
February	-0.5	-1.2	1.8
March	3.7	4.8	3.7
April	8.0	3.7	9.0
May	11.8	2.6	9.7
June	16.1	15.4	19.1
July	17.2	12.5	25.6
August	18.3	14.5	14.2
September	14.4	11.0	9.2
October	9.3	10.5	6.4
November	6.0	7.8	0.0
December	1.0	2.0	-2.5

Table 5. Monthly data of humidity in the three selected sites

	Spain (Madrid)	Morocco (Sefrou)	Iran (Khorramabad)
	Humidity (%)	Humidity (%)	Humidity (%)
January	77.6	63.3	61.0
February	59.8	61.9	65.9
March	59.0	59.2	55.8
April	62.9	72.2	57.0
May	54.9	50.7	53.8
June	38.5	47.7	19.5
July	30.7	39.0	15.0
August	32.1	34.6	17.5
September	41.3	52.6	18.5
October	60.4	66.0	27.3
November	80.3	75.0	40.5
December	80.9	71.8	52.1

Table 6. Monthly data of wind speed in the three selected sites

	Spain (Madrid)	Morocco (Sefrou)	Iran (Khorramabad)
	Wind speed (Km/h)	Wind speed (Km/h)	Wind speed (Km/h)
January	7.1	8.9	7.2
February	11.1	11.2	7.4
March	9.7	9.5	8.9
April	12.3	10.8	8.1
May	9.8	5.5	8.0
June	11.3	5.7	9.5
July	11.5	6.6	13.6
August	11.0	6.4	8.7
September	9.2	11.7	7.5
October	8.2	11.5	7.4
November	8.1	13.5	5.9
December	8.3	10.2	4.7

$$\text{Normalized vector activity in January} = \frac{[\text{Vector activity in January}]}{[\text{Month with maximal vector activity}]} \quad (1)$$

Statistical Analysis

The temporal association between the climatic data and the monthly phlebotomine vector was carried out using Pearson coefficient that allows measuring the direction and strength of correlation between the variables using Equation 2:

$$r = \frac{1}{n-1} \sum \frac{(x_i - \bar{X})(y_i - \bar{Y})}{s_x s_y}$$

In this equation, X and Y refer to the correlated parameters, Sx and Sy, the standard deviations, and $\bar{X}$ and $\bar{Y}$, the mains. The Arab processor in social statistics software (APSS 1.0) is the tool used to conduct the statistical analysis.

RESULTS

Climate and *P. sergenti* Seasonality

Three represerntative sites of the North Africa, South Europe, and West Asia regions were studied to investigate the association of some climatic variables with the seasonal activity of *P. sergenti*. The monthly normalized density of this vector in these three sites from Spain (Madrid), Morocco (Sefrou), and Iran (Khorramabad) were presented (Figure 2a) along with some climatic variables, like the monthly mean Maximum (Figure 2b) and Minimum temperatures in °C (Figure 2c), monthly mean humidity in % (Figure 2d), and monthly mean wind speed in Km/h (Figure 2e). In Spain, Morocco and Iran, the highest *P. sergenti* density occurs in July and/or August when the mean minimum and maximum temperatures were in the range of 17 and 33 °C respectively, mean humidity around 30 to 32%, and the mean wind speed around 11 km/h (Figure 2). Overall, this data highlighted that the highest activity of the vector is linked with the highest maximum and minimum temperatures and the lowest humidity, at least in the sites that were studied in the selected litterature in all sites.

The results showed a high activity of the vector in July in the Moroccan site with a bimodal distribution, in July and August in Spain (monomodal distribution, and in August in Iran (Monomodal repartition).

Figure 2. Phlebotomus sergenti density available climatic variables from three sites representive of the three regions explored in this study; (a) monthly P. sergenti normalized activity; (b and c) mean monthly maximum and minimum temperatures in °C; (d) mean monthly humidity in %; and (e) mean monthly wind speed in Km/h
Source: the authors

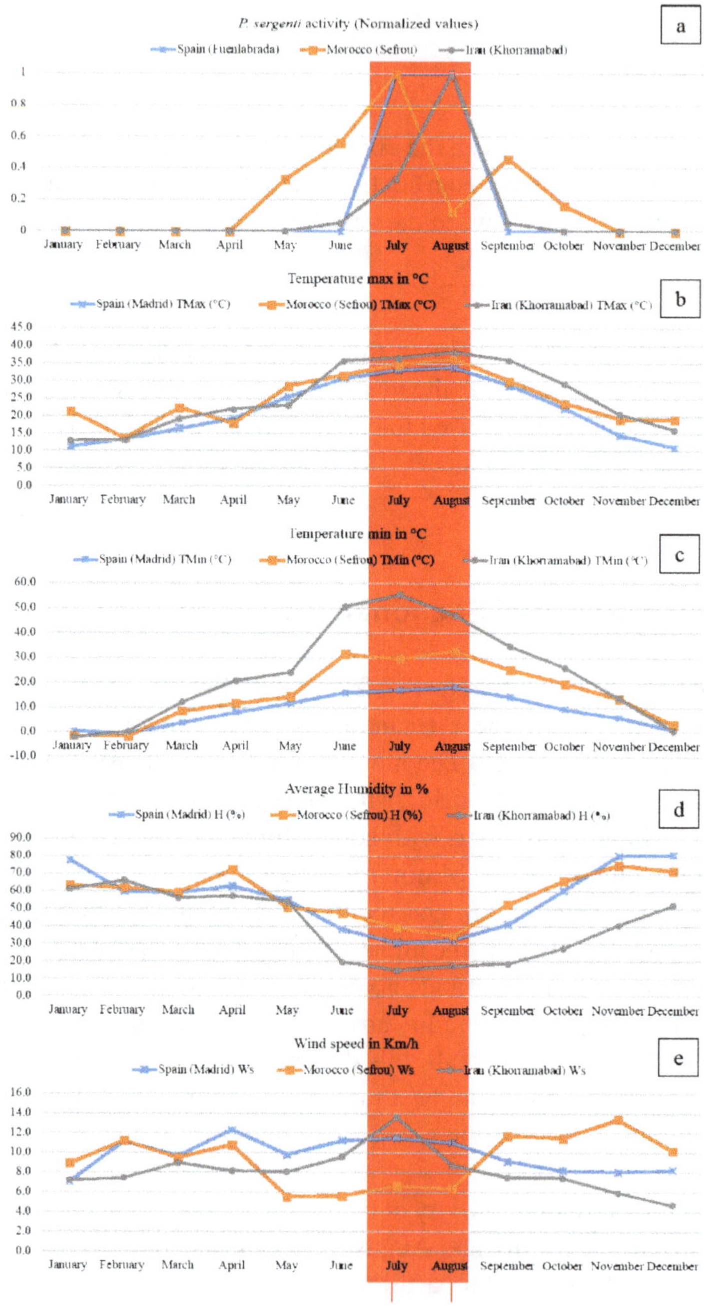

Table 7. Association between the P. sergenti abundance and the climatic parameters. Pearson correlation coefficient at a=0.05 (n=12)

	Maximal temperature	Minimal temperature	Humidity	Wind speed
P. sergenti in Madrid (Spain)	0.64	0.61	-0.66	0.4
	Strong (+)	Strong (+)	Strong (-)	Average
P. sergenti in Sefrou (Morocco	0.73	0.27	-0.67	-0.51
	Strong (+)	Low (+)	Strong (-)	Average (-)
P. sergenti in Khorramabad (Iran)	0.58	0.46	-0.52	-0.35
	Average (+)	Average (+)	Average (-)	Low (-)

Source: the authors

Maximum and minimum temperatures (°C), monthly humidity (%), and monthly wind speed (Km/h), are the climatic parameters correlated with the data on monthly activity of the sand fly vector in three representative countries. The monthly climatic data was standardized and correlated with the normalized monthly abundance of the *P. sergenti* vector. The correlations were carried out using Pearson correlation coefficient at a=0.05. In Madrid (Spain), the findings show a strong positive correlation between the abundance of the vector, the maximal temperature (0.64), and minimal temperature (0.61) (Table 1). A negative strong correlation was observed between the vector activity and humidity (-0.66) while an average positive association was found with the wind speed variable (0.4). Concerning the Moroccan site, the density of the vector has a strong positive correlation with maximal temperature (0.73), and low positive correlation with minimal temperature (0.27). However, the abundance of the vector has negative strong relationship with humidity (-0.67) and negative average correlation with the wind speed (-0.51). In Iran selected site, a positive average correlation was found between the vector activity and the maximal temperature (0.58), positive low correlation with minimal temperature (0.46). However, a negative average correlation was recorded between the vector activity and the humidity (-0.52) in this site while a negative low association was found with the wind speed (-0.35).

Generally, maximum temperatures and humidity have respectivelly strong positive and strong negative correlation with the vector density in two sites (Madrid and Sefrou) and have an average correlation in the third site (Khorramabad). This allows to conclude that the maximum temperatures and humidity can be considered as the most relevant indicators of the vector abundance. However, the minimal temperature showed a significant difference in the statistical relationship (low, average, and strong in each of the three selected sites). For the wind speed variable, the results depict a low to average association with the *P. sergenti* vector (Table 8).

Table 8. Association between the P. sergenti abundance in the three selected sites. Pearson correlation coefficient at a=0.05 (n=12)

	Morocco	*Iran*	*Spain*
Morocco	-	*0.18*	*0.50*
Iran		-	*0.87*
Spain			-

Source: the authors

Most of studies on *P. sergenti* seasonality were performed in countries located between the 30 and 40° of latitude and a large interval of longitude. The highest seasonal density for *P. sergenti* is recorded during hot and dry seasons, and in most of country during July and August, in the latitude 32° (see Table 2).

DISCUSSION

The findings suggest using the maximal temperature and the humidity as high risk factors of the anthroponotic cutaneous leishmaniasis (ACL) disease. The highest phlebotomine density including *P. perniciosus* (73.69%) and *P. sergenti* (6.58%) is recorded in the months with the highest average temperature (25.6°C) and lowest relative humidity (60%) (Branco et al., 2013). The maximum density of sand flies including *Phlebotomus tobbi* (65.6%), *P. perfiliewi galilaeus* (31%), *P. papatasi* (2%), *P. sergenti* (0.1%) were found in months with average monthly temperatures of 23.7–31.2 °C and relative humidity 46.6–66.8% (Kasap et al., 2009). This last study recorded no significant associations between the density of these vectors, temperature (r = 0. 5, p= 0.3) and relative humidity (r = - 0.19, p= 0.7).

Knowing the seasonal activity of arthropod vectors is of paramount importance for the management of leishmaniasis. *P. (Par.) sergenti* that is the proven primary vector of *L. tropica* has high activity in some places can indicate a high risk of leishmaniasis transmission. This may suggest that the findings may be used in control programs on the leishmaniasis transmission. The findings show that the seasonal dynamic of this vector followed, mono or bimodal activity. Likewise, in Morocco, a monomodal distribution of *P.(Par.) sergenti* activity is recorded while *P. (Phl.) papatasi* seasonal density displays a bimodal ones (Boussaa et al., 2005). In the

North West area of Libya (Al Rabta area), high *P. (Par.) sergenti* density spans from August to October (Dokhan et al., 2016). However, in Iran, Abedi-Astaneh et al., (2015) found a bimodal activity of *P. sergenti* in the mountainous region, frequent indoor and outdoor. Reza & Mansour (2006) and Doha & Samy, (2010) identified a higher density of *P. sergenti* outdoors than indoors. In contrast, Veysi et al., (2012) and Yaghoobi-Ershadi et al., (2015) recorded this species only indoors. Regarding the biotope preference in Morocco, *Phlebotomus sergenti* was the most abundant both indoors and outdoors, and the highest density is recorded in outdoor in June (Gijón-Robles et al., 2021). In a global study (Karmaoui, 2020), a high activity of *P. papatasi* was reported indoors in May and June and high abundance outdoors in August with a high preference to indoor than to outdoor. Otherwise, Somwang et al., (2021) advanced that the location of the trap at the entrance of the cave has greater effect to collect the vectors. None or a few *P. sergenti* were found in the plains, including the coastal zone (Kassiri et al., 2013) while in mountainous area from Khorramabad (Iran) the vector was found by Kayedi et al., (2016) and from Qom province by Abedi-Astaneh et al., (2015). However, Reza & Mansour (2006) reported the presence of *P. sergenti* in urban areas in Shiraz City and Boussaa et al., (2005) in Marrakech city. Consequently, the presence and density of the vectors depends on climatic parameters, geographic location, habitat type, the collection method, and the place of traps. The sand fly collection methods (the traps) are the principal tools used to study the abundance of sand fly vectors in a selected site or biotope. In the literature there are various types such as Light traps (LT) and sticky traps (ST) (Sawalha et al., (2017), CO_2 traps, mouth aspirators, human landing collection, animal-baited traps (Kasap et al., (2009), and Knock-down spray catch method (Ali et al., (2016). Studies dealing with the comparison between these sand fly collectors were found in the literature. Ritchie et al., (2008) conducted a comparison between the Centers for Disease Control and Prevention (CDC) light trap and updraft light traps. In this study, the authors found that CDC updraft trap and "the Northern Australia Quarantine Strategy (NAQS) Mozzie Trap" catch notably more mosquitoes than the CDC light trap. In a recent study, Somwang et al., (2021) compared the efficacy of Encephalitis Vector Survey trap (EVS) that showed a lowest performance with CDC Updraft Blacklight trap (CDC UB), Laika trap (LK), and Centers for Disease Control miniature light trap (CDC LT) that have similar performance for sand fly collection. In this study Laika Trap 4.0 was the most appropriate due to less attractive effect to non-target insects, reduced power consumption, and its light weight.

Compared to *P. (Phl.) papatasi*, Karmaoui (2020) noted that high density at a global scale was observed generally in the latitude 33° in August and September. Generally, the two vectors, *P. papatasi* and *P. (Par) sergenti* share close latitudes and the same month, August. Many studies have explored the omnipresence of these vectors in countries not considered in our study. In a research effort using the molecular diagnosis of sandflies Hussain et al., (2017) recorded that these two vectors are abundant in Kohat from Khyber Pakhtunkhwa (Pakistan). Isenring et al., (2018) reported expansive existence of these sandflies in Iraq and Syria. In many villages of Şanlıurfa in southeast Turkey, these two vectors were found to be the most frequent species (Demir and Karakuş, 2015).

The distribution of the vector between the latitudes 30-40° and the significant presence in a large longitude dimension were also found for the *P. papatasi,* as reported by Karmaoui (2020). Attitudinally, *P. sergenti* was found in a mountainous area (Seyedi Rashti & Nadim 1992). However, Doha & Samy, (2010) reported that *P. sergenti* was found in all altitudes, whereas, *P. papatasi* was recorded to the 0-200 m and 800-1200 m, and *P. sergenti* with high density at low elevation. The elevational zonation of *P. sergenti* is influenced by climate variables such as humidity, precipitation, and temperature.

The association between *P. sergenti* and *L. tropica* was reported in many countries in several studies. In Khyber Pakhtunkhwa province (Pakistan), a molecular identification of *L. tropica* was performed by Hussain et al., (2017) in *P. sergenti.* In South-eastern (Iran), the first report of *L. tropica* from *P. sergenti* using high resolution melting curve method was conducted by Aghaei et al., (2014). In Turkey, the first molecular detection of *L. tropica* within naturally infected *P. sergenti* in areas with ACL was investigated by Demir and Karakuş, (2015). However, the *P. sergenti* was identified as *Leishmania killicki* host in south of Algeria (Ghardaïa) (Boubidi et al., 2011). According to Jaouadi et al., (2012) *L. killicki* DNA was detected in *P. sergenti* which indicates the likely role of vector of this parasite while *L. major* DNA was identified in *P. papatasi.*

This disease caused by *L. tropica* is considered as a major public health. *The high P. (Par.) sergenti* activity would be linked with an increased risk of leishmaniasis transmission; our study provides information that can be used for control programs on ACL transmission. The current study has some limitations including the fact that it is based only on three studies. This means that the studied vector is present in many other countries and places not explored in the current study. This raises the need to a large collaboration in the region where the leishmaniasis is present to include more sites in several countries. This can be supervised by an international coalition on vector-borne-diseases with the aims to standardize the methods and construct a database about the diseases cases, parasites, vectors, and reservoirs. Studies about the correlation between climatic parameters and the vectors are required to show the behavior of this vector with the change in climate. Predictive studies about the vector localization and expansion are of high interest since they can be used fight against the disease in the areas with high possible risk.

CONCLUSION

The current study highlights the seasonal repartition of the *Phlebotomus (Paraphlebotomus) sergenti,* the principal vector that transmits anthroponotic cutaneous leishmaniasis (ACL). The identification of high density in some regions may indicate a high risk of pathogen transmission. In this chapter, an analysis of association between the monthly repartition of *P. sergenti* and some climatic parameters. The findings indicate that in selected sites, the *P. sergenti* peaks in the hot and dry season, mainly in July and August. The analysis depicts that the high activity of this vector can indicate a high risk of leishmaniasis transmission. The current study suggests that the findings may be used in control programs on the leishmaniasis transmission.

REFERENCES

Afshar, A. A., Parizi, M. H., Sharifi, I., Gorouhi, M. A., Sharifi, F., & Baafi, B. (2017). Evaluation of the ecological characteristics in the vector of anthroponotic cutaneous leishmaniasis in a new focus of Mohammad Abad, Kerman, southeast of Iran. *Asian Pacific Journal of Tropical Disease*, *7*(2), 84–87. doi:10.12980/apjtd.7.2017D6-354

Aghaei, A. A., Rassi, Y., Sharifi, I., Vatandoost, H., Mollaie, H. R., Oshaghi, M. A., Abai, M. R., & Rafizadeh, S. (2014). First report on natural Leishmania infection of Phlebotomus sergenti due Leishmania tropica by high resolution melting curve method in South-eastern Iran. *Asian Pacific Journal of Tropical Medicine*, *7*(2), 93–96. doi:10.1016/S1995-7645(14)60002-X PMID:24461520

Ajaoud, M., Es-sette, N., Hamdi, S., El-Idrissi, A. L., Riyad, M., & Lemrani, M. (2013). Detection and molecular typing of Leishmania tropica from Phlebotomus sergenti and lesions of cutaneous lesihmaniasis in an emerging focus of Morocco. *Parasites & Vectors*, *6*(1), 217. doi:10.1186/1756-3305-6-217 PMID:23890256

Ali, N., Ullah, A., Wahid, S., Khisroon, M., & Rasheed, S. B. (2016). Distribution, species composition and relative abundances of sandflies in North Waziristan Agency, Pakistan. *Medical and Veterinary Entomology*, *30*(1), 89–94. doi:10.1111/mve.12141 PMID:26582763

Alten, B., Maia, C., Afonso, M. O., Campino, L., Jiménez, M., González, E., Molina, R., Bañuls, A. L., Prudhomme, J., Vergnes, B., Toty, C., Cassan, C., Rahola, N., Thierry, M., Sereno, D., Bongiorno, G., Bianchi, R., Khoury, C., Tsirigotakis, N., ... Gradoni, L. (2016). Seasonal Dynamics of Phlebotomine Sand Fly Species Proven Vectors of Mediterranean Leishmaniasis Caused by Leishmania infantum. *PLoS Neglected Tropical Diseases*, *10*(2), e0004458. doi:10.1371/journal.pntd.0004458 PMID:26900688

Belen, A., & Alten, B. (2011). Seasonal dynamics and altitudinal distributions of sand fly (Diptera: Psychodidae) populations in a cutaneous leishmaniasis endemic area of the Cukurova region of Turkey. *Journal of Vector Ecology*, *36*(s1), S87–S94. Advance online publication. doi:10.1111/j.1948-7134.2011.00116.x PMID:21366785

Boubidi, S., Benallal, K., Boudrissa, A., Bouiba, L., Bouchareb, B., Garni, R., Bouratbine, A., Ravel, C., Dvorak, V., Votypka, J., Volf, P., & Harrat, Z. (2011). Phlebotomus sergenti (Parrot, 1917) identified as Leishmania killicki host in Ghardaia, south Algeria. *Microbes and Infection*, *13*(7), 691–696. doi:10.1016/j.micinf.2011.02.008 PMID:21382502

Branco, S., Alves-Pires, C., Maia, C., Cortes, S., Cristovão, J. M. S., Gonçalves, L., Campino, L., & Afonso, M. O. (2013). Entomological and ecological studies in a new potential zoonotic leishmaniasis focus in Torres Novas municipality, Central Region, Portugal. *Acta Tropica*, *125*(3), 339–348. doi:10.1016/j.actatropica.2012.12.008 PMID:23262215

Demir, S., & Karakuş, M. (2015). Natural Leishmania infection of Phlebotomus sergenti (Diptera: Phlebotominae) in an endemic focus of cutaneous leishmaniasis in Şanlıurfa, Turkey. *Acta Tropica*, *149*, 45–48. doi:10.1016/j.actatropica.2015.05.010 PMID:25997884

Doha, S. A., & Samy, A. M. (2010). Bionomics of phlebotomine sand flies (Diptera: Psychodidae) in the province of Al-Baha, Saudi Arabia. *Memorias do Instituto Oswaldo Cruz*, *105*(7), 850–856. doi:10.1590/S0074-02762010000700002 PMID:21120352

Hussain, M., Munir, S., Jamal, M. A., Ayaz, S., Akhoundi, M., & Mohamed, K. (2017). Epidemic outbreak of anthroponotic cutaneous leishmaniasis in Kohat District, Khyber Pakhtunkhwa, Pakistan. *Acta Tropica*, *172*, 147–155. doi:10.1016/j.actatropica.2017.04.035 PMID:28476600

Isenring, E., Fehr, J., Gültekin, N., & Schlagenhauf, P. (2018). Infectious disease profiles of Syrian and Eritrean migrants presenting in Europe: A systematic review. *Travel Medicine and Infectious Disease*, *25*, 65–76. doi:10.1016/j.tmaid.2018.04.014 PMID:29702253

Jaouadi, K., Depaquit, J., Haouas, N., Chaara, D., Gorcii, M., Chargui, N., Dedet, J.-P., Pratlong, F., Boubabous, R., & Babba, H. (2012). Twenty-four new human cases of cutaneous leishmaniasis due to Leishmania killicki in Metlaoui, southwestern Tunisia. Probable role of Phlebotomus sergenti in the transmission. *Acta Tropica*, *122*(3), 276–283. doi:10.1016/j.actatropica.2012.01.014 PMID:22306359

Karmaoui, A. (2020). Seasonal Distribution of Phlebotomus papatasi, Vector of Zoonotic Cutaneous Leishmaniasis. *Acta Parasit.* doi:10.2478/s11686-020-00201-6

Karmaoui, A., Sereno, D., Maia, C., Campino, L., El Jaafari, S., Taybi, A., & Hajj, L. (2022). A conceptual model for understanding the zoonotic cutaneous leishmaniasis transmission risk in the Moroccan pre-Saharan area. *Parasite Epidemiology and Control*, *00243*, e00243. Advance online publication. doi:10.1016/j.parepi.2022.e00243 PMID:35243048

Kasap, Ö. E., Belen, A., Kaynas, S., Simsek, F. M., Biler, L., Ata, N., & Alten, B. (2009). *Activity patterns of sand fly (Diptera: Psychodidae) species and comparative performance of different traps in an endemic cutaneous leishmaniasis focus in Cukurova Plain*. Southern Anatolia. doi:10.2754/avb200978020327

Kayedi, M. H., Abdali, N., & Rostami, R. (2016). Seasonal activity of sand flies (Diptera Psychodidae) in Khorramabad County, Lorestan Province, Iran. *Nature and Science*, *14*(1).

Lisi, O., D'Urso, V., Vaccalluzzo, V., Bongiorno, G., Khoury, C., Severini, F., Di Muccio, T., Gramiccia, M., Gradoni, L., & Maroli, M. (2014). Persistence of phlebotomine Leishmania vectors in urban sites of Catania (Sicily, Italy). *Parasites & Vectors*, *7*(1), 560. doi:10.118613071-014-0560-0 PMID:25487039

Ritchie, S. A., Zborowski, P., Banks, D., Walsh, I., & Davis, J. (2008). Efficacy of novel updraft traps for collection of mosquitoes in Cairns, Australia. *Journal of the American Mosquito Control Association*, *24*(4), 520–527. doi:10.2987/5698.1 PMID:19181059

Somwang, P., Khositharattanakool, P., Pathawong, N., Pongsiri, A., Davidson, S. A., & Ponlawat, A. (2021). Field evaluation of four commercial light traps, trap placement, and effect of carbon dioxide on phlebotomine sand fly collection in northern Thailand. *Acta Tropica*, *220*, 105953. doi:10.1016/j.actatropica.2021.105953 PMID:33979638

Talbi, F. Z., El Ouali Lalami, A., Janati Idrissi, A., Sebti, F., & Faraj, C. (2015). Leishmaniasis in central Morocco: Seasonal fluctuations of phlebotomine sand fly in Aichoun locality, from Sefrou province. *Pathology Research International, 2015*, 1–4. doi:10.1155/2015/438749

WHO. (2022). *Leishmaniasis*. 8 January 2022. https://www.who.int/news-room/fact-sheets/detail/leishmanias is

KEY TERMS AND DEFINITIONS

ArcGIS: Refers to a set of geographical and statistical programs that constitute a system that allows the manipulation of geographical information by ensuring the functions of collection, management, organization, and analysis of data. ArcGIS software is used in several fields and disciplines such as geography, geology, climatology, ecology, epidemiology, urban planning, and topography by many sectors including government, business, media, education.

Climate: Is the statistic distribution of the atmospheric parameters including precipitation and temperature, wind speed, and relative humidity in a defined area and period.

Leishmaniasis: Is a vector borne disease caused by a protozoan pathogen and considered in the neglected tropical diseases list.

Normalization: In the current chapter normalization or standardization is a mathematical method to homogenize data to avoid the differences in the variables units.

***P. papatasi*:** Is the primary vector the zoonotic cutaneous leishmaniasis disease (ZCL).

***P. sergenti*:** Is the primary vector of the anthroponotic cutaneous leishmaniasis disease (ACL).

Pearson Correlation: or Pearson's r, frequently used in statistic to measure of linear relationship (association or correlation) between two groups of variables or data, called also Pearson correlation coefficient (PCC) or the bivariate correlation.

Chapter 6
Coleopteran Species in Arid Environments:
Ecological Importance and Damages in the Pre-Sahara of Morocco

Ahmed Karmaoui
https://orcid.org/0000-0003-3881-4029
Moulay Ismail University of Meknes, Morocco & Moroccan Center for Culture and Science, Morocco

Guy Chavanon
Mohammed I University, Ouajda, Morocco

Adil Moumane
https://orcid.org/0000-0003-0296-2679
Ibn Tofail University, Morocco

ABSTRACT

Draa oases are declared as a part of the biosphere reserve of Moroccan oases. These oases are agro-ecosystem dominated by the palm and fed by Draa Wadi, the longest in Morocco. Outside the oasis, the acacia trees and many pre-Saharan herbaceous species expand along the flooded areas and at the edges of temporary waterways. This desert vegetation creates a microclimate that favors the proliferation of a vast invertebrate diversity including Coleopteran species (beetles) that support the soil and flora-fauna ecosystem services. These insects play a very important role in the health of this environment. Indeed, they can perform the functions of ecosystems such as pollination, purification, biological control, and soil fertilization. In this chapter, the authors provide new information and update the list of the Coleopteran species in the Draa Valley. Thirty-one species and 15 families were observed through several entomological trips.

DOI: 10.4018/978-1-6684-4824-3.ch006

INTRODUCTION

In terms of number and diversity, Coleoptera is the most important group of insects representing about a fourth of all living organisms with significant ecological interactions (Peris, 2020). This order is very abundant in the forest ecosystem representing about 40% of insects and 400,000 species exist worldwide (Hong *et al.*, 2017). Several coleopteran species provide valuable ecosystem services including forage fouling, nitrogen volatilization (break down dung to usable form of nitrogen), pest control (decompose dung may reduce parasites) (Losey & Vaughan, 2006), and soil fertilization, which support many ecological functions. For example, subfamily Scarabaeinae (Coleoptera: Scarabaeidae), through manipulating feces assures the bioturbation, parasite suppression, the seed dispersal, nutrient cycling, and plant growth enhancement (Nichols *et al.*, 2008). For example, Carabids are frequently used as indicators (Parisi *et al.*, 2005; Rainio *et al.*, 2003) to monitor the influence of different types of control (Porhajašová & Babošová, 2022; Legrand *et al.*, 2011). In fact, they are used as indicators in habitat alteration (Rainio *et al.*, 2003). In managed area, carabids are less rich and of smaller body size (Blake *et al.*, 1996) and decrease in isolated and fragmented zones (de Vries, 1994). They are affected by the elements like soil, water, and elevation (Eyre & Luff, 1990), are sensitive (decreased) to forest cutting (Niemela *et al.*, 1993), and to growing urbanization (Do & Choi, 2022). The importance of the coleopteran species resides in one of the phases of development whether it is the adult or larval form or in both forms. For example, several groups of coleopteran like scarabeids and carabides live the larval stage in the litter while in the adult stage use soil as a food source (Jeffery *et al.*, 2010). Globally, the coleopteran species are found in several ecosystems and climatic zones including the arid areas such as the Middle Draa Valley (MDV) (Southeastern of Morocco). In this area, the knowledge on coleopteran species is not updated. These valleys belong to the biosphere reserve of Moroccan oases declared by UNESCO in 2000. The main ecosystems are oases dominated by the palm and fed by Draa Wadi and surrounded by steppe environments. These ecosystems create microclimate that favors the proliferation of various families of beetles. In this chapter, several field trips were carried out between 2010 and 2020 to identify the coleopteran species in this region.

MATERIAL AND METHODS

Study Area

Saharan ecosystems are characterized by arborescent trees mainly, *Acacia raddiana, A. ehrenbergiana,* and *Maerua crassifolia.* For other groups, this region homes several species like *Hamada* and *Nucuclaria, Nerium oleander,* and *Chenopodium ambrosoide.* In this Saharan zone, Franchimont & Saadaoui (2001) recorded more than 500 coleopterans, 100 Hymenopters, 50 Lepidopters, 50 Dipters, and 40 Orthopters. In addition to this richness, the Middle Draa Valley (MDV) encompases five sites of ecological interest, namely two RAMSAR sites, a national park and two observatories of the Sahara and the Sahel (The observatory of Oued Mird and Fezouata). These sites meet in the area of the Southern Moroccan Oasis Biosphere Reserve (RBOSM). For the MDV (Figure 1), the periphery of the oasis ecosystems is represented by the same vegetal species and a rich diversity of fauna. The fauna is represented by four amphibians, 33 reptiles, 102 birds, 44 mammals, and 131 plant species grouped in 41 families (HCEFLCD, 2007). A lack of inventory of invertebrate species in this valley was noted, which suggest to update the list of recorded Coleopteran species.

Figure 1. Localization of the study area, the Middle Draa Valley in the southeastern region of Morocco. A, the delimitation of the study area; B, map showing the altitudes of the region
Source: Authors

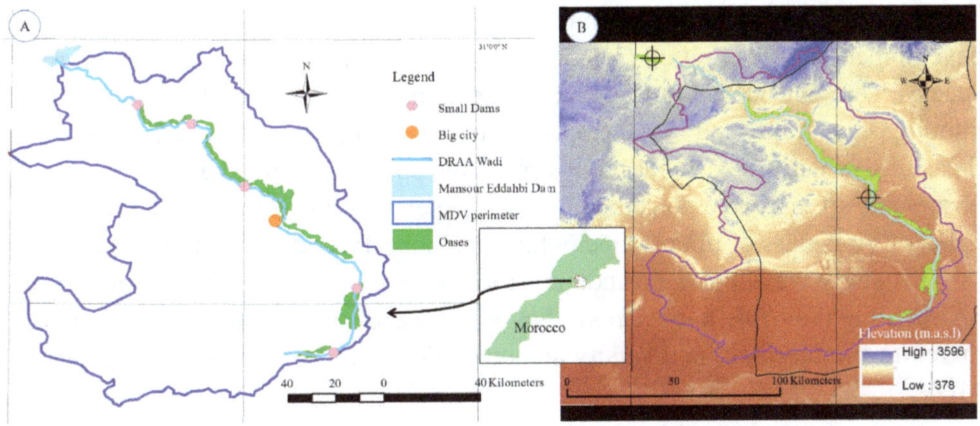

The MDV is located in the southeast of Morocco. It is crossed by several temporary streams. It is characterized by arid climate, with average annual rainfall of less than 100 mm and a sub-desert vegetation. The oasis agroecosystems in this arid valley constitutes a primordial source of livelihood for centuries. Desertification, drought, and floods are the main impact factors that affect these environments and the associated ecosystem services (Karmaoui *et al.,* 2021; Moumane *et al.,* 2021). In this area, in the last decade, little is known about the diversity of invertebrate including the coleopteran species. To our knowledge there is no recent studies dealing with entomological studies, which suggests investigating this group of animals.

Methodology

The field trips were carried out between 2010 and 2020. Several sites were visited as shown in Figure 2. This map shows the exact location of the species found and compiled through the bibliographic study on the Draa Valley. The species found on land were not harvested, but they were photographed. For some species found in the valley (see the photos in figures 3 and 4), sometimes, only the family name is giving because of the non-possibility of species determinations. The following stages were followed in this study:

1. Review of the Coleoptera in MDV
2. Update the list by fields in Draa
3. Determine the ecosystem services associate with each species

The study was carried out in the MDV, an arid area including oasis and steppe ecosystems. The diversified habitats along with climatic conditions favor the development of a diversified Coleoptera species. Both the beneficial and harmful species were considered in this study. The sampling was carried out using sight hunting, light traps, and aerial trap. The micro-biotope of stones was omitted in this study, which suggest that the list observed species is not exhaustive. Only the observable species were considered, the collected species were not observed under a magnifying glass or microscope, which can pose a misidentification issue.

This study contributes to updating the presence of coleopteran diversity and highlighting their role in the ecosystem of oasis and surrounding environments. It therefore summarizes the diversity and ecological status of coleopteran species in MDV.

RESULTS

Identification of Coleopteran Species in the MDV

Several species were found and photographed through the field trips were carried out between 2010 and 2020. A list of 31 potential species of 15 families is provided (Figure 2 and Table 1). The collected species include *Julodis aequinoctialis* (Buprestidae), three indeterminate species of Tenebrionidae, *Tentyriini* (Tenebrionidae), *Sclerum sp.* (Tenebrionidae), *Adesmia biskreensis asperrima* (Tenebrionidae), *Pimelia angulata ssp.* (Tenebrionidae), *Blaps sp.* (Tenebrionidae), *Erodius sp.* (Tenebrionidae), *Attagenus trifasciatus* (Dermestidae), Elateridae, *Paussus thomsonii* (Carabidae), *Chlaenius spoliatus* (Carabidae), *Cleonini* (Curculionidae), *Sinoxylon sexdentatum* (Bostrichidae), *Melolonthini* (Scarabaeidae), *Pentodon sp.* (Scarabaeidae), *Formicomini* (Anthicidae), *Stromatium auratum* (Cerambycidae), *Melyrini* (Melyridae), *Eretes sticticus (s. lat.)* (Dytiscidae); Alleculinae (Tenebrionidae), *Meladema coriacea* (Dytiscidae), *Anthrenus verbasci* (Dermestidae), *Tropinota squalida pilosa* (Scarabaeidae), *Coccinella septempunctata* (Coccinellidae), *Clytrini* (Chrysomelidae), *Scarabaeus sacer* (Scarabaeidae), and two indeterminate species of Mylabrini (Meloidae).

The records from the MDV confirmed the presence of 15 families including Tenebrionidae, Scarabaeidae, Carabidae, Dermestidae, Dytiscidae, Meloidae, Buprestidae, Elateridae, Curculionidae, Bostrichidae, Anthicidae, Cerambycidae, Melyridae, Coccinellidae, and Chrysomelidae. This classification of the observed colepteran species yielded 32% of Tenebrionidae, 13% of Scarabaeidae, and 7% for Carabidae, Dytiscidae, and Meloidae (Figure 3).

Figure 2. Species found in the Middle Draa Valley. 1, Julodis aequinoctialis (Buprestidae); 2, Tenebrionidae; 3, Tenebrionidae; 4, Tentyriini (Tenebrionidae); 5, Tenebrionidae; 6, Sclerum sp. (Tenebrionidae); 7, Adesmia biskreensis asperrima (Tenebrionidae); 8, Pimelia angulata ssp. (Tenebrionidae); 9, Blaps sp. (Tenebrionidae); 10, Erodius sp. (Tenebrionidae); 11, Attagenus trifasciatus (Dermestidae); 12, Elateridae; 13, Paussus thomsonii (Carabidae);14, Chlaenius spoliatus (Carabidae); 15, Cleonini (Curculionidae); 16, Sinoxylon sexdentatum (Bostrichidae); 17, Melolonthini (Scarabaeidae); 18, Pentodon sp. (Scarabaeidae); 19, Formicomini (Anthicidae); 20, Stromatium auratum (Cerambycidae); 21, Melyrini (Melyridae); 22, Eretes sticticus (s. lat.) (Dytiscidae); 23, Alleculinae (Tenebrionidae); 24, Meladema coriacea (Dytiscidae); 25, Anthrenus verbasci (Dermestidae) ; 26, Tropinota squalida pilosa (Scarabaeidae); 27, Coccinella septempunctata (Coccinellidae); 28, Clytrini (Chrysomelidae) ; 29, Scarabaeus sacer (Scarabaeidae); 30, Julodis aequinoctialis (Buprestidae) ; 31, Mylabrini (Meloidae) ; and 32, Mylabrini (Meloidae)

Source: Authors

Table 1. Brief description, habitat, and ecosystem services associated with the coleopteran species observed in the Middle Draa Valley (see Figure 2); Source: Authors

N°	Species/family	Description	Habitat	Ecosystem service/ deservice
1 and 30	*Julodis aequinoctialis* (Buprestidae)	Fusiform with glossy iridescent colors and live on bark and dead wood	Desert ecosystem	Phytophagous, the larva lives in the roots
2	Tenebrionidae	Cylindrical larvae, Hard and convex integuments usually black,	Desert ecosystem	Detritivorous, Feed on dry or decomposed plant matter
3	Tenebrionidae	Hard and black integuments	Desert ecosystem	Feed on dry or decomposed plant matter
4	*Tentyriini* (Tenebrionidae)	Hard and black integuments	Desert ecosystem and forests	Plant scavengers and dead insect omnivores
5	*Tenebrionidae*	Small brown species	Desert ecosystem	adults and larvae generally predaceous,
6	*Sclerum sp.* (Tenebrionidae)	Rigid cuticle, Brown-black	Arid and semi-arid ecosystem	Omnivorous and saprophagous, dead insects and dry and decomposed food
7	*Adesmia biskreensis asperrima* (Tenebrionidae)	25 mm, black body with long legs and head narrower. Antennae short frequent in arid areas.	Observed in two sites in Mountains and bare land in desert ecosystem,	detritivorous, nutrient cycling using litter in hot and dry periods
8	*Pimelia angulata ssp.* (Tenebrionidae)	Black and void body, gibbous, with the head narrower and short than the abdomen. Antennae short and moniliform at their extremety;	Sand dunes in desert ecosystem Diurnal species	Detritivorous
9	*Blaps sp.* (Tenebrionidae)	Black color and flightless with pointed abdomen present in arid environments and lives in home gardens; Invaded gardens of a village in Zagora in summer of 2022	Desert ecosystem In wet soil, and under stones	Decomposing waste, dead insect, and plant matter
10	*Erodius sp.* (Tenebrionidae)	15 mm, black body.	Observed on sand near banks of temporary river of desert ecosystem	Detritivorous (Desert detritus cycling and prey for several predators)
11	*Attagenus trifasciatus* (Dermestidae)	Dark brown color pest insects measuring 3-5 mm with an oval shape with a hairy body. The adults are able to fly and the larva is hairy cylindrical.	Present near the houses of Beni Zouli Village. Larvae at home (carpets and woolen goods) Adult feeds on nectar	The larvae attacks woolen goods and grains causing extensive damage to carpets
12	Elateridae	Brown beetles, unusual click mechanism. *Nocturnal adults.*	Observed at night near a house attracted by light.	Phytophagous, Larvae in the soil
13	*Paussus thomsonii* (Carabidae)	Brown small beetle Called also ant nest beetles. It measures 8-10 mm with caracterestic-flattened antennae.	Observed at night near a house attracted by light. Myrmecophiles, living within the nests of ants	Predatory on ants

continued on following page

Table 1. continued

N°	Species/family	Description	Habitat	Ecosystem service/ deservice
14	*Chlaenius spoliatus* (Carabidae)	A ground carabidae collected from wet banks of rivers of Draa temporary river.	They shelter under stones or plant debris.	Both adults and larvae hunt various invertebrates.
15	*Cleonini* (Curculionidae)	Flightless with a head extended by a snout, chewing mouthparts, sucking-type rostrum. Antennas bent at right angles. White larvae, legless and curved.	Surrounding area of the oasis	Phytophagous, cause significant damage to various crops.
16	*Sinoxylon sexdentatum* (Bostrichidae)	Generally, the head is downwardly oriented and hidden by the thorax. biting mouthparts and a hard cuticle	Acacia trees	Xylophagous or feed on cereal grains
17	*Melolonthini* (Scarabaeidae)	Brown beetles, the ventral parts of femur and tibiotarsus fossorial setae.	Observed at night near a house (arid ecosystem) attracted by light.	Adults feed on the foliage and flowers. Larvae attacks plant roots.
18	*Pentodon sp.* (Scarabaeidae)	Oval-shaped beetle and black color with a maximum length of 1 cm to 3 cm with a short head glued to the body with two antennae in the form of pincers. Its abdomen composed of several fragments to which are attached strong wings (elytra) that protect the membranous wings.	On the soil and under the stones	Larvae and adults are herbivores
19	*Formicomini* (Anthicidae)	Anthicinae black bown beetle with a long antennae and a small body that resembles to ants. The prothorax much narrower than the elytra.	Attracted by light near houses of a village 17 Km Northwest Zagora Present on plant debris	Predator adult found also on the flowers
20	*Stromatium auratum* (Cerambycidae)	Measures 15-30 mm the body and parts (Wings, legs, antennae) with clear brown color.	The larva lives in dead wood,	Xylophage larvae
21	*Melyrini* (Melyridae)	Called soft-wing flower beetles. Small colepteran Species with elongated and oval red and black body.	In a village 17 Km Northwest Zagora	Adults and larvae are predatory
22	*Eretes sticticus (s. lat.)* (Dytiscidae)	15-20 mm, dark brown and oval body	Present in wet areas and temporary water Collected at light trap near wet site	Predator beetle
23	Alleculidae (Tenebrionidae)	Black beetle with hard and convex integument	Desert ecosystem	Floricolous adult
24	*Meladema coriacea* (Dytiscidae)	Water shiny black beetle (about 20 mm), oval body adapted for aquatic life.	Mediterranean aquatic beetles	Predator
25	*Anthrenus verbasci* (Dermestidae)	Harmful small beetle (1.8 to 3.4mm) covered with scales of brown color with shades.	Near the rural houses	Phytophagous adult and detritivorus larvae

continued on following page

Table 1. continued

N°	Species/family	Description	Habitat	Ecosystem service/ deservice
26	*Tropinota squalida pilosa* (Scarabaeidae)	Dark-brown beetle (10–15 mm), long tawny hair with two series of white spots on elytra.	Mediterranean distribution, associated with the vegetation in the wetland ecosystem of the study area	Adult feeds on flowers and larvae on roots.
27	*Coccinella septempunctata* (Coccinellidae)	Red beetle (5-8mm) with 7 black spots on the elytra. The elytral groove with two triangular white spots and a white spot behind the eyes.	Palaearctic distribution	Aphid predator
28	*Clytrini (Chrysomelidae)*	Orange small beetle with black spots.	Lives on the leaves of trees or herbaceous. Present in wetlands of arid area.	Phytophagous
29	*Scarabaeus sacer* (Scarabaeidae)	Black oval beetle. Edge of the head, flat, is crenellated with six serrations	Mediterranean distribution	Coprophagous adult and larvae
30	*Mylabrini* (Meloidae)	Orange beetle with black spots.	Found on flowers in the arid region	Pollinator, phytophagous adults and feed on leaves and flowers. The larvae are insect parasites
31	Mylabrini (Meloidae)	Black beetle with rectangular head and horizontally elongated red spots.	Lives on the leaves of trees or herbaceous. Present in wetlands of arid area.	Phytophagous adults. The larvae are insect parasites

DISCUSSION

The oasis is an agro-ecosystem that supports a rich biodiversity that is of exceptional importance for the maintenance of certain ecosystem goods and services. These oases provide food, fiber and water. A large number of plant species, including a large number of species used in the local food, are pollinated by animals. Without flying insects as beet, and bees, there will be few fruits and vegetables to consume. This supports high biodiversity, and then a better productivity of biomass. At the sol and subsoil level, biodiversity constitutes a vast genetic reserve including a large number of species that produce and accumulate all the elements necessary for the development (nitrogen, phosphorus…) of the main support of life (the soil). The oasis is characterized by favorable conditions for the biodiversity due the micro humid climate under the palm trees system. The fauna is represented by four amphibians, 33 reptiles, 102 birds, 44 mammals, and 131 plant species grouped in 41 families (HCEFLCD, 2007). Invertebrate species including beetles are also present but not well studied. In fact, a lack of recent inventory of beetles species in this valley was

Figure 3. Percentage of the coleopteran species classified by family of the Middle Draa Valley
Source: Authors

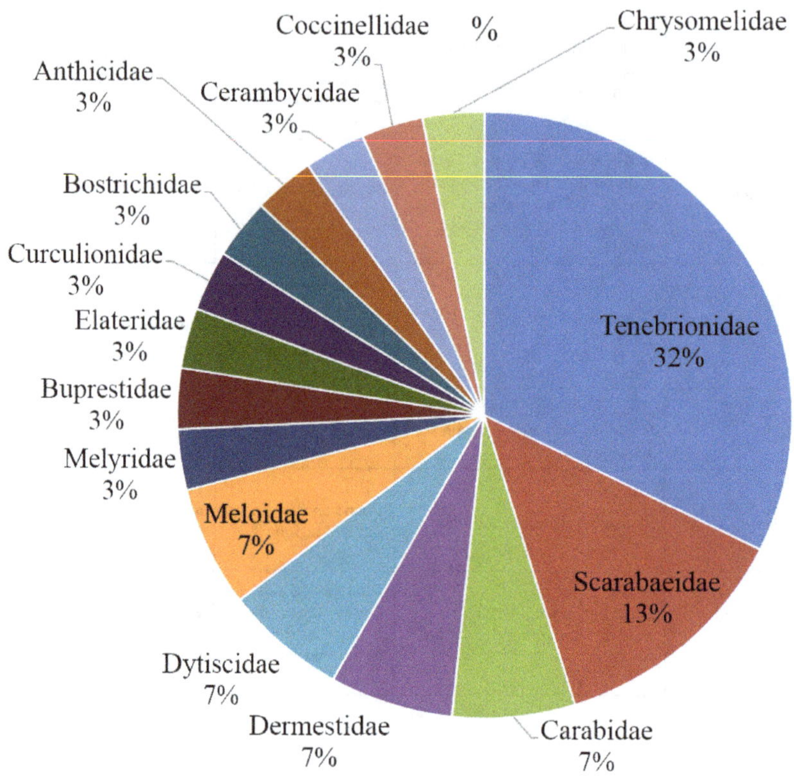

noted, which suggest to draw and update the list of recorded coleopteran species. The choice of this group of animals was based on the ecological importance on soil and underground and in several biotopes. It was reported that have many roles as being organic matter recyclers, herbivores, pollinators, and prey (Peris, 2020) in fly control and suppression of parasites (Bergstrom *et al.,* 1976), and in seed dispersal (Andresen, 2002). They can be used as ecological indicators (Porhajašová & Babošová, 2022), and used in research to explore the impact of environmental disturbances on ecosystem structure because is a sensitive group to habitat perturbations (Cajaiba *et al.,* 2017).

Beetles support the processes of soil fertility (Cheli *et al.,* 2022), which provides several ecosystem services including provisioning, regulating, and supporting services. Particularly, dung beetles are important ecosystem service suppliers, making them significant ecological indicator to monitor the impact of land-use change on biodiversity (Tocco *et al.,* 2020).

Figure 4. Habitats of the beetles found in the MDV. A, mountains; B, Draa temporary river; C, Acacia formation, D, Oasis (Palm grove); E, temporary lake; and F, Sand dunes

The findings revealed the presence of 31 species classified into 15 families including Tenebrionidae, Scarabaeidae, Carabidae, Dermestidae, Dytiscidae, Meloidae, Buprestidae, Elateridae, Curculionidae, Bostrichidae, Anthicidae, Cerambycidae, Melyridae, Coccinellidae, and Chrysomelidae. This classification of the observed colepteran species yielded 32% of Tenebrionidae, 13% of Scarabaeidae, and 7% for

Carabidae, Dytiscidae, and Meloidae. Tenebrionidae with 30 000 described species (Bouchard *et al.,* 2021) are the most frequent macrodetritivores in several arid areas at global scale and are responsible for most of the nutrient cycling using litter in hot and dry periods (Cheli *et al.,* 2022) and considered prey for various predators (Bartholomew *et al.,* 2016). However, the Scarabaeidae encompasses 91% of all scarabaeoids with 27,800 species worldwide (Ratcliffe & Jameson, 2004) including both harmful (phytophagous) and beneficial (coprophagous) beetles (Chandra *et al.,* 2012). The Carabidae is the largest group of Coleopteran that includes about 33,000 species (Hong *et al.,* 2017). It was observed that the majority of observations were carried out in summer, which reflects the seasonal activity of several species while some were present throughout the year. The family Carabidae prefer mostly moist soil sites, but can be adapted to various environments and climatic areas (Choi *et al.,* 2022). Beetles can live and feed in various biotopes and have an essential role in the majority of terrestrial ecosystem (On ground or plants, in water, or on dung, in mountain and deserts) (Peris, 2020). However, the presence and density of the carabidae community for example in agroecosystems (sugar beet, maize, alphaalpha, wheat, and potatoes, etc.) are impacted by several factors including hydrological and pedological conditions, microclimate, presence of pests and diseases, and agrotechnical practices (Porhajašová & Babošová, 2022).

These factors vary from a site to another and from a biotope to another. This is the case for the species observed in the study area, which may be present in different characteristic habitats of this area (Figure 4).

Importance of Some Coleopteran Species Found in the MDV

Pollination

Crop pollination is an important ecosystem service to support the global food security (Potts *et al.,* 2016a). 75% of the global food crops is based on the pollination (Ganuza *et al.,* 2022; Klein *et al.,* 2007) with an annual economic value of US$235 to US$577 billion (Ganuza *et al.,* 2022; Potts *et al.,* 2016b). It is widely accepted that pollinating insects make it possible for many plants to grow. Beetles are among the most important pollinating insects of basal flowering plants and considered the earliest pollinators of angiosperms since the Cretaceous (Peris, 2020). Indeed, the flowers attract by their colors these beetles, which will transport the pollen (grains with male genetic heritage of the plant) from the plant or male organ to the plant or female organ. This results in an egg that will perpetuate the species. Several pollinating species were found in the study area associated with the oasis agroecosystem (Figure 5).

Figure 5. Exemples d'espèces pollinisatrices observées dans les oasis de la vallée de Draa. a-b, Mylabrini; c, Alleculinae; d, Mylabrini; e, Tropinota squalida pilosa; f, Julodis aequinoctialis
Source: Authors

Biologic Control

The relationship between the Coccinellidae predators and aphids was recognized centuries ago (Obrycki *et al.*, 2009). In the study area Madkouri, (1973) reported the use of Coccinellidae to combat the *Parlatoria blanchardi* that infests the date Palm scattered in the whole MDV. The first inventory of entomophages highlighted the presence of four species: *Chilocorus bipustulalus* L,. *Pharoscymnus sp., Exochomus nigripennis* Erichson., and *Cybocephalus sp.* (Madkouri, 1973). *Coccinella septempunctata* is a beetle that belongs to the family Coccinellidae with a red color feed on aphids and are therefore used in biological control (Figure 6). It was used as the principal agent to combat all kinds of aphids in both greenhouse and farmland (Cheng *et al.*, 2022). Biddinger *et al.*, (2009) reported the role of some Coccinellidae such as Stethorini in biological control as predators of mites (primarlly Tetranychidae). However, any change of ecological conditions may affect the behavior of Coccinellidae. Islam *et al.* (2022) investigating predation and development of *Harmonia axyridis* foraging on *Acyrthosiphon pisum* at different temperatures (15, 20, 25, and 30 °C) and revealed that the net predation rate decreased with increasing temperatures. Otherwise, the change in habitat has an effect on the behavior of *Harmonia axyridis* Pallas and *Hippodamia variegata* Goeze (The dispersal behavior). The density of these two species was higher in the unfragmented habitat because the fragmentation and habitat loss negatively influence natural enemies searching (Liu *et al.,* 2021).

Figure 6. Morphology of the Coccinella septempunctata (Coccinellidae) observed in Middle Draa Valley
Source: Authors

Xylophagous and Polyphagous Colepteran Species

Globally, beetles of the Bostrichidae family groups nine subfamilies and 580 species, destructive of the dry and dead wood while some species attack dry grain, and others degrade living trees (Park *et al.,* 2015). In the study area, this family is pest of *Acacia radiana*, especially on weakened or dead branches (Figure 7). Both adults and larvae were observed in dead branches found this current year in the study area.

Figure 7. Sinoxylon sexdentatum (Bostrichidae)
Source: Authors

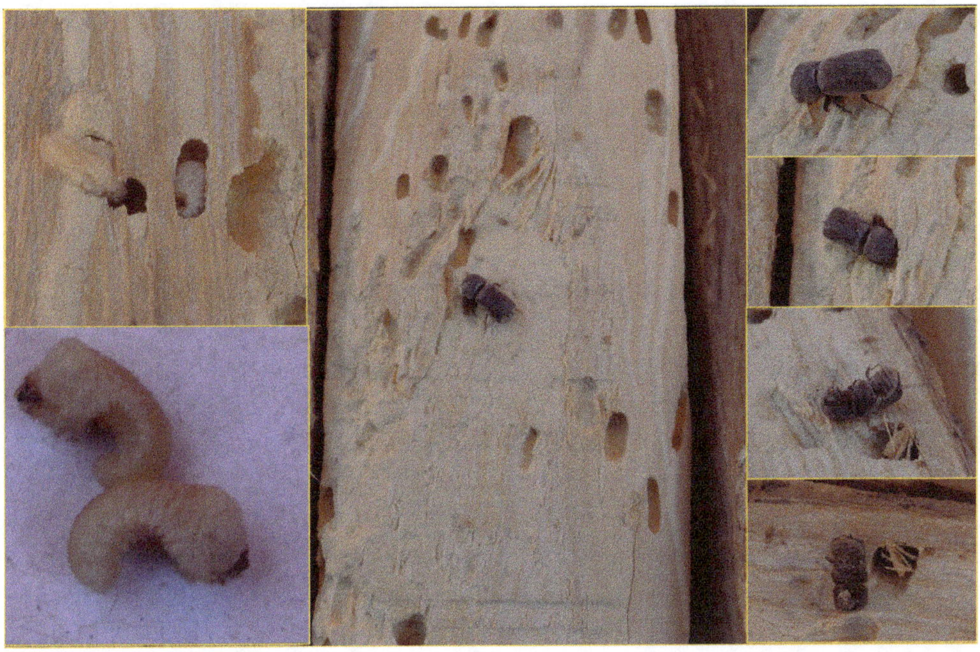

Figure 8. Morphology of the Chlaenius spoliatus (Carabidae) observed in Middle Draa Valley
Source: Authors

Carnivorous Species

Chlaenius spoliatus (Carabidae)

Carabidae is the largest group out of Coleopteran representing about 33,000 species worldwide (Hong *et al.,* 2017). It is well recognized as carnivorous and cosmopolitan group (Choi *et al.,* 2022). In the study area, this ground Carabidae was collected from wet banks of the Draa temporary river (Figure 8). They shelter under stones or plant debris. Both adults and larvae hunt various invertebrates.

Decomposers (Necrophagous and Coprophagous Species)

Scavenger animals are species capable of feeding on organic matter by decomposing corpses (Figure 9). These decomposers are essential to the functioning of ecosystems, as they recycle energy and nutrients from dead matter. Many insects are involved in this process including beetles. It is estimated that about 7,000 species of coprophagous dung beetles interrelate with a wide range parasite in ways that both reduce and increase transmission (Nichols *et al.,* 2017). The dung beetles also support various ecological functions such as recycling of animal feces. However, these species are threatened by deforestation, the loss of traditional grazing practices, the accumulation of the residues (veterinary products and herbicides), which induced a gradual decline in the numbers of species (Lumaret *et al.,* 2022). In regard to the necrophage aspect, a dead animal attracts several necrophagous insects including beetles that accelerate the corpse decomposition. In a recent study conducted on rabbit corpse and revealed that the most numerous beetles are Dermestidae, Staphylinidae, Silphidae, Scarabaeidae, Cleridae, Trogidae, and Histeridae) (Guerroudj & Berchi, 2019).

Figure 9. Decomposition of corpses. a, hare corpse; and b, fox corpse from the study area
Source: Authors

Scarabaeus Sacer (Scarabaeidae)

Scarabaeus sacer (Figure 10) is a black oval beetle belongs to the Genus *Scarabaeus* Linnaeus, 1758 and Scarabaeidae family distributed in the Mediterranean countries. It is a coprophagous species. This dung beetle uses vertebrates feces for food and reproduction (Mohtasebi *et al.,* 2021).

Figure 10. Morphology of the Scarabaeus sacer observed in Middle Draa Valley
Source: Authors

a) Crops devastating species

Melolonthini (Scarabaeidae) (Figure 11)

Melolonthini belongs to the Melolonthinae subfamily largest taxa of scarab beetles (Scarabaeidae), that cause considerable damage through the larval attack of several crops and pastures (Jia *et al.,* 2020). The larvae are of scarabaeiform.

Figure 11. Morphology of the Scarabaeidae Melolonthini observed in Middle Draa Valley
Source: Authors

Several studies have recorded the presence of many species of Coleoptera in MDV. Platia, (2008) descibes *Aeoloides houschkai (Elateridae)* from Zagora 30 km SE, Zaouia el-Barrahnia, Draa river while Morris & Schön, (2010) mentioned the presence of *Corimalia martini* Brisout, 1883, (Brentidae), associated with Tamaricaceae in 30 Km N of Zagora. Bennas & Sàinz (2006), noted *Meladema coriacea* Laporte, 1835 (Dytiscidae) in the north Zagora, after Guignot (1946) and Gomy, (2007) describes *Teretrius (Teretrius) angelae* (Histeridae) from 45 km S. Zagora. Gerstmeier, (2002) reported the presence of *Trichodes audouini-dubreuili* Reymond, 1956 (Cleridae), between Tagmoute and Tisgui-Ida-ou- Ballou. Háva, (2004) recorded *Attagenus lineatus* (Pic, 1894) (Dermestidae) and d Bezděk, & Baselga, (2015) *Oulema duftschmidi* (Redtenbacher, 1874) (Chrysomelidae). Gomy *et al.* (2016) found 9 species of *Saprinus* (Histeridae) in different sites of Draa valley, as well as the following Histeridae species *Teretrius (Teretrius) pulex* Fairmaire 1877,

Chalcionellus tunisius (Marseul, 1875), *Pholioxenus rutilus* (Erichson, 1834), and *Hypocaccus (Hypocaccus) fochi* (Auzat, 1921). However, Yélamos, (1992) observed three species of Saprinus: *Saprinus (s. str.) ornatus* Erichson, 1834, *Saprinus (s. str) chalcites* (Illiger, 1807), *Saprinus (s. str) moyses* Marseul, 1862. Jaskula, *et al.,* (2015) found in this region three species of Cicindelidae: *Lophyra (Lophyra) flexuosa flexuosa* (Fabricius, 1787), *Cicindela (Cicindela) campestris atlantis* Surcouf, 1933, and *Megacephala (Grammognatha) euphratica* (Dejean, 1822).

The reviewed and observed species in this arid area shows a considerable diversity including benefit and damaging species. The findings revealed that the observed species provide several ecosystem services and deservices including the beetles: phytophagous, detritivorous, decomposers, scavengers, predaceous, omnivorous, saprophagous, xylophagous, herbivores, floricolous, coprophagous, pollinator, and insect parasites. These species are present in many micro and macro ecosystems of this region and their abundance depend to the density of the vegetation cover, the presence of wetlands including the waterways, to extreme events such as droughts, floods, and fires and to climate change (CC) (Primarily, the change in temperature and precipitations). It is widely recognized that CC is a major factor of insect pest population change.

Climate Change and Beetles' Activity

Like other insects, the coleopteran species are affected by the environmental conditions including the climatic factors such as temperature, precipitations, and relative humidity. It was reported that the carabidae community for example are impacted by several factors including microclimate (Porhajašová & Babošová, 2022). Ganuza *et al.,* (2022) found that higher annual temperature increases the gamma diversity of beetles. It was also found that the climatic factors affect the biocontrol agents used in the pest management strategies because temperature controls biological rates from individuals to populations (Islam *et al.,* 2022).

In fact, insects including coleopteran species are sensitive, and have a quick response to temperature change, making them ideal bioindicators to monitor the impacts of climate change (CC) (Colares *et al.,* 2021). Otherwise, the beetle pests are numerous and are responsible for considerable losses in agriculture and forestry. For example, Bark beetles damaged about 35 million m^3 of wood per year in the period 1950–2000 (Schelhaas *et al.,* 2003). Climate change is affecting vegetation, which increases the vulnerability to insect pest species. Drought stress induced by CC can contribute to the fragility of forests and lead to bark beetle infestations (Trubin *et al.,* 2022). The species, *Dendroctonus frontalis* Zimmermann (Southern pine beetle among the most destructive pest insects in Central and North America) was found to be positively associated with warmer temperature that enhance the outbreaks and expansion (Aoki *et al.*, 2022). McKee *et al.*, (2022) concluded that the effects of the species *Dendroctonus simplex* (the eastern larch beetle) on tamarack forests are expected to be multiplied as CC continues.

CONCLUSION

Beetles are insects that play a very important role for the health of the environment. The flying beetles pollinate flowers that determine fruit production. A second essential role is that of the purification of ecosystems by decomposition of organic matter. In fact, these insects degrade leaf litter fallen on land by fertilizing the soil. A large number of beetles feed and degrade animal dung. The presence of this beetle diversity is a food source for other mostly vertebrate animals such as birds and reptiles and amphibians.

The Coleopteran diversity of this region was not well studied out according to our information. This chapter provides new observation of several species inside and outside the oasis agroecosystem of the region (the southeastern of Morocco). This work allows to update coleopteran species of the oasis system of MDV. 15 families and 31 species were observed. The contrast in climate between the interior and exterior of the palm grove provides a rich biodiversity, which supports a large number of ecosystem services. This richness reflects the importance of this arid area in terms of biodiversity, which can be used to awareness local population and government for the importance of oasis ecosystem.

ACKNOWLEDGMENT

The authors are grateful to Ibn El Mahdi Eljirari Middle School for the technical assistance.

REFERENCES

Andresen, E. (2002). Dung beetles in a Central Amazonian rainforest and their ecological role as secondary seed dispersers. *Ecological Entomology*, *27*(3), 257–270. doi:10.1046/j.1365-2311.2002.00408.x

Aoki, C. F., Munro, H. L., & Gandhi, K. J. (2022). Responses and modeling of southern pine beetle and its host pines to climate change. In Bark Beetle Management, Ecology, and Climate Change (pp. 55-85). Academic Press. doi:10.1016/B978-0-12-822145-7.00015-5

Bartholomew, A., Hafezi, S. A., & Karimi, S. (2016). Effects of habitat complexity on the abundance, species richness and size of darkling beetles (Tenebrionidae) in artificial vegetation. *Journal of Arid Environments*, *129*, 35–41. doi:10.1016/j.jaridenv.2016.02.004

Bennas, N., & Sàinz-Cantero, C. E. (2006). Coléoptères aquatiques Adephaga de la chaîne rifaine marocaine (Coleoptera Gyrinidae, Haliplidae, Noteridae, Hygrobiidae, Dytiscidae). *Memorie della Societa Entomologica Italiana*, *85*(1), 31–73. doi:10.4081/memorieSEI.2006.31

Bergstrom, B. C., Maki, L. R., & Werner, B. A. (1976). Small dung beetles as biological control agents: Laboratory studies of beetle action on trichostrongylid eggs in sheep and cattle feces. *Proceedings of the Helminthological Society of Washington*, *43*, 171–174.

Bezděk, J., & Baselga, A. (2015). Revision of western Palaearctic species of the *Oulema melanopus* group, with description of two new species from Europe (Coleoptera: Chrysomelidae: Criocerinae). *Acta Entomologica Musei Nationalis Pragae*, *55*(1), 273–304. https://www.biotaxa.org/AEMNP/article/view/12440

Biddinger, D. J., Weber, D. C., & Hull, L. A. (2009). Coccinellidae as predators of mites: *Stethorini* in biological control. *Biological Control*, *51*(2), 268–283. doi:10.1016/j.biocontrol.2009.05.014

Blake, S., Foster, G. N., Fisher, G. E., & Ligertwood, G. L. (1996). Effects of management practices on the carabid faunas of newly established wildflower meadows in southern Scotland. *Annales Zoologici Fennici*, *33*, 139–147. https://www.jstor.org/stable/23735413

Bouchard, P., Bousquet, Y., Aalbu, R. L., Alonso-Zarazaga, M. A., Merkl, O., & Davies, A. E. (2021). Review of genus-group names in the family Tenebrionidae (Insecta, Coleoptera). *ZooKeys*, *1050*, 1–633. doi:10.3897/zookeys.1050.64217 PMID:34385881

Brander, L. (2013). *Guidance manual on value transfer methods for ecosystem services. United Nations Environment Programme*. UNEP.

Cajaiba, R. L., Périco, E., Dalzochio, M. S., da Silva, W. B., Bastos, R., Cabral, J. A., & Santos, M. (2017). Does the composition of Scarabaeidae (Coleoptera) communities reflect the extent of land use changes in the Brazilian Amazon? *Ecological Indicators*, *74*, 285–294. doi:10.1016/j.ecolind.2016.11.018

Chandra, K., Gupta, D., Uniyal, V. P., Bharadwaj, M., & Sanyal, A. K. (2012). Studies on scarabaeid beetles (Coleoptera) of Govind wildlife sanctuary, Garhwal, Uttarakhand, India. *Biological Forum: An International Journal*, *4*(1), 48–54.

Cheli, G. H., Bosco, T., & Flores, G. E. (2022). The role of *Nyctelia circumundata* (Coleoptera: Tenebrionidae) on litter fragmentation processes and soil fertility in northeastern arid patagonia. *Geoderma*, *415*, 115770. doi:10.1016/j.geoderma.2022.115770

Cheng, Z., Wang, D., Han, S., Zuo, C., & He, Y. (2022). Transcriptome analysis in the thiamethoxam resistance of seven-spot ladybird beetle, *Coccinella septempunctata* (Coleoptera: Coccinellidae). *Ecotoxicology and Environmental Safety*, *230*, 113144. doi:10.1016/j.ecoenv.2021.113144 PMID:34998260

Choi, J. B., Choi, E. Y., Kwon, T., Jeong, C. H., Oh, S. N., Han, G. Y., Kim, M. C., & Park, J. K. (2022). A checklist of the family Carabidae (Coleoptera: Adephaga: Caraboidea) from Cambodia. *Journal of Asia-Pacific Biodiversity*, *15*(2), 180–195. doi:10.1016/j.japb.2021.12.007

Colares, C., Roza, A. S., Mermudes, J. R., Silveira, L. F., Khattar, G., Mayhew, P. J., Monteiro, R. F., Nunes, M. F., & Macedo, M. V. (2021). Elevational specialization and the monitoring of the effects of climate change in insects: Beetles in a Brazilian rainforest mountain. *Ecological Indicators*, *120*, 106888. doi:10.1016/j.ecolind.2020.106888

Do, Y., & Choi, M. B. (2022). Effects of urbanization on carabid beetles in an urban riparian area. *Entomological Research*, *52*(6), 281–294. doi:10.1111/1748-5967.12601

de Vries, H. H. (1994). Size of habitat and presence of ground beetle species. In K. Desender, M. Dufrêne, & J. P. Maelfait (Eds.), *Carabid Beetles. Ecology and Evolution* (pp. 253–259). Kluwer Academic Publishers. doi:10.1007/978-94-017-0968-2_39

Eyre, M. D., & Luff, M. L. (1990). A preliminary classification of European grassland habitats using carabid beetles. In N. E. Stork (Ed.), *The Role of Ground Beetles in Ecological and Environmental Studies* (pp. 227–236). Intercept.

Franchimont, J., & Saadaoui, E. M. (2001). *National Study on Biodiversity Synthesis Report. Kingdom of Morocco Ministery of Territorial Planing, Water and Environment.* Department of Environment.

Ganuza, C., Redlich, S., Uhler, J., Tobisch, C., Rojas-Botero, S., Peters, M. K., Zhang, J., Benjamin, C. S., Englmeier, J., Ewald, J., Fricke, U., Haensel, M., Kollmann, J., Riebl, R., Uphus, L., Müller, J., & Steffan-Dewenter, I. (2022). Interactive effects of climate and land use on pollinator diversity differ among taxa and scales. *Science Advances*, *8*(18), eabm9359. Advance online publication. doi:10.1126ciadv.abm9359 PMID:35544641

Gerstmeier, R. (2002). *Trichodes audouini-dubreuili* Reymond—Rediscovery and Redescription of an Obscure Species (Coleoptera: Cleridae). *Coleopterists Bulletin*, *56*(2), 273–278. doi:10.1649/0010-065X(2002)056[0273:TADRRA]2.0.CO;2

Gomy, Y., de Castro-Arrazola, I., Sanchez-Piñero, F., & Labrique, H. (2016). Contribution à la connaissance des Histeridae du Maroc (quatrième note)(Coleoptera). *Entomologiste*, *72*(1), 7–32.

Gomy, Y. (2007). Description d'une nouvelle espèce de *Teretrius* Erichson, 1834 du Maroc (Coleoptera, Histeridae). *Bulletin Mensuel de la Societe Linneenne de Lyon*, *76*(5), 113–119. doi:10.3406/linly.2007.13656

Guerroudj, F. Z., & Berchi, S. (2019). Composition of the Necrophagous Fauna on a Rabbit Corpse in Constantine, Northeastern Algeria. *Energy Procedia*, *157*, 1083–1088. doi:10.1016/j.egypro.2018.11.275

Háva, J. (2004). Notes on Dermestidae (Coleoptera) with new synonyms and description of three new species. *Annales Historico-Naturales Musei Nationalis Hungarici*, *96*, 81–95.

HCEFLCD. (2007). *Etude d'aménagement du Parc national d'Iriqui.* Rapport de synthèse. HCEFLCD-DREF SO. 2007.

Hong, E., Kim, Y., Jeong, J. C., Kang, S. H., Jung, J. K., & Suk, S. W. (2017). Community structure and distribution of ground beetles (Coleoptera: Carabidae) in Sobaeksan National Park, Korea. *Journal of Ecology and Environment*, *41*(1), 1–7. doi:10.118641610-017-0036-1

Islam, Y., Güncan, A., Fan, Y., Zhou, X., Naeem, A., & Shah, F. M. (2022). Age-stage, two-sex life table and predation parameters of *Harmonia axyridis* Pallas (Coleoptera: Coccinellidae), reared on *Acyrthosiphon pisum* (Harris)(Hemiptera: Aphididae), at four different temperatures. *Crop Protection (Guildford, Surrey)*, *106029*. Advance online publication. doi:10.1016/j.cropro.2022.106029

Jaskula, R., Rewicz, T., & Kwiatkowski, K. (2015). Tiger beetle fauna (Coleoptera: Carabidae, Cicindelinae) of Morocco: distribution, phenology and list of taxa. *Entomologica Fennica*, *26*(3), 132–155. doi:10.33338/ef.52825

Jeffery, S., Gardi, C., Jones, A., Montanarella, L., Marmo, L., Miko, L., Ritz, K., Peres, G., Römbke, J., & van der Putten, W. H. (Eds.). (2010). *European Atlas of Soil Biodiversity*. Publications Office of the European Union.

Jia, Z. C., Lu, C., Zhou, R. J., & Jiang, L. (2020). Comparative morphology between the white grubs *Pseudosymmachia tumidifrons* and *Brahmina faldermanni* (Coleoptera: Scarabaeidae: Melolonthinae) using scanning electron microscopy. *Zoologischer Anzeiger*, *289*, 8–17. doi:10.1016/j.jcz.2020.08.008

Karmaoui, A., El Jaafari, S., Chaachouay, H., & Hajji, L. (2021). The socio-ecological system of the pre-Sahara zone of Morocco: A conceptual framework to analyse the impact of drought and desertification. *GeoJournal*, 1–14. doi:10.100710708-021-10546-8

Klein, A. M., Vaissière, B. E., Cane, J. H., Steffan-Dewenter, I., Cunningham, S. A., Kremen, C., & Tscharntke, T. (2007). Importance of pollinators in changing landscapes for world crops. *Proceedings. Biological Sciences*, *274*(1608), 303–313. doi:10.1098/rspb.2006.3721 PMID:17164193

Legrand, A., Gaucherel, C., Baudry, J., & Meynard, J. M. (2011). Long-term effects of organic, conventional, and integrated crop systems on Carabids. *Agronomy for Sustainable Development*, *31*(3), 515–524. doi:10.100713593-011-0007-3

Liu, J., Zhao, Z., Yan, Y., Ali, A., Ahmed, Z., He, D., Yu, M., Hang, J., Perven, M., Nazir, T., Tayyib, M., & Zhao, Z. (2021). Effect of alfalfa habitat change on dispersal behavior of *Harmonia axyridis Pallas* and *Hippodamia variegata Goeze* (Coleoptera: Coccinellidae). *Journal of Asia-Pacific Entomology*, *24*(4), 997–1003. doi:10.1016/j.aspen.2021.09.002

Losey, J. E., & Vaughan, M. (2006). The economic value of ecological services provided by insects. *Bioscience*, *56*(4), 311–323. doi:10.1641/0006-3568(2006)56[311:TEVOES]2.0.CO;2

Lumaret, J. P., Kadiri, N., & Martinez Morales, I. (2022). The Global Decline of Dung Beetles. Imperiled: The Encyclopedia of Conservation, 553-562. doi:10.1016/B978-0-12-821139-7.00018-0

Madkouri, M. (1973). Etude Préliminaire sur la Bio écologie de *Parlatoria blanchardi* Targ (Hom. Diaspididae) dans la Vallée du Draa. *Al-Awamia*, *48*, 39–72.

McKee, F. R., Windmuller-Campione, M. A., Althoff, E. R., Reinikainen, M. R., Dubuque, P. A., & Aukema, B. H. (2022). Eastern larch beetle, a changing climate, and impacts to northern tamarack forests. In Bark Beetle Management, Ecology, and Climate Change (pp. 261-300). Academic Press. doi:10.1016/B978-0-12-822145-7.00001-5

Mohtasebi, S., Teimouri, A., Afshar, M. J. A., Mobedi, I., Abbasian, H., Totonchian, N., & Mowlavi, G. (2021). First report of *Spirocerca lupi* larva in dung beetles (*Scarabaeus armeniacus*) in the central region of Iran: A morphological and molecular identification. *Comparative Immunology, Microbiology and Infectious Diseases, 77*, 101671. doi:10.1016/j.cimid.2021.101671 PMID:34049036

Morris, M. G., & Schön, K. (2010). *Corimalia martini* (Brisout, 1883)(Insecta, Coleoptera, Nanophyidae), a weevil new to the Canary Islands from Fuerteventura. *Museum Scientiarum Naturalium Nivariense, 38*(Vieraea 38), 83–90. doi:10.31939/vieraea.2010.38.06

Moumane, A., El Ghazali, F. E., Al Karkouri, J., Delorme, J., Batchi, M., Chafiki, D., & Karmaoui, A. (2021). Monitoring spatiotemporal variation of groundwater level and salinity under land use change using integrated field measurements, GIS, geostatistical, and remote-sensing approach: Case study of the Feija aquifer, Middle Draa watershed, Moroccan Sahara. *Environmental Monitoring and Assessment, 193*(12), 1–21. doi:10.100710661-021-09581-2 PMID:34735624

Nichols, E., Alarcón, V., Forgie, S., Gomez-Puerta, L. A., & Jones, M. S. (2017). Coprophagous insects and the ecology of infectious diseases of wildlife. *ILAR Journal, 58*(3), 336–342. doi:10.1093/ilar/ilx022 PMID:29036417

Nichols, E., Spector, S., Louzada, J., Larsen, T., Amezquita, S., Favila, M. E., & Network, T. S. R. (2008). Ecological functions and ecosystem services provided by Scarabaeinae dung beetles. *Biological Conservation, 141*(6), 1461–1474. doi:10.1016/j.biocon.2008.04.011

Obrycki, J. J., Harwood, J. D., Kring, T. J., & O'Neil, R. J. (2009). Aphidophagy by Coccinellidae: Application of biological control in agroecosystems. *Biological Control, 51*(2), 244–254. doi:10.1016/j.biocontrol.2009.05.009

Parisi, V., Menta, C., Gardi, C., Jacomini, C., & Mozzanica, E. (2005). Microarthropod communities as a tool to assess soil quality and biodiversity: A new approach in Italy. *Agriculture, Ecosystems & Environment, 105*(1-2), 323–333. doi:10.1016/j.agee.2004.02.002

Park, S., Lee, S., & Hong, K. J. (2015). Review of the family Bostrichidae (Coleoptera) of Korea. *Journal of Asia-Pacific Biodiversity*, *8*(4), 298–304. doi:10.1016/j.japb.2015.10.015

Platia, G. (2008). Descriptions of new species of click-beetles from the Palearctic region and Taiwan with chorological and synonymical notes. *Quaderno di Studi e Notizie di Storia Naturale della Romagna*, *27*, 187–215.

Peris, D. (2020). Coleoptera in amber from Cretaceous resiniferous forests. *Cretaceous Research*, *113*, 104484. doi:10.1016/j.cretres.2020.104484

Potts, S. G., Ngo, H. T., Biesmeijer, J. C., Breeze, T. D., Dicks, L. V., Garibaldi, L. A., Hill, R., Settele, J., & Vanbergen, A. (2016a). *The assessment report of the Intergovernmental Science-Policy Platform on Biodiversity and Ecosystem Services on pollinators, pollination and food production.* Bonn, Germany, Secretariat of the Intergovernmental Science-Policy Platform on Biodiversity and Ecosystem Services. Available on https://nora.nerc.ac.uk/id/eprint/519227/

Potts, S. G., Imperatriz-Fonseca, V., Ngo, H. T., Aizen, M. A., Biesmeijer, J. C., Breeze, T. D., Dicks, L. V., Garibaldi, L. A., Hill, R., Settele, J., & Vanbergen, A. J. (2016b). Safeguarding pollinators and their values to human well-being. *Nature*, *540*(7632), 220–229. doi:10.1038/nature20588 PMID:27894123

Porhajašová, J. I., & Babošová, M. (2022). Impact of arable farming management on the biodiversity of Carabidae (Coleoptera). *Saudi Journal of Biological Sciences*, *29*(9), 103371. doi:10.1016/j.sjbs.2022.103371 PMID:35865322

Rainio, J., & Niemelä, J. (2003). Ground beetles (Coleoptera: Carabidae) as bioindicators. *Biodiversity and Conservation*, *12*(3), 487–506. doi:10.1023/A:1022412617568

Ratcliffe, B. C., & Jameson, M. L. (2004). The Revised classification for Scarabaeoidea: What the Hell is going on? *Papers in Entomology*, 25. http://www-museum.unl.edu/research/entomology/Scarabs-Newsletter.htm

Schelhaas, M. J., Nabuurs, G. J., & Schuck, A. (2003). Natural disturbances in the European forests in the 19th and 20th centuries. *Global Change Biology*, *9*(11), 1620–1633. doi:10.1046/j.1365-2486.2003.00684.x

Tocco, C., Midgley, J. M., & Villet, M. H. (2020). Intermediate disturbance promotes diversity and the conservation of dung beetles (Scarabaeoidea: Scarabaeidae and Aphodiidae) in the Eastern Cape, South Africa. *Basic and Applied Ecology*, *49*, 45–56. doi:10.1016/j.baae.2020.10.005

Trubin, A., Mezei, P., Zabihi, K., Surový, P., & Jakuš, R. (2022). Northernmost European spruce bark beetle *Ips typographus* outbreak: Modelling tree mortality using remote sensing and climate data. *Forest Ecology and Management*, *505*, 119829. doi:10.1016/j.foreco.2021.119829

Yélamos, T. (1992). Nuevos datos sobre los Histéridos de Marruecos. *Zapateri*, *12*, 55–64.

Section 4

Climate Change and Honeybees

Chapter 7
Climate Change Impact on Honeybees (Apis spp.) and Their Pollination Services:
Climate Change and Honeybees Services

Mohamed Ouknin
https://orcid.org/0000-0001-5251-3033
Moulay Ismail University, Morocco

Hassan Alahyane
https://orcid.org/0000-0002-4105-8327
Cadi Ayyad University, Morocco

Abdelaziz Mounir
Cadi Ayyad University, Morocco

Lhou Majidi
Moulay Ismail University, Morocco

ABSTRACT

Honeybees participate in the pollination of many crops and wild species and serve as bioindicators of environmental health, and therefore play an essential role in agricultural production. While it is currently facing serious threats worldwide, its disappearance will impact the balance of ecosystems. Global economic value of crop pollination services has been estimated at 153 billion euros per year, representing 9.5% of global agricultural food value. As bees are ectothermic species, the ambient temperature determines their activity, and therefore, climate change, with the increase in temperatures experienced in recent years, can influence their biology, distribution, and behavior. Climate change affects negatively bees by impacting their

DOI: 10.4018/978-1-6684-4824-3.ch007

food resources. Increased temperature induces a differential response of insects and plants that generate a temporal and spatial variation in the biological cycle of the species concerned, with serious demographic consequences. This results in an alteration of the synchronization between plants and insects.

INTRODUCTION

The European honeybee, *Apis mellifera*, is considered the main pollinator of about 87% of the world's agricultural crops (McGregor, 1976; Watanabe, 1994). It was found in a wide geographical area, including northern Europe, sub-Saharan Africa, and Central Asia, known by an important environmental diversity, such as the mists of the United Kingdom, the Alps, and African desert oases (Ruttner, 1988). Each year, the economic value of pollinator species, including honeybee has been estimated at tens of billion dollars (Reddy et al., 2012). Bees play a crucial role in biodiversity conservation by pollinating many plant varieties essential to humans (Gallai et al., 2009; Potts et al., 2010).

During the last few years, honeybee populations have suffered several threats which have led to the collapse of colonies (Flores et al., 2019; Klein et al., 2018). As a result, climate change has the potential to affect the health and vitality of bee colonies at various levels. Developing pathogens and predators can pose direct risks to honeybees. It can also alter the behavior and physiology of honeybee colonies and affect the quality of the floral environment and decrease the harvesting capacity and development of the colony (Le Conte & Navajas, 2008). Furthermore, climate change may create a new range for bees and new competitive relationships between species, as well as between their parasites and pathogens (Figure 1).

Recently, beekeeping in Morocco has been affected by this phenomenon, with approximately 25,000 hives lost and causing serious economic damage to beekeepers, estimated at two million Dirhams (MAD) (ONSSA, 2022). Beekeepers will also have to change the way they practice beekeeping. They will focus on moving their hives to new feeding areas and importing foreign breeds to test their value in new environments to make them appropriate and sustainable for environmental conditions in each region.

Nowadays, the incremental effects of climate change are increasingly quantified and modeled around the world. As a result, the understanding of their impacts on pollination and pollinators is better interpreted and has evolved significantly (Settele et al., 2016). Increases in the frequency, magnitude, and intensity of some extreme weather events (heat waves or drought periods) are likely to accelerate bees declines (Brown et al., 2016). It is becoming clear that such climate events would rapidly lead to extinctions of wild plants and associated pollinators (Oliver et al., 2015;

Figure 1. Potential effects of climate change on health and vitality of honeybee colonies

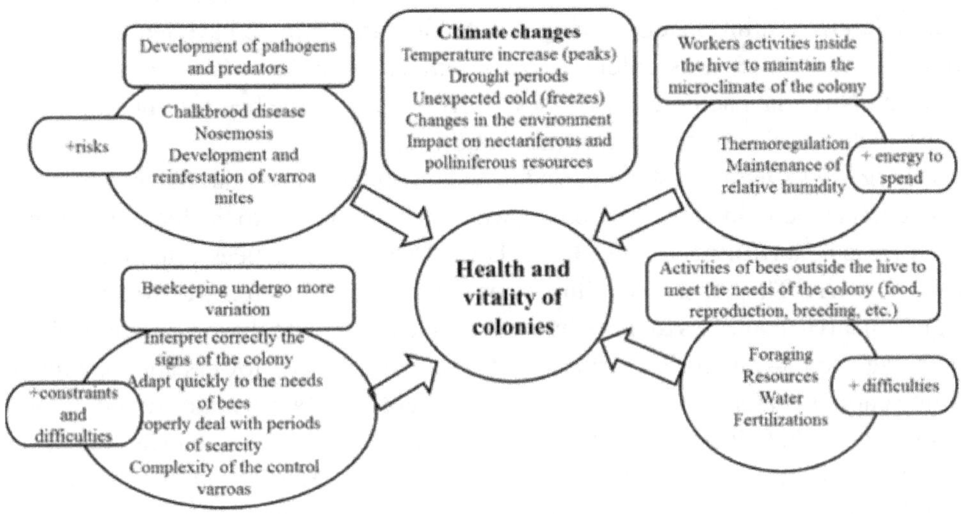

Rasmont & Iserbyt, 2012). Thus, it would be interesting to expand our knowledge in this area, especially regarding the relative importance of extreme climate events in order to improve the management of animal pollination-dependent habitats. In fact, bees are intrinsically linked to the spatial-temporal distribution of floral resources (Kremen et al., 2007; Potts et al., 2003, 2006). Climate change therefore has the potential to restructure or simplify networks of plant-pollinator interactions (Schweiger et al., 2010).

In this chapter, we outline the impacts of climate change on honeybees and their pollination services. We also demonstrate the climate change interactions with floral hosts of bees and their possible effect on food availability.

CLIMATE CHANGE IMPACTS ON HONEYBEE'S BEHAVIOUR, PHYSIOLOGY, AND DISTRIBUTION

The potential impact of climate change is felt in the behavior, physiology, and distribution of bees, as well as in the evolution of the interaction between bees and diseases. Changing weather conditions will inevitably affect the survival of these archetypes or honeybee species closely associated with their environment. Migration and changes in their life cycle and behavior could help them succeed in new biotopes. Climate change could disrupt the relationship between flowers and pollinators, especially bees, and pollinators will need to be protected to continue their pollination function, which is so important for the economy and ecological balance; a recent molecular study has confirmed the importance of the evolutionary branches of *A. mellifera*. Anthropogenic changes in habitat and climate have led to various types of disorders such as colony collapse, attack by mites - *Varroa destructor*, protozoa - Nosema, bacteria and viruses. Varroa weakens the bee's immune system and promotes viral growth (Chen et al., 2006).

Apis mellifera, has the potential to adapt to hot climates. For example, *A. mellifera sahariensis* is found in the Sahara oases, where it has adjusted to the local flowering and extreme heat (Ruttner, 1988). In the United States, honeybees can thrive in the Arizona Desert. The survival conditions of these bees are a supply of water, which they use in large quantities to raise their larvae and to regulate the temperature of the colony between 34 and 35°C. In an arid environment, desert flowers are not able to provide enough water for bees, which die. Temperature plays a major role in insect and plant growth. Climate change can then induce temporal discrepancies between flowering and the emergence of associated pollinators (Hegland et al., 2009). For example, a meta-analysis by Parmesan (2007) showed that the emergence or arrival of migratory pollinator species was three times faster than the first blooms of the herbaceous layer. Thus, there is a risk that this phenological decoupling will lead to a reduction in the availability of floral resources for pollinators and, at the same time, a reduction in the pollination of associated host plants. According to climate change predictions, desert regions will become even drier, resulting in the disappearance of oases and their bees. It is highly unlikely that *A. mellifera sahariensis* will naturally migrate to more favorable desert areas, as oases are very isolated and not conducive to long-distance migration or swarming. It is thus essential to consider conservation measures to transfer this bee towards zones favorable to its development, at the risk of losing this ecotype so precious for the world biodiversity. Among other things, climate change has the potential to restrict or expand the ranges of species. Pollinators with a lower migratory capacity will tend to see their range shrink, but for some species that are able to keep pace with climate change, their range may expand (Thomas et al., 2006). Climate predictions are already available for the spatiotemporal distributions of bumblebees in North America and Europe and European butterflies (Kerr et al., 2015; Rasmont et al., 2015; Settele et al., 2008).

Climate change may affect the honeybee development cycle. It is generally accepted that each race of honeybees develops at its own rate (Louveaux et al., 1966). Therefore, any type of climate change or movement of a honeybee race from one geographic region to another will therefore necessarily have measurable consequences. In cooler regions, bees spend the winter clustered in a tight ball and use their honey reserves to provide energy for survival until spring. The ability of the bee to store energy reserves and manage colony development exerts considerable adaptive pressure. In spring, when the weather becomes milder, the queen starts to lay eggs and the colony grows and increases the size of the working population. There can be a multi-week cold wave during which bees cannot harvest. The large size of the honeybee population results in such a rapid depletion of reserves that the colony can starve to death. This can easily happen to hybrid bees (crosses of several races by bee breeders), which grow very quickly in the spring. In contrast, local ecotypes, better adapted to the environmental conditions, are more cautious and develop more slowly in the spring, until after the cold snap they reproduce very quickly. They thus avoid endangering the survival of the colony. A distinction must therefore be made between local ecotypes, which must adapt their development and reserves to the climate, and hybrid bees selected by bee breeders. Thus, hybrids have not been bred to build up food reserves, the queen does not adapt her egg laying, and the workers do not adapt their larval rearing, so the bees cannot survive without the help of a beekeeper who provides an unlimited supply of sugar solution (Louveaux et al., 1966). The variability of honeybee life history traits with temperature and environment shows such genetic plasticity and variability that it could lead to the selection of developmental cycles adapted to new climatic conditions. Among the

Figure 2. Climate change impacts on bee species

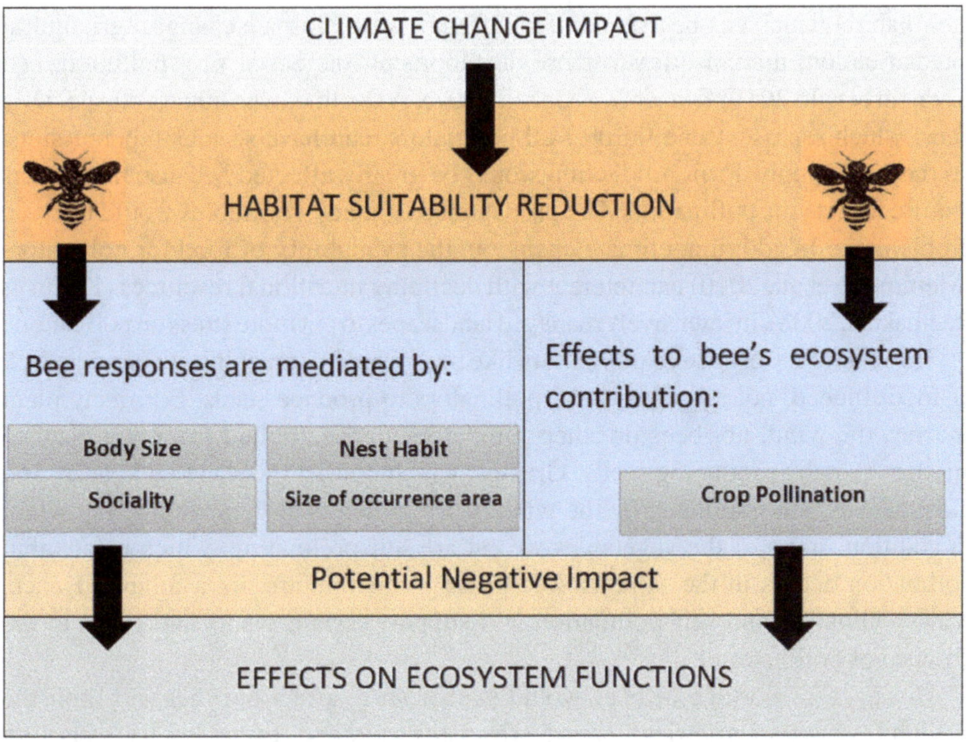

effects of climate change on bee species is a reduction in habitat suitability. Bee responses to climate change can be complex, with potential negative effects on ecosystem functions (Figure 2).

Floral chemical ecology is also involved in plant-pollinator relationships through complex odor clusters associated with nectar and pollen. Variations in temperature and elevated atmospheric carbon dioxide concentration are thought to impact the chemical ecology of plants and the composition of nectar and pollen produced (Farré Armengol et al., 2014; Parachnowitsch & Manson, 2015). For example, butterflies feeding on nectar with high concentrations of amino acids lay more eggs and thus increase their fitness than those that collected nectar in lower concentrations of amino acids (Mevi-Schutz & Erhardt, 2005). Hoover et al. (2012) also reported that the joint effects of increased temperatures and atmospheric CO_2 might affect flower phenology and morphology, as well as nectar composition, by this way altering its attractiveness to pollinators. This research topic remains little explored and scientific investigations are ongoing.

GLOBAL DECLINE OF HONEYBEES

The honeybees decline and other pollinators are mainly due to the degradation of their habitat which can be explained in different ways: climate change, agricultural intensification, human infrastructure development, increased fire, pollution, *etc.* (Drossart et al., 2019; Kluser & Peduzzi, 2007). A decline in pollinators, especially bees, which represent one third of all pollinators, can have serious repercussions on the food supply. Fruit production would be greatly affected because most fruits require insects for pollination. The propagation of many vegetables would become problematic. In addition, climate change in the availability of food for pollinators (Memmotte et al., 2010) can interact with declining nutritional resources (Kleijn & Raemakers, 2008), in intensively managed landscapes to put more stress on pollinators.

However, the absence of pollinators like bees would not result in a complete lack of food. Indeed, not all plants need pollinators to produce seeds. For many plant species, the wind, not bees or other pollinators, carries pollen from one plant to another, thereby producing seeds. Grasses, which consist of wheat and barley and most cereals, are pollinated by the wind. Corn pollen is also carried by the wind. In addition, some of the vegetables we eat are self-pollinating, which means that pollination occurs in the same flower when pollinators are not available. Hence, wind pollination and self-pollination will support certain plants and crops in the absence of pollinators.

However, a shortage of bees would significantly affect our diet and limit the availability of various nutrients provided by many fruits and vegetables. By restricting alfalfa production, the lack of pollinators would also be detrimental to meat and dairy production. Consequently, a reduction in the number of bees would have a negative impact on human nutrition. Human health depends on the health of bees.

Bees and other pollinators are under threat. Current rates of extirpation range from 100 to 1,000 times higher than normal. Insects will probably account for the majority of the future biodiversity loss, with 40% of invertebrate pollinator species - especially bees and butterflies - threatened with extinction. Though to a lesser extent, vertebrate pollinators (16.5%) are also under threat of global extinction. Reddy et al. (2012), reports that over the last 25 years, the rate of decline worldwide varies from 35 to 65% (Table 1).

Table 1. Global honeybee populations decline (Reddy et al., 2012)

Country	Decline (%)	Duration
Germany	57	Last 15 yrs
UK	61	Last 10 yrs
USA	>50	Last 20 yrs
Poland	>35	Last 15 yrs
India	>40	Last 25 yrs
Brazil	>53	Last 15 yrs
Netherland	58-65	Last 25 yrs
China	>50	Last 20 yrs

One of the concerns associated with the local decline of pollinators in an area is the possible impact on the plants that depend on them. In fact, the International

Figure 3. IUCN red list status of vertebrate pollinators across regions

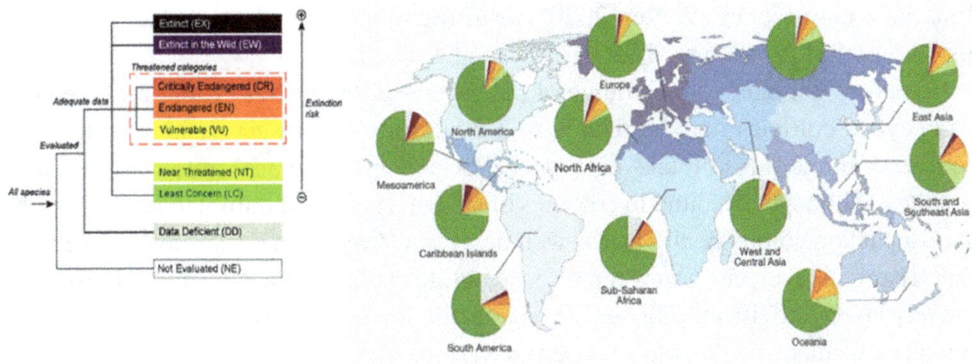

Union for Conservation of Nature (IUCN) predicts that 20,000 species of flowering plants will disappear in the coming decades. While declining pollinator numbers are not the only cause of these plant extinctions, it is recognized that some plant-pollinator associations are truly at risk. The loss of flowering plants, as much as the loss of pollinators, is likely to result in a cascading loss in both groups.

CLIMATE CHANGE AND DISEASE AFFECT BEE POPULATIONS

Agricultural production has been affected in the last decade by climate change. It favors the installation of several pests and diseases in agriculture that affect the honeybee populations, the most important pollinator of agricultural crops in the worldwide (Free, 1993; Klein et al., 2007). Honeybee populations are susceptible to many pathogenic agents, parasites, pests, and predators. The impacts of climate change include the spread and virulence of diseases and pests. Furthermore, it could favor the transfer of the haplotypes of honeybee populations (Le Conte & Navajas, 2008), movements of the honeybees of different species and races could bring them into contact with pathogens with which they have never co-evolved. One such example is *V. destructor* and *A. mellifera*. In a few decades, two highly homogenous haplotypes of this parasite were sufficient to invade almost the entire range of *A. mellifera* (Solignec et al., 2005). Consequently, high temperatures may increase the resistance or sensitivity of bees to pathogens due to physiological stress and changes in immune response (Ferguson et al., 2016; Laughton et al., 2017; Zaragoza-Trello et al., 2021). Similarly, as with human diseases (Ogden, 2017), the pathogen vector through pollinator expansion can introduce new infections in indigenous pollinator communities. Climate change induced plant extinctions and can also affect flower-mediated pathogen transmission by altering the community structure, which becomes dominated by plant species with different potential to act as hubs for pathogen transmission (Adler et al., 2020). In addition, pollinating forms may disappear or produce fewer flowers or lower quality floral rewards under altered climatic circumstances (Walter, 2020), resulting in nutritional stress and increased host susceptibility (Pike et al., 2019).

Bees were threatened by a range of pathogens and parasites (Table 2). Damage to honeybee colonies may be related to certain mite and insect pests, resulting in the loss of numerous colonies each year. For example, tracheal mites are internal parasites of honeybees found in colonies of several species of bumblebees, although their impact on colony health is less well designed. American Foulbrood, caused by a bacterium, is a serious honeybee disease, while chalk foulbrood, a fungal disease, severely affects alfalfa cutting bees. The intestinal parasite Nosema, a protozoan, as well as new emerging diseases, caused by various viruses, has a negative impact on

Table 2. Main parasites and pathogens of honeybees

Classification	Pathogenic characteristics	Species	Disease cause	Susceptible developmental stages	References
Bacteria	Infection lasts a long time, spreads rapidly and causes severe damage.	*Spiroplasma apis* *Spiroplasma melliferum*	Spiroplasmosis Spiroplasmosis	Adults Adults	Mouches et al., 1983 Meeus, et al., 2012
		Paenibacillus larvae	American foulbrood disease	Larvae	Genersch, 2010
		Melissococcus plutonius	European foulbrood disease	Larvae	Forsgren et al., 2005
Fungi	Frequently associated with humidity, rain, and unstable temperature	*Ascosphaera apis* *Aspergillus flavus*	Chalkbrood disease *Aspergillus flavus* disease	Larvae Different life stages	Chen et al., 2018 Chen, 2013
Parasite	Nutrient ingestion in honeybees	*Nosema ceranae* *Nosema apis*	Microsporidiosis Microsporidiosis	Adults Adults	Fries et al., 1991 Sak et al., 2004
		Tropilaelaps	Acariasis of bees	Larvae	Chantawannakul et al., 2018
		Varroa destructor	Acariasis of bees	Larvae, pupae and adults	Rondeau et al., 2018
Virus	Occurs without obvious symptoms until it is activated under certain environmental conditions, resulting in high pathogenicity.	Black queen cell virus Deformed wing virus	Black queen cell disease Deformed wing disease	Larvae and pupae Different life stages	Zhang et al., 2012 Bailey & Ball, 1991
		Israeli acute paralysis virus	Israeli acute paralysis disease	Different life stages	Chen et al., 2014
		Sac brood virus	Sacbrood bee disease	Different life stages	Thai et al., 2018 ; Park et al., 2016
		Chronic bee paralysis virus	Chronic bee paralysis disease	Adults	Olivier et al., 2008
		Kashmir bee virus	Kashmir bee disease	Different life stages	Milićević, et al., 2018
		Acute bee paralysis virus	Acute bee paralysis disease	Different life stages	Ullah et al., 2021
		Tobacco ringspot virus	Unclear	At least adult worker bees	Li et al., 2014

the health of honeybees and bumblebees. The impact of these parasites and pathogens can be amplified by other factors such as poor nutrition, a result of degradation or loss of habitat.

DOES CLIMATE CHANGE INFLUENCE THE CROP PRODUCTION VALUE?

The negative effects of pollinator declines caused by climate change include direct economic losses due to reduced crop yields, as well as broader impacts on agricultural activity due to reduced productivity of supporting ecosystems. Although there is concern that the latter impacts are very important, the appropriate causal chains-from reduced animal pollination to population dynamics of wild plant species, changes in food web structure, ecosystem health, and provision of ecosystem services to agriculture-have not yet been systematically developed. Perhaps in recognition of the daunting task, literature on the societal impacts of pollinator decline has focused on the direct implications for agricultural production and global food security. However, the overall economic impact of potential yield losses due to pollination deficiency remains unclear. Kevan (2001) concludes from modeling that there is considerable evidence to suggest that pollinator decline affects agricultural productivity.

The most important role of bees is not only the production of honey and other products of the hive, but also the pollination of plants. Bees visit plants to find their food, nectar, and pollen. In addition to the food, they get from their visit, bees pollinate several crops. Klein et al. (2007) found that 87 crops, accounting for 70% of the 124 main crops used directly for human consumption worldwide, rely on pollinators. Insects (bees) pollination are both an ecosystem service and a widely used production practice among farmers worldwide crop production. It is an ecosystem service in that wild pollinators, including wild bees, contribute significantly to the pollination of a wide range of crops (Greenleaf & Kremen, 2006; Kremen et al., 2002; Winfree et al., 2008). Honeybees are the most cost-effective pollinators. An estimated 35% of human food consumption is directly or indirectly related to insect pollination (Delaplane & Mayer, 2000). In the United States, honeybees contribute an estimated $117 billion annually (Costanza et al., 1997); approximately 84% of cultivated plant species are linked to the activity of these insects (Williams, 1994). *Apis mellifera* is the most cost-effective agricultural pollinator in the world (Klein et al., 2007). According to Agriculture and Agri-Food Canada and the Canadian Association of Professional Apiculturists, the value of bee pollination in Canada could be as high as $1 to $5.5 billion or more per year (Mukezangango & Page, 2017). In Europe, pollinators (honeybees, etc.) contribute at least 22 billion euros per year to the European agricultural industry, and they provide pollination for many crops and wild plants (European Commission, 2021).

Overall, the total economic value of crop pollination has been estimated at €153 billion per annum (Gallai et al., 2009). Among the main pollinating crops, vegetables and fruits are worth around €50 billion each, followed by edible oilseeds, stimulants (coffee, cocoa, etc.), nuts and spices. Honeybees play a central role in agriculture as pollinators and their global economic contribution to food production is estimated to be between $235 and $285 billion per year (Lautenbach et al., 2012). They are valuable to the ecosystem because they pollinate more than 90% of insect-pollinated plants and, as generalists, are essential for buffeting pollination systems (Potts et al., 2010).

IMPACT OF CLIMATE CHANGE ON NECTAR AND POLLEN CHEMISTRY

According to Hoover et al. (2012) the nectar chemistry (e.g. sugars and amino acids) was modified by differences in the concentration of its compounds. The combined impact of CO_2 and nitrogen provoked a change in the ratio of molecules present in nectar, with the amount of sucrose becoming less compared to other sugars, glucose, and fructose (Hoover et al., 2012). In fact, it has been shown that high night temperatures lead to a decrease in the percentage of sucrose present in the nectar, again favoring glucose and fructose for the species *Trifolium repens* (Jakobsen & Kristjansson, 1994).

In addition, a reduced starch concentration in pollen grains, due to exposure to high heat shortly before anthesis (Firon et al., 2006). Similarly, the total sugar concentration in mature pollen can be reduced. It can be noted that other studies have shown the sucrose concentration of pollen increased under the same conditions, with however probable relation with the augmentation of CO_2 concentration in the air (Aloni et al., 2001). The starch concentration was also reduced, a decrease in metabolism possibly explaining all these results. Other important molecules could be disrupted by the effects of climate change, such as volatile organic compounds, molecules that play a role in attracting honeybees (Farré-Armengol et al., 2013). The alteration of the volatile compounds emitted could also affect their physiological and ecological functions (Yuan et al., 2009).

Finally, with regard to the quality of pollen in terms of proteins and amino acids, some studies have previously demonstrated the harmful effect that high temperatures could be induced to these molecules, therefore to the quality of honey. High temperatures applied for a prolonged period can indeed lead to a reduced concentration of soluble proteins (Tang et al., 2008). Similar findings have also been established for an amino acid such as proline (Mutters et al., 1989; Tang et al., 2008).

ADAPTATION AND MITIGATION MEASURES TO MANAGE POLLINATORS AND BIODIVERSITY IN A CHANGING CLIMATE

Adaptation to climate change can be defined as a continuous process of implementing strategies to manage current risks and reduce future risks from the impacts of climate change (Howden et al., 2007). Climate change makes pest infestations more unpredictable and increased their distribution area. They affect directly production yields, the interactions between plants and pests, but their mechanisms remain unclear (Gregory et al., 2009). The adaptive capacity of beekeeping sector will depend on many factors such as sociological, biological and economy. The beekeeper ability to adapt their pest management practices will depend on their financial and social resources (Sutherst et al., 2011). To reduce the risks of appearance and spreading of invasive pests and new diseases in beekeeping sectors, the potential adaptation strategies must also be identified to mitigate the negative effects of existing pests. Integrated pest management, genetic variability and monitoring pest populations are the most commonly mentioned strategies (Raza et al., 2019).

Potential for Adaptation: Genetic Variability

Apis mellifera considered as one of the great species that shown a highly adaptive capacity. It's found all over the world and in diverse types of climates. This capacity to adapt to different ecosystems may allow *A. mellifera* to have an important genetic variability that can be used as a great source to minimize the effect of climate change. It seems to have more adaptive potential than its Asian cousins, which have been subject to little transhumance and have low yields. The human being, with the *A. mellifera* species have co-evolved for many centuries, will certainly be decisive in helping honeybees to survive in hostile environments and in preserving the biodiversity of these species.

Integrated Pest Management Practices

Integrated pest management defined as various measures preventive or direct that aimed to control harmful species of animals (mainly mites and insects) and pathogens taking into account the environmental approach. The decision to use the control practices must be taken by referring to scientifically validated thresholds (Boller et al., 2004). In fact, integrated pest management programs on beekeeping have focused on decisions based on databases indicated how many pests can be supported before economic yield losses occur, also known as economic thresholds. When the intervention decision is made based on the appropriate data often the pest management has yielded good results. The intervention done via various integrated pest management techniques.

The climate change induced a rapid development of insects when the temperatures increased, that participate to the faster development of pest populations, consequently the augmentation of hives damage at earlier than currently expected. Therefore, treatment based on chemical products led to the persistence of their residues in the environment and the appearance of resistance to many synthetic products in the pest populations (Vadez et al., 2012). Furthermore, it will be necessary to develop the new and efficient biological control agents to be used in management practices to control the pests of honeybees.

Efficient control of the honeybee parasites should be focus on the development of alternative pest management strategies and new formulations of pesticides as well as products with repellent and attractant activities. Recently, pesticides based on essential oils have become great alternative ways to control different types of pests and can have an important impact in the development of integrated pest management programs. For example, Alahyane et al. (2021) in their work, showed the acaricidal effects of some biopesticides on the honeybee ectoparasite mite, *V. destructor*. The results of this study indicate that different concentration of environmentally friendly pesticides (essential oils) in terms of their fumigant and repellent activities. As a result, they showed that the high concentration of applied essential oils killed and repelled individuals of treated species significantly. These results could be used as a good basis for the development of new pesticides for sustainable and environmentally friendly pest management.

MONITORING ABUNDANCE AND DISTRIBUTION

The monitoring of pest populations on honeybees allow to collect the baseline data through which we can determinate the long-term effect of climate change on population dynamics of pest species (Yamamura et al., 2006). This data constituted an important tool to predict the future evolution of population dynamics (Andrew & Hill, 2017). Long term monitoring of the pest populations of honeybees may provide some valuable information in biological response to climate change (Halsch et al., 2021; Heeb et al., 2019). The climate changes induced several changes in the dynamic of diseases, vectors and host populations at different levels and which need to monitor. Honeybees attacked by new invasive species in the worldwide, stimulated by climate change inducing the changes in the geographical distribution of pests. Effective management and monitoring are necessary to limit the potential effects of invasive species and therefore to prevent them from becoming a species with the economic importance in the invaded regions (Hellmann et al., 2008).

Mitigating the impact of pests and diseases on bees will need an integrated understanding of pest interactions or host-pathogen and the role of alternate hosts and vectors in disease epidemiology. Beekeeping surveillance programs remain essential to control the spread of diseases that caused by the movement of colonies and their products (Moritz et al., 2010). Interventions such as improved bee husbandry and innovative disease treatments (e.g., inoculating bees' colonies with lactic acid bacteria to inhibit gut pathogens or molecular technology to treat diseases) could help limit the virulence of parasites and pathogens (Moritz et al., 2010).

ECOSYSTEM SERVICES OF HONEYBEES

Wild bees play a fundamental ecological and economic role in agroecosystems. They considered as the most economically important pollinating insect (Kremen et al., 2007), because of their high flower constancy and reliability (Garibaldi et al., 2013; Grüter & Ratnieks, 2011; Rader et al., 2016). These shows that bees have the ability to actively search for flowers and are able to identify the species they prefer to visit. Many studies reported that the honeybee populations decline and have raised an alarm over their ecosystem services and their conservation (Cresswell et al., 2012; Goulson et al., 2008; Potts et al., 2010). This requires urgent measures that allow to conserve honeybee breeds and ecotypes in order to limit the loss of bee biodiversity. The decline explained by the anthropic activities such as deforestation, overuse of chemical products in agricultural lands, and especially by climatic change (González-Varo et al., 2013; Le Conte & Navajas, 2008).

FUTURE PERSPECTIVES

Honeybees are unique insects that produce food for humans, as well as substances used as raw materials and pharmaceuticals. In natural ecosystems, this insect represents an important pollinator, contributing to the preservation of plant biodiversity, while pollination in agro-ecosystems can promote plant production. In addition, bees are key bioindicators of environmental pollution and can provide valuable information on the impact of human activities, allowing the implementation of measures to mitigate risks to human and ecosystem health (Papa et al., 2022).

Misuse of agricultural pesticides is known to be detrimental to the development of honeybee colonies. Similarly, it is imperative to improve the understanding of the relationship between pesticide use and wildlife. The development of a field tool to measure the impact of pesticides on natural populations would allow for better protection of these species. A better management of these phytosanitary products would allow to improve the natural populations of pollinators which contribute to the increase of agricultural yields. Therefore, it is essential to reduce pollinator mortality resulting from excessive pesticide use, to study the lethal and sublethal effects of chemical products (insecticides, herbicides) on wild pollinators, and to conduct research on the long-term impact of pesticides on natural populations (Bertrand, 2001).

Future research needs to cover a wide range of landscapes, climates, and local circumstances to gain a robust overview of how pathogen dynamics will respond to anthropogenic pressures at different spatiotemporal scales. The interaction of global change pressures may reorganize pollinator assemblages and species interactions in ways that increase the risk of pathogen emergence and epidemics within populations or communities, which may compromise policies to protect pollinators and pollination services.

With respect to further characterization of more radical margins of adjustment, future research could also explore the role of technology-based and conservation-based mitigation strategies. Technology-based strategies include the development of management regimes for more effective control of pollinator pests and pathogens, more efficient mechanized pollen collectors, and plant cultivars that rely less on animal pollination, while conservation-based mitigation strategies include habitat conservation. A finer understanding of substitution and mitigation alternatives will greatly improve our understanding of grower decisions and enhance our ability to characterize the risks associated with declining pollinators and their services. Therefore, beekeeper education programs are essential for bee conservation and maintaining ecological balance. Taking appropriate steps for community involvement can accelerate biodiversity conservation efforts and improve the quality of life for bees and their services.

CONCLUSION

In this chapter, we have described the impacts of climate change on honeybees and their services. Also, we have demonstrated the interactions of climate change with honeybee floral hosts and their potential effect on food availability. Climate change is a multifaceted global phenomenon with serious implications for the distribution and abundance of a wide range of ecosystems and organisms, including pollinators and plants. Honeybees play a key role in providing a variety of ecosystem services. They are also considered the primary pollinator of approximately 87% of plant production and face multiple threats related to global environmental change. The global economic value of crop pollination services was estimated at 153 billion euros per year, representing about 9.5% of the total value of global agricultural food production. Honeybee populations have suffered several threats in recent years due to climate change, resulting in colony collapse, decreased colony density, and the development of pathogens and pests that can pose direct risks to bees and their services. To address this issue, it has become paramount to follow mitigation strategies and sustainable management practices to ensure the health of bees as a driving force in the functioning of natural ecosystem services. Beekeepers will also need to change rearing practices to help bee populations to adapt to climate change, shifting from ecotypes or local breeds to more suitable species. This means that measures must be taken to conserve honeybee breeds and archetypes to limit the loss of bee biodiversity. One attractive technique is sperm cryopreservation. Finally, a clear demonstration of the multiple threats to pollinators, especially bees, must be factored into the joint decisions of government agencies, non-governmental organizations, and the agrochemical, food, and other industries to ensure that the plant-pollinator relationship serves a productive and sustainable natural ecosystem.

REFERENCES

Adler, L. S., Barber, N. A., Biller, O. M., & Irwin, R. E. (2020). Flowering plant composition shapes pathogen infection intensity and reproduction in bumble bee colonies. *Proceedings of the National Academy of Sciences of the United States of America, 117*(21), 11559–11565. doi:10.1073/pnas.2000074117 PMID:32393622

Alahyane, H., Ouknin, M., Aboussaid, H., El Messoussi, S., Costa, J., & Majidi, L. (2022). Biological activities of essential oils from Moroccan plants against the honey bee ectoparasitic mite, *Varroa destructor. International Journal of Acarology, 48*(1), 50–56. doi:10.1080/01647954.2021.2015436

Aloni, B., Peet, M., Pharr, M., & Karni, L. (2001). The effect of high temperature and high atmospheric CO_2 on carbohydrate changes in bell pepper (*Capsicum annuum*) pollen in relation to its germination. *Physiologia Plantarum, 112*(4), 505–512. doi:10.1034/j.1399-3054.2001.1120407.x PMID:11473710

Andrew, N. R., & Hill, S. J. (2017). Effect of climate change on insect pest management. In M. Coll & E. Wajnberg (Eds.), *Environmental Pest Management: Challenges for Agronomists, Ecologists, Economists and Policymakers* (1st ed., pp. 197–215). John Wiley and Sons Ltd. doi:10.1002/9781119255574.ch9

Bailey, L., & Ball, B. V. (1991). *Index-Honey Bee Pathology* (2nd ed.). Academic press.

Bailey, L., & Fernando, E. F. W. (1972). Effects of sacbrood virus on adult honey-bees. *Annals of Applied Biology*, *72*(1), 27–35. doi:10.1111/j.1744-7348.1972.tb01268.x

Bertrand, J. (2001). *Agriculture et biodiversité: un partenariat à valoriser*. Educagri éditions.

Böhme, F., Bischoff, G., Zebitz, C. P., Rosenkranz, P., & Wallner, K. (2018). Pesticide residue survey of pollen loads collected by honeybees (*Apis mellifera*) in daily intervals at three agricultural sites in South Germany. *PLoS One*, *13*(7), e0199995. doi:10.1371/journal.pone.0199995 PMID:29979756

Boller, E. F., Avilla, J., Joerg, E., Malavolta, C., Wijnands, F. G., & Esbjerg, P. (2004). *Integrated Production: Principles and Technical Guidelines*. IOBC/WPRS. Available online: https://www.iobcwprs.org/ip_ipm/01_IOBC_Principles_and_Tech_Guidelines_2004.pdf

Brown, M. J., Dicks, L. V., Paxton, R. J., Baldock, K. C., Barron, A. B., Chauzat, M. P., Freitas, B. M., Goulson, D., Jepsen, S., Kremen, C., Li, J., Neumann, P., Pattemore, D. E., Potts, S. G., Schweiger, O., Seymour, C. L., & Stout, J. C. (2016). A horizon scan of future threats and opportunities for pollinators and pollination. *PeerJ*, *4*, e2249. doi:10.7717/peerj.2249 PMID:27602260

Chantawannakul, P., Ramsey, S., Khongphinitbunjong, K., & Phokasem, P. (2018). Tropilaelaps mite: An emerging threat to European honey bee. *Current Opinion in Insect Science*, *26*, 69–75. doi:10.1016/j.cois.2018.01.012 PMID:29764663

Chen, D., Guo, R., Xiong, C., Zheng, Y., Hou, C., & Fu, Z. (2018). Morphological and molecular identification of chalkbrood disease pathogen *Ascosphaera apis* in *Apis cerana cerana*. *Journal of Apicultural Research*, *57*(4), 516–521. doi:10.1080/00218839.2018.1475943

Chen, X. Y. (2013). Harm and control of chalkbrood disease and Aspergillus flavus disease. *Mifeng Zazhi*, *33*, 24.

Chen, Y. P., Evans, J., & Feldlaufer, M. (2006). Horizontal and vertical transmission of viruses in the honeybee *Apis mellifera*. *Journal of Invertebrate Pathology*, *92*(3), 152–159. doi:10.1016/j.jip.2006.03.010 PMID:16793058

Chen, Y. P., Pettis, J. S., Corona, M., Chen, W. P., Li, C. J., Spivak, M., Visscher, P. K., DeGrandi-Hoffman, G., Boncristiani, H., Zhao, Y., Vanengelsdorp, D., Delaplane, K., Solter, L., Drummond, F., Kramer, M., Lipkin, W. I., Palacios, G., Hamilton, M. C., Smith, B., ... Evans, J. D. (2014). Israeli acute paralysis virus: Epidemiology, pathogenesis and implications for honey bee health. *PLoS Pathogens*, *10*(7), e1004261. doi:10.1371/journal.ppat.1004261 PMID:25079600

Costanza, R., d'Arge, R., de Groot, R., Farber, S., Grasso, M., Hannon, B., Limburg, K., Naeem, S., O'Neill, R. V., Paruelo, J., Raskin, R. G., Sutton, P., & van den Belt, M. (1997). The value of the world's ecosystem services and natural capital. *Nature, 387*(6630), 253–260. doi:10.1038/387253a0

Cresswell, J. E., Desneux, N., & VanEngelsdorp, D. (2012). Dietary traces of neonicotinoid pesticides as a cause of population declines in honey bees: An evaluation by Hill's epidemiological criteria. *Pest Management Science, 68*(6), 819–827. doi:10.1002/ps.3290 PMID:22488890

Delaplane, K. S., & Mayer, D. F. (2000). *Crop pollination by bees*. CAB. doi:10.1079/9780851994482.0000

Drossart, M., Rasmont, P., Vanormelingen, P., Dufrêne, M., Folschweiller, M., Pauly, A., Vereecken N. J., Vray, S., Zambra, E., D'Haeseleer, J., & Michez, D. (2019). Belgian red list of bees. *Multidisciplinary assessment of BELgian wild BEE decline to adapt mitigation management policy*, 3–140.

European Commission Agriculture & Rural Development. (2020). *Honey Market Presentation*. Available online: https://ec.europa.eu/info/sites/info/files/food-farming fish eries/animals_and_animal_products/documents/market-presentat ion honey_autumn2020_en.pdf

European Commission. (n.d.). *Honey Bees*. Available online: https://ec.europa.eu/food/animals/live_animals/bees_en

Farré-Armengol, G., Filella, I., Llusia, J., & Peñuelas, J. (2013). Floral volatile organic compounds: Between attraction and deterrence of visitors under global change. *Perspectives in Plant Ecology, Evolution and Systematics, 15*(1), 56–67. doi:10.1016/j.ppees.2012.12.002

Farré-Armengol, G., Filella, I., Llusià, J., Niinemets, Ü., & Peñuelas, J. (2014). Changes in floral bouquets from compound-specific responses to increasing temperatures. *Global Change Biology, 20*(12), 3660–3669. doi:10.1111/gcb.12628 PMID:24817412

Ferguson, L. V., Heinrichs, D. E., & Sinclair, B. J. (2016). Paradoxical acclimation responses in the thermal performance of insect immunity. *Oecologia, 181*(1), 77–85. doi:10.100700442-015-3529-6 PMID:26846428

Firon, N., Shaked, R., Peet, M. M., Pharr, D. M., Zamski, E., Rosenfeld, K., Althan, L., & Pressman, E. (2006). Pollen grains of heat tolerant tomato cultivars retain higher carbohydrate concentration under heat stress conditions. *Scientia Horticulturae*, *109*(3), 212–217. doi:10.1016/j.scienta.2006.03.007

Flores, J. M., Gil-Lebrero, S., Gámiz, V., Rodríguez, M. I., Ortiz, M. A., & Quiles, F. J. (2019). Effect of the climate change on honey bee colonies in a temperate Mediterranean zone assessed through remote hive weight monitoring system in conjunction with exhaustive colonies assessment. *The Science of the Total Environment*, *653*, 1111–1119. doi:10.1016/j.scitotenv.2018.11.004 PMID:30759551

Forsgren, E., Lundhagen, A. C., Imdorf, A., & Fries, I. (2005). Distribution of *Melissococcus plutonius* in honeybee colonies with and without symptoms of European foulbrood. *Microbial Ecology*, *50*(3), 369–374. doi:10.100700248-004-0188-2 PMID:16254762

Free, J. B. (1993). *Insect Pollination of Crops*. Academic Press.

Fries, I. (1991). Treatment of sealed honey bee brood with formic acid for control of Varroa jacobsoni. *American Bee Journal*.

Gallai, N., Salles, J. M., & Vaissiere, B. E. (2009). Évaluation de la contribution économique du service de la pollinisation à l'agriculture européenne. *Bulletin Technique Apicole*, *36*(2), 110–116.

Garibaldi, L. A., Steffan-Dewenter, I., Winfree, R., Aizen, M. A., Bommarco, R., Cunningham, S. A., Kremen, C., Carvalheiro, L. G., Harder, L. D., Afik, O., Bartomeus, I., Benjamin, F., Boreux, V., Cariveau, D., Chacoff, N. P., Dudenhoffer, J. H., Freitas, B. M., Ghazoul, J., Greenleaf, S., ... Klein, A. M. (2013). Wild pollinators enhance fruit set of crops regardless of honey bee abundance. *Science*, *339*(6127), 1608–1611. doi:10.1126cience.1230200 PMID:23449997

Genersch, E. (2010). American Foulbrood in honeybees and its causative agent, *Paenibacillus larvae*. *Journal of Invertebrate Pathology*, *103*, S10–S19. doi:10.1016/j.jip.2009.06.015 PMID:19909971

Gilliam, M. (1990). Chalkbrood disease of honey bees, *Apis mellifera*, caused by the fungus, *Ascosphaera apis*: A review of past and current research. *Proceedings and abstracts, Vth International Colloquium on Invertebrate Pathology and Microbial Control*, 398-402.

González-Varo, J. P., Biesmeijer, J. C., Bommarco, R., Potts, S. G., Schweiger, O., Smith, H. G., Steffan-Dewenter, I., Szentgyorgyi, H., Woyciechowski, M., & Vila, M. (2013). Combined effects of global change pressures on animal-mediated pollination. *Trends in Ecology & Evolution*, *28*(9), 524–530. doi:10.1016/j.tree.2013.05.008 PMID:23746938

Goulson, D., Lye, G. C., & Darvill, B. (2008). Decline and conservation of bumble bees. *Annual Review of Entomology*, *53*(1), 191–208. doi:10.1146/annurev.ento.53.103106.093454 PMID:17803456

Greenleaf, S. S., & Kremen, C. (2006). Wild bees enhance honey bees' pollination of hybrid sunflower. *Proceedings of the National Academy of Sciences of the United States of America*, *103*(37), 13890–13895. doi:10.1073/pnas.0600929103 PMID:16940358

Gregory, P. J., Johnson, S. N., Newton, A. C., & Ingram, J. S. (2009). Integrating pests and pathogens into the climate change/food security debate. *Journal of Experimental Botany*, *60*(10), 2827–2838. doi:10.1093/jxb/erp080 PMID:19380424

Grüter, C., & Ratnieks, F. L. (2011). Flower constancy in insect pollinators: Adaptive foraging behaviour or cognitive limitation. *Communicative & Integrative Biology*, *4*(6), 633–636. doi:10.4161/cib.16972 PMID:22446521

Halsch, C. A., Shapiro, A. M., Fordyce, J. A., Nice, C. C., Thorne, J. H., Waetjen, D. P., & Forister, M. L. (2021). Insects and recent climate change. *Proceedings of the National Academy of Sciences of the United States of America*, *118*(2), e2002543117. doi:10.1073/pnas.2002543117 PMID:33431560

Heeb, L., Jenner, E., & Cock, M. J. (2019). Climate-smart pest management: Building resilience of farms and landscapes to changing pest threats. *Journal of Pest Science*, *92*(3), 951–969. doi:10.100710340-019-01083-y

Hegland, S. J., Nielsen, A., Lázaro, A., Bjerknes, A. L., & Totland, Ø. (2009). How does climate warming affect plant-pollinator interactions. *Ecology Letters*, *12*(2), 184–195. doi:10.1111/j.1461-0248.2008.01269.x PMID:19049509

Hellmann, J. J., Byers, J. E., Bierwagen, B. G., & Dukes, J. S. (2008). Five potential consequences of climate change for invasive species. *Conservation Biology*, *22*(3), 534–543. doi:10.1111/j.1523-1739.2008.00951.x PMID:18577082

Hoover, S. E., Ladley, J. J., Shchepetkina, A. A., Tisch, M., Gieseg, S. P., & Tylianakis, J. M. (2012). Warming, CO_2, and nitrogen deposition interactively affect a plant-pollinator mutualism. *Ecology Letters*, *15*(3), 227–234. doi:10.1111/j.1461-0248.2011.01729.x PMID:22221802

Howden, S. M., Soussana, J. F., Tubiello, F. N., Chhetri, N., Dunlop, M., & Meinke, H. (2007). Adapting agriculture to climate change. *Proceedings of the National Academy of Sciences of the United States of America, 104*(50), 19691–19696. doi:10.1073/pnas.0701890104 PMID:18077402

Jakobsen, H. B., & Kritjánsson, K. (1994). Influence of temperature and floret age on nectar secretion in *Trifolium repens* L. *Annals of Botany, 74*(4), 327–334. doi:10.1006/anbo.1994.1125

Kerr, J. T., Pindar, A., Galpern, P., Packer, L., Potts, S. G., Roberts, S. M., Rasmont, P., Schweiger, O., Colla, S. R., Richardson, L. L., Wagner, D. L., Gall, L. F., Sikes, D. S., & Pantoja, A. (2015). Climate change impacts on bumblebees converge across continents. *Science, 349*(6244), 177–180. doi:10.1126cience.aaa7031 PMID:26160945

Kevan, P. G. (2001). Pollination: a plinth, pedestal, and pillar for terrestrial productivity. The why, how, and where of pollination protection, conservation, and promotion. Bees and Crop Pollination: Crisis, Crossroads, Conservation, 7–68.

Kleijn, D., & Raemakers, I. (2008). A retrospective analysis of pollenhost plant use by stable and declining bumble bee species. *Ecology, 89*(7), 1811–1823. doi:10.1890/07-1275.1 PMID:18705369

Klein, A. M., Boreux, V., Fornoff, F., Mupepele, A. C., & Pufal, G. (2018). Relevance of wild and managed bees for human well-being. *Current Opinion in Insect Science, 26*, 82–88. doi:10.1016/j.cois.2018.02.011 PMID:29764666

Klein, A. M., Vaissière, B. E., Cane, J. H., Steffan-Dewenter, I., Cunnigham, S. A., Kremen, C., & Tscharntke, T. (2007). Importance of pollinators in changing landscapes for world crops. *Proceedings of the Royal Society, 274*(1608), 303–313. doi:10.1098/rspb.2006.3721 PMID:17164193

Klein, A., Vaissiere, B. E., Cane, J. H., Steffan, D., Cunningham, S. A., Kremen, C., & Tscharntke, T. (2007). Importance of pollinators in changing landscapes for world crops. *Proceeding of the Royal Society B: Biological Sciences, 274*, 303–313.

Kluser, S., & Peduzzi, P. (2007). *Global pollinator decline: A literature review.* Academic Press.

Kremen, C., Williams, N. M., & Thorp, R. W. (2002). Crop pollination from native bees at risk from agricultural intensification. *Proceedings of the National Academy of Sciences of the United States of America, 99*(26), 16812–16816. doi:10.1073/pnas.262413599 PMID:12486221

Kremen, C., Williams, N. M., Aizen, M. A., Gemmill-Herren, B., LeBuhn, G., Minckley, R., & Ricketts, T. H. (2007). Pollination and other ecosystem services produced by mobile organisms: A conceptual framework for the effects of land-use change. *Ecology Letters*, *10*(4), 299–314. doi:10.1111/j.1461-0248.2007.01018.x PMID:17355569

Kremen, C., Williams, N. M., Aizen, M. A., Gemmill-Herren, B., LeBuhn, G., Minckley, R., & Ricketts, T. H. (2007). Pollination and other ecosystem services produced by mobile organisms: A conceptual framework for the effects of land-use change. *Ecology Letters*, *10*(4), 299–314. doi:10.1111/j.1461-0248.2007.01018.x PMID:17355569

Laughton, A. M., O'Connor, C. O., & Knell, R. J. (2017). Responses to a warming world: Integrating life history, immune investment, and pathogen resistance in a model insect species. *Ecology and Evolution*, *7*(22), 9699–9710. doi:10.1002/ece3.3506 PMID:29188001

Lautenbach, S., Seppelt, R., Liebscher, J., & Dormann, C. F. (2012). Spatial and temporal trends of global pollination benefit. *PLoS One*, *7*(4), e35954. doi:10.1371/journal.pone.0035954 PMID:22563427

Le Conte, Y., & Navajas, M. (2008). Climate change: Impact on honey bee populations and diseases. *Revue Scientifique et Technique (International Office of Epizootics)*, *27*, 499–510. PMID:18819674

Li, J. L., Cornman, R. S., Evans, J. D., Pettis, J. S., Zhao, Y., Murphy, C., Peng, W. J., Wu, J., Hamilton, M., Boncristiani, H. F. Jr, Zhou, L., Hammond, J., & Chen, Y. P. (2014). Systemic spread and propagation of a plant-pathogenic virus in European honeybees, *Apis mellifera. mBio*, *5*(1), e00898–e13. doi:10.1128/mBio.00898-13 PMID:24449751

Louveaux, J., Albisetti, M., Delangue, M., & Theurkauff, J. (1966). Les modalités de l'adaptation des abeilles (*Apis mellifica* L.) au milieu naturel. *Anné Abeille*, *9*(4), 323–350. doi:10.1051/apido:19660402

McGregor, S. E. (1976). Insect pollination of cultivated crop plants (No. 496). *Agricultural Research Service, US Department of Agriculture.*, *496*, 93–98.

Meeus, I., Vercruysse, V., & Smagghe, G. (2012). Molecular detection of *Spiroplasma apis* and *Spiroplasma melliferum* in bees. *Journal of Invertebrate Pathology*, *109*(1), 172–174. doi:10.1016/j.jip.2011.11.006 PMID:22138255

Memmott, J., Carvell, C., Pywell, R. F., & Craze, P. G. (2010). The potential impact of global warming on the efficacy of field margins sown for the conservation of bumble-bees. *Philosophical Transactions of the Royal Society of London. Series B, Biological Sciences*, *365*(1549), 2071–2079. doi:10.1098/rstb.2010.0015 PMID:20513715

Mevi-Schutz, J., & Erhardt, A. (2005). Amino acids in nectar enhance butterfly fecundity: A long-awaited link. *American Naturalist*, *165*(4), 411–419. doi:10.1086/429150 PMID:15791533

Milićević, V., Radojičić, S., Kureljušić, J., Šekler, M., Nešić, K., Veljović, L., Maksimović Zorić, J., & Radosavljević, V. (2018). Molecular detection of black queen cell virus and Kashmir bee virus in honey. *AMB Express*, *8*(1), 1–5. doi:10.118613568-018-0655-7 PMID:30088183

Moritz, R. F., De Miranda, J., Fries, I., Le Conte, Y., Neumann, P., & Paxton, R. J. (2010). Research strategies to improve honeybee health in Europe. *Apidologie*, *41*(3), 227–242. doi:10.1051/apido/2010010

Mouches, C., Bové, J. M., Tully, J. G., Rose, D. L., McCoy, R. E., Carle-Junca, P., Garnier, M., & Saillard, C. (1983). Spiroplasma apis, a new species from the honey-bee Apis mellifera. In Annales de l'Institut Pasteur/Microbiologie (Vol. 134, No. 3, pp. 383-397). Elsevier Masson.

Mukezangango, J., & Page, S. (2017). *Statistical overview of the Canadian honey and bee industry and the economic contribution of honey bee pollination*. Horticulture and Cross Sectoral Division. Agriculture and Agri-Food Canada.

Mutters, R. G., Ferreira, L. G. R., & Hall, A. E. (1989). Proline content of the anthers and pollen of heat-tolerant and heat-sensitive cowpea subjected to different temperatures. *Crop Science*, *29*(6), 1497–1500. doi:10.2135/cropsci1989.0011183X002900060036x

Ogden, N. H. (2017). Climate change and vector-borne diseases of public health significance. *FEMS Microbiology Letters*, *364*(19). Advance online publication. doi:10.1093/femsle/fnx186 PMID:28957457

Oliver, T. H., Marshall, H. H., Morecroft, M. D., Brereton, T., Prudhomme, C., & Huntingford, C. (2015). Interacting effects of climate change and habitat fragmentation on drought-sensitive butterflies. *Nature Climate Change*, *5*(10), 941–945. doi:10.1038/nclimate2746

Olivier, V., Blanchard, P., Chaouch, S., Lallemand, P., Schurr, F., Celle, O., Dubois, E., Tordo, N., Thiéry, R., Houlgatte, R., & Ribiere, M. (2008). Molecular characterisation and phylogenetic analysis of chronic bee paralysis virus, a honey bee virus. *Virus Research, 132*(1-2), 59–68. doi:10.1016/j.virusres.2007.10.014 PMID:18079012

ONSSA. (2022). *La santé des colonnies d'abeilles au Maroc*. http://www.onssa.gov.ma

Papa, G., Maier, R., Durazzo, A., Lucarini, M., Karabagias, I. K., Plutino, M., Bianchetto, E., Aromolo, R., Pignatti, G., Ambrogio, A., Pellecchia, M., & Negri, I. (2022). The Honey Bee *Apis mellifera*: An Insect at the Interface between Human and Ecosystem Health. *Biology (Basel), 11*(2), 233. doi:10.3390/biology11020233 PMID:35205099

Parachnowitsch, A. L., & Manson, J. S. (2015). The chemical ecology of plant-pollinator interactions: Recent advances and future directions. *Current Opinion in Insect Science, 8*, 41–46. doi:10.1016/j.cois.2015.02.005 PMID:32846674

Park, C., Kang, H. S., Jeong, J., Kang, I., Choi, K., Yoo, M. S., & Chae, C. (2016). In-situ hybridization for the detection of sacbrood virus in infected larvae of the honey bee (*Apis cerana*). *Journal of Comparative Pathology, 154*(2-3), 258–262. doi:10.1016/j.jcpa.2015.12.003 PMID:26852344

Parmesan, C. (2007). Influences of species, latitudes and methodologies on estimates of phenological response to global warming. *Global Change Biology, 13*(9), 1860–1872. doi:10.1111/j.1365-2486.2007.01404.x

Pike, V. L., Lythgoe, K. A., & King, K. C. (2019). On the diverse and opposing effects of nutrition on pathogen virulence. *Proceedings of the Royal Society B, 286*(1906), 20191220. 10.1098/rspb.2019.1220

Potts, S. G., Biesmeijer, J. C., Kremen, C., Neumann, P., Schweiger, O., & Kunin, W. E. (2010). Global pollinator declines: Trends, impacts and drivers. *Trends in Ecology & Evolution, 25*(6), 345–353. doi:10.1016/j.tree.2010.01.007 PMID:20188434

Potts, S. G., Vulliamy, B., Dafni, A., Ne'eman, G., & Willmer, P. (2003). Linking bees and flowers: How do floral communities structure pollinator communities. *Ecology, 84*(10), 2628–2642. doi:10.1890/02-0136

Rader, R., Bartomeus, I., Garibaldi, L. A., Garratt, M. P., Howlett, B. G., Winfree, R., ... Woyciechowski, M. (2016). Non-bee insects are important contributors to global crop pollination. *Proceedings of the National Academy of Sciences of the United States of America, 113*(1), 146–151. doi:10.1073/pnas.1517092112 PMID:26621730

Rasmont, P., & Iserbyt, S. (2012). The Bumblebees Scarcity Syndrome: Are heat waves leading to local extinctions of bumblebees (Hymenoptera: Apidae: Bombus)? *Annales de la Société Entomologique de France*, *48*(3–4), 275–280. doi:10.1080/00379271.2012.10697776

Rasmont, P., Franzén, M., Lecocq, T., Harpke, A., Roberts, S. P., Biesmeijer, J. C., & Schweiger, O. (2015). *Climatic risk and distribution atlas of European bumblebees* (Vol. 10). Pensoft Publishers.

Raza, A., Razzaq, A., Mehmood, S. S., Zou, X., Zhang, X., Lv, Y., & Xu, J. (2019). Impact of climate change on crops adaptation and strategies to tackle its outcome: A review. *Plants*, *8*(2), 34. doi:10.3390/plants8020034 PMID:30704089

Reddy, P. R., Verghese, A., & Rajan, V. V. (2012). Potential impact of climate change on honeybees (*Apis spp.*) and their pollination services. *Pest Management in Horticultural Ecosystems*, *18*(2), 121–127.

Rondeau, S., Giovenazzo, P., & Fournier, V. (2018). Risk assessment and predation potential of Stratiolaelaps scimitus (Acari: Laelapidae) to control *Varroa destructor* (Acari: Varroidae) in honey bees. *PLoS One*, *13*(12), e0208812. doi:10.1371/journal.pone.0208812 PMID:30532269

Ruttner, F. (1988). *Biogeography and taxonomy of honeybees*. Springler. doi:10.1007/978-3-642-72649-1

Sak, B., Sakova, K., & Ditrich, O. (2004). Effects of a novel anti-exospore monoclonal antibody on microsporidial development in vitro. *Parasitology Research*, *92*, 74–80. doi:10.100700436-003-0988-1 PMID:14610668

Schweiger, O., Biesmeijer, J. C., Bommarco, R., Hickler, T., Hulme, P. E., Klotz, S., & Settele, J. (2010). Multiple stressors on biotic interactions: How climate change and alien species interact to affect pollination. *Biological Reviews of the Cambridge Philosophical Society*, *85*(4), 777–795. doi:10.1111/j.1469-185X.2010.00125.x PMID:20184567

Settele, J., Bishop, J., & Potts, S. G. (2016). Climate change impacts on pollination. *Nature Plants*, *2*(7), 1–3. doi:10.1038/nplants.2016.92 PMID:27364126

Settele, J., Kudrna, O., Harpke, A., Kühn, I., Van Swaay, C., Verovnik, R., & Schweiger, O. (2008). Climatic risk atlas of European butterflies (Vol. 1). Sofia: Pensoft. *BioRisk*, *1*, 1–712. doi:10.3897/biorisk.1

Solignac, M., Cornuet, J. M., Vautrin, D., Le Conte, Y., Anderson, D., Evans, J., & Navajas, M. (2005). The invasive Korea and Japan types of *Varroa destructor*, ectoparasitic mites of the Western honeybee (*Apis mellifera*), are two partly isolated clones. *Proceedings. Biological Sciences*, *272*(1561), 411–419. doi:10.1098/rspb.2004.2853 PMID:15734696

Sutherst, R. W., Constable, F., Finlay, K. J., Harrington, R., Luck, J., & Zalucki, M. P. (2011). Adapting to crop pest and pathogen risks under a changing climate. *Wiley Interdisciplinary Reviews: Climate Change*, *2*(2), 220–237. doi:10.1002/wcc.102

Tang, R. S., Zheng, J. C., Jin, Z. Q., Zhang, D. D., Huang, Y. H., & Chen, L. G. (2008). Possible correlation between high temperature-induced floret sterility and endogenous levels of IAA, GAs and ABA in rice (*Oryza sativa* L.). *Plant Growth Regulation*, *54*(1), 37–43. doi:10.100710725-007-9225-8

Thai, P. H., Huyen, N. T., Van Toan, T., & Jung, C. (2018). *Apis cerana* beekeeping and sacbrood disease management in Vietnam. *Journal of Apiculture*, *33*(4), 269–275. doi:10.17519/apiculture.2018.11.33.4.269

Thomas, C. D., Franco, A. M. A., & Hill, J. K. (2006). Range retractions and extinction in the face of climate warming. *Trends in Ecology & Evolution*, *21*(8), 415–416. doi:10.1016/j.tree.2006.05.012 PMID:16757062

Ullah, A., Gajger, I. T., Majoros, A., Dar, S. A., Khan, S., Shah, A. H., Khabir, M. N., Hussain, R., Khan, H. U., Hameed, M., & Anjum, S. I. (2021). Viral impacts on honey bee populations: A review. *Saudi Journal of Biological Sciences*, *28*(1), 523–530. doi:10.1016/j.sjbs.2020.10.037 PMID:33424335

Vadez, V., Berger, J. D., Warkentin, T., Asseng, S., Ratnakumar, P., Poorna Chandra Rao, K., Gaur, P. M., Munier-Jolain, N., Larmure, A., Voisin, A.-S., ... Zaman, M. A. (2012). Adaptation of grain legumes to climate change: A review. *Agronomy for Sustainable Development*, *32*(1), 31–44. doi:10.100713593-011-0020-6

Walter, J. (2020). Dryness, wetness and temporary flooding reduce floral resources of plant communities with adverse consequences for pollinator attraction. *Journal of Ecology*, *108*(4), 1453–1464. doi:10.1111/1365-2745.13364

Watanabe, M. E. (1994). Pollination worries rise as honey bees decline. *Science*, *265*(5176), 1170. doi:10.1126cience.265.5176.1170 PMID:17787573

Williams, I. H. (1994). The dependence of crop production within the European Union on pollination by honey bees. *Agricultural Zoology Reviews*, *6*, 229–257.

Winfree, R., Williams, N. M., Gaines, H., Ascher, J. S., & Kremen, C. (2008). Wild bee pollinators provide the majority of crop visitation across land-use gradients in New Jersey and Pennsylvania, USA. *Journal of Applied Ecology*, *45*(3), 793–802. doi:10.1111/j.1365-2664.2007.01418.x

Yamamura, K., Yokozawa, M., Nishimori, M., Ueda, Y., & Yokosuka, T. (2006). How to analyze long-term insect population dynamics under climate change: 50-year data of three insect pests in paddy fields. *Population Ecology*, *48*(1), 31–48. doi:10.100710144-005-0239-7

Zaragoza-Trello, C., Vilà, M., Botías, C., & Bartomeus, I. (2021). Interactions among global change pressures act in a non-additive way on bumblebee individuals and colonies. *Functional Ecology*, *35*(2), 420–434. doi:10.1111/1365-2435.13703

Zhang, X., He, S. Y., Evans, J. D., Pettis, J. S., Yin, G. F., & Chen, Y. P. (2012). New evidence that deformed wing virus and black queen cell virus are multi-host pathogens. *Journal Invertebrate Pathology, 109*, 156–159. doi: .09.010 doi:10.1016/j.jip.2011

Chapter 8
Honey Bees as Environmental Biomonitors and Effects of Climate Change on Their Population

Debraj Biswal
https://orcid.org/0000-0002-7317-6774
Government General Degree College at Mangalkote, India

ABSTRACT

Honey bees (Apis sp.) have gained research attention lately because of their role in pollination, food production, and biodiversity conservation. Additionally, they can serve as excellent environmental biomonitors because they can concentrate pollutants in their products like honey and beeswax. Interestingly, honey bee colonies are resilient to stressors while the individuals are sensitive to them. Consequently, the negative impacts of stressors on their population tends to go unnoticed. However, research shows their population is declining rapidly. Pollution, intensification of agriculture, habitat loss, fragmented landscapes, reduction of floral heterogeneity, and climate change have been reasoned as the major factors for this observation. Their ectothermic nature makes them sensitive to climatic conditions which may have long-term effects on the environment. However, this field has been poorly investigated. Further research endeavours can provide sufficient information to plan conservation strategies for honey bee populations and ensure future food security.

DOI: 10.4018/978-1-6684-4824-3.ch008

INTRODUCTION

The honey bees (*Apis* spp.) are well known for the ecosystem services they provide in the form of pollination. According to reports, about 84% of the species of cultivated plants along with 35% of agricultural crops depend on the activity of insect pollinators including honey bees (Klein et al., 2007). Additionally, the honey bees help to pollinate several species of entomophilous wild plants (Michener, 2000). Thus, they are equally important for the agricultural industry as well as for the conservation of biodiversity. The honey bees benefit from their mutualistic association with plants by obtaining nutrient-rich nectar and/or pollen from the flowers. In fact, the major source of protein for the bee colony is pollen which nourishes both the brood as well as the adult bees (Flores et al., 2019). Recently, some studies have reported severe decline in the population of honey bees (Reddy et al., 2012). This threatens the agricultural sector as well as the global biodiversity.

Investigations into the reasons for the vanishing population of honey bees reveal pollution, intensification of agriculture, introduction of new pesticides, habitat loss, fragmented landscapes, reduction of floral biodiversity and climate change as the major factors (Vasiliev and Greenwood, 2021). Additionally, invasion by predators, pathogens and parasites also cause destruction to their population (Abou-Shaara, 2016). Despite these findings, the issue of eroding population of honey bees has been neglected for long because honey bee colonies are resilient to stressors even though the individuals are sensitive to them enabling many of the negative impacts to go unnoticed (Cunningham et al., 2022). Interestingly, the honey bees can accumulate pollutants without collapsing. They do so by concentrating the pollutants in their products like honey, beeswax, stored pollen, etc. (Cunningham et al., 2022). As such, analysis of these bee products can help to gain qualitative and quantitative information about the environmental condition of the adjoining areas. Thus, the honey bees have emerged as excellent candidates for monitoring environmental quality (Cunningham et al., 2022).

Among all the factors affecting honey bees, it is expected that climate change will have serious repercussions in the future. Their sensitivity to climatic factors arise from their ectothermic nature. The existence of different honey bee ecotypes can aggravate the problems further because they may face challenges in adjusting themselves to the new climate regimes (Le Conte and Navajas, 2008). It may eventually lead to their colony collapse (Flores et al., 2019). Studies show that alterations in climatic conditions will have direct effects on their physiology, behaviour, development and distribution (Reddy et al., 2012). Additionally, the effects of climate change on plant phenology and their natural enemies (i.e. predators, parasites and pathogens) can also exert indirect effects on honey bee population (Reddy et al., 2012). Under such circumstances, the honey bees may not only be deprived of nutritional resources but also be exposed to new and emerging agents of infection (Reddy et al., 2012).

Despite these findings, the field lacks extensive research. Further studies in this sector can help the agricultural practitioners, the environmentalists and the policy makers plan conservation strategies for honey bees (*Apis* spp.). This will not only form a protective measure against the future food security but also an important tool for the conservation of biodiversity and maintenance of ecological balance. Based on this background, the current chapter tries to explore the diversity and importance of honey bees (*Apis* spp.) together with the means by which climate change can exert negative impacts on their population. Last but not the least, it also tries to suggest mitigation strategies for the emerging problems.

HONEY BEES

There are ten different species of honey bees belonging to the genus, *Apis* that are scattered worldwide (Arias and Sheppard, 2005). Research shows that *Apis* originated in Asia with nine species prevalent in this region. Of these, the major species include *A. cerana*, *A. laboriosa*, *A. florea*, *A. dorsata*, *A. andreniformis*, *A. nigrocincta*, *A. nuluensis* and *A. koschevnikovi* (Franck et al., 2000). The species, *Apis mellifera*, however, has its origin in Africa with subsequent colonization events in Eurasia followed by their exports to other continents. This honey bee species is commonly called the European honey bee and is the most widely distributed species that has been recorded from different climate regimes existing worldwide (Le Conte and Navajas, 2008). The European races of the species, *Apis mellifera* has gained importance because they are very efficient in harvesting honey with intricate strategies for storage of pollen. This helps them to maintain a steady food supply during the harsh winter season of the temperate zones (Martin and Medina, 2004). On the other hand, the Asian races of *Apis mellifera* are not very good harvesters because of the tropical climate to which they are adapted. The warm tropics provide them with plentiful resource supply (in the form of flowers) all through the year, making them less dependent on honey and pollen reserves (Ruttner, 1988). *Apis mellifera* is considered to have diverged from a closely related species, *Apis cerana* approximately eight million years ago (Ruttner, 1988). *A. cerana* is prevalent in Japan and the adjoining areas. It has high similarities with *A. mellifera* with respect to its genetics, morphology and behaviour (Le Conte and Navajas, 2008). All the four major honey bee species, namely, *A. mellifera*, *A. cerana*, *A. dorsata* and *A. florea* have been reported from India (Reddy et al., 2012).

The most widely distributed honey bee species, *Apis mellifera*, has been subdivided into several subspecies from time to time. The taxonomic approaches

Table 1. Evolutionary lineages of different honey bee (Apis mellifera) subspecies together with their geographical distribution (Based on Ilyasov et al., 2020)

Serial no.	*Apis mellifera* subspecies	Geographical distribution	Evolutionary lineage
1	*Apis mellifera unicolor*	Africa	**A**
2	*A. m. sahariensis*		
3	*A. m. scutellata*		
4	*A. m. adansonii*		
5	*A. m. litorea*		
6	*A. m. monticola*		
7	*A. m. capensis*		
8	*A. m. intermissa*		
9	*A. m. simensis*		
10	*A. m. ruttneri*	The Middle East and Western Asia	
11	*A. m. meda*	The Middle East and Western Asia	**A** (sublineage **Z**)
12	*A. m. syriaca*		
13	*A. m. anatoliaca*		
14	*A. m. mellifera*	The Middle East and Western Asia	**M**
15	*A. m. sinisxinyuan*		
16	*A. m. iberiensis*	Europe	
17	*A. m. caucasia*	The Middle East and Western Asia	**C**
18	*A. m. rodopica*	Europe	
19	*A. m. ligustica*		
20	*A. m. cecropia*		
21	*A. m. cypria*		
22	*A. m. carnica*		
23	*A. m. macedonica*		
24	*A. m. carpathica*		
25	*A. m. siciliana*		
26	*A. m. adami*		
27	*A. m. lamarckii*	Africa	**O**
28	*A. m. pomonella*	The Middle East and Western Asia	
29	*A. m. remipes*		
30	*A. m. jemenitica*	Africa	**Y**
31	*A. m. sossimai*	Europe	**C/O** (undefined lineage)
32	*A. m. artemisia*		
33	*A. m. taurica*		

undertaken by researchers include morphometric studies, allozyme based analyses, mitochondrial DNA (mtDNA) based techniques and nuclear DNA (nDNA) based methods (Ilyasov et al., 2020). This has often led to several confusions that exist even today. Surprisingly, the morphometric descriptions of some honey bees are not backed by molecular evidence (Arias and Sheppard, 1996). Literature shows that the honey bee subspecies have been described as geographic races leading to the issues of allocating them proper geographic locations and describing their genetic distribution in Europe, Africa and the Near East (Ruttner, 1988; Franck et al., 2000; Dukku, 2016). The description of evolutionary lineages of honey bees also lacks important clarifications and has been revised several times. Initially, twenty *A. mellifera* subspecies were divided into three lineages – 'A', 'M' and 'C' based on morphometric studies (Ruttner et al., 1978). Subsequently, the morphometric studies were combined together with molecular studies to describe twenty five honey bee subspecies under four lineages – 'A', 'M', 'C' and 'O' (Ruttner, 1988). The latest studies describe about thirty three *A. mellifera* subspecies under five evolutionary lineages – 'A', 'M', 'C', 'O' and 'Y' (Ilyasov et al., 2020) (Table 1). The sublineage 'Z' (belonging to the lineage 'A') and the undefined lineage – 'C/O' have also been described in these studies (Ilyasov et al., 2020) (Table 1). These 33 subspecies are distributed in Europe (with 13 subspecies), Africa (with 11 subspecies) and the Middle East and Western Asia (with 09 subspecies) (Ilyasov et al., 2020) (Table 1). However, this does not seem to be universal because the subspecies can be assigned to different evolutionary lineages depending on the technique used during their taxonomic studies (Ilyasov et al., 2020). For instance, morphometric studies place *A. m. caucasia* in the lineage 'O' (Ruttner, 1988) while those based on mt DNA or allozyme place them in the lineage 'C' (Smith, 1991; Franck et al., 2000). According to Ilyasov et al. (2020), the anomalies in *A. mellifera* taxonomy arise due to their genetic plasticity and the ease of gene flow among the subspecies because of the absence of strict mechanisms of reproductive isolations among them. This also highlights their tendencies of genetic homogenizations under adverse conditions which will be discussed later in this chapter.

HONEY BEES AS ENVIRONMENTAL BIOMONITORS

Honey bees usually travel a long way to collect pollen, nectar, resin and other important resources for maintaining their colony. Consequently, the pollutants present in the environment can adhere to the body of honey bees during their foraging activities (Cunningham et al., 2022). The primary sources from where they pick up the contaminants include plants, soil, air and water. The pollutants can also be ingested by the honey bees while collecting pollen or nectar or even while sipping water (Cunningham et al., 2022). In a similar manner, the pathogens present in their food sources or on adjacent surfaces can also be accumulated by the honey bees. Furthermore, they can concentrate the accrued contaminants in their products such as beeswax and honey (Zarić et al., 2017). These products can, thereby, act as environmental biomonitors apart from the honey bees themselves (Cunningham et al., 2022). Thus, their importance as biomonitors are manifold.

For this purpose, the honey bees, mostly the foragers, are sampled from the hives either at their entrance or from their outer frames. Their products like honey, wax and/or stored pollen can also be collected and analysed by various techniques [such as gas–liquid chromatography, atomic absorption spectrometry, high throughput sequencing and polymerase chain reaction (PCR)] from the hive frames. Moreover, pollen traps may be utilized to sample the freshly collected pollen by the forager bees (Cunningham et al., 2022). Their roles in monitoring some of the major environmental contaminants have been discussed below.

Pesticides

The importance of studying pesticides lies in the fact that they can persist in the environment for a long time and continue to exert toxic effects on organisms coming close to them (de Oliviera et al., 2016). Moreover, biomonitoring the environment for the presence of pesticides has been found to be informative for both the honey bees and other insect pollinators because of the overlaps in their foraging habits (Bishop et al., 2020). Therefore, analysing the pesticides in an area by using honey bees can give an idea about the pesticide-related potential threats to the pollinators of the area (Cunningham et al., 2022).

The honey bees usually die when they are exposed to lethal doses of pesticides. However, they can withstand the sub-lethal doses of pesticides and can, thereby, serve as biomonitors of pesticides (both individually and as a colony) (Cunningham et al., 2022). Insecticides and pesticides have been reported to have negative effects on colony survival (Wu et al., 2011). Additionally, in sub-lethal doses they can impede the memory of honey bees and their foraging behaviour, apart from decreasing their immunity (DesJardins et al., 2021).

The presence and concentration of the pesticides in honey bee products largely depend on their nature as well as their transfer ratios (Niell et al., 2017). Pesticides that are volatile usually have poor transfer ratio (i.e. their transfer from various surfaces to the bee products) (Niell et al., 2017). Additionally, the concentrating capacities of the pesticides vary with the bee products. Beeswax, for example, are efficient in concentrating agrochemicals while pollen exhibits the most diverse types of accumulated pesticides (Calatayud-Vernich et al., 2018). Therefore, the diversity and concentration of pesticides in stored pollen can be studied to understand the spatial variations of pesticide contamination of the environment (Calatayud-Vernich et al., 2018). Usually, the pesticide amounts are higher in pollen from bee hives in areas of intensive farming practices compared to grasslands and rural areas (Calatayud-Vernich et al., 2018).

Heavy Metals

Heavy metals are a matter of concern because most of them [e.g. cadmium (Cd), chromium (Cr), copper (Cu), mercury (Hg) and lead (Pb)] are toxic and have serious health implications in humans (Jyothi, 2020). The major sources of heavy metal pollution include smelting of ores, exhausts from vehicles, agrochemical industries and ignition of fossil fuels. They can settle on the surfaces of nearby vegetation or be incorporated into one or more plant products. They can also be dispersed into the air as particulate matters or pollute water bodies nearby (Zarić et al., 2017). The honey bees can be exposed to these heavy metal sources in the environment while foraging. Eventually, they may stick to their body hairs making their way into the hives and get accumulated in the stored pollen (or bee bread), honey, beeswax and propolis. Since this has been found to be non-lethal to the colony they can be used for environmental monitoring (Cunningham et al., 2022).

Ruschioni et al. (2013) showed that the concentration of heavy metals was higher in the forager bees compared to the honey samples. Greater exposure of the bees to pollutants during their foraging activities was put forward as the reason for such observation. Many other workers like Álvarez-Ayuso and Abad-Valle (2017) and Smith and Weis (2020) reported similarly and suggested that the amounts of heavy metal contaminants in bees and their products vary from one site to the other. Moreover, the amounts of heavy metals in the bees and their products have also been shown to vary with the weather changes (Zarić et al., 2017). For example, chromium (Cr) was found to be higher in months without rainfall (Ruschioni et al., 2013). Furthermore, the influence of anthropogenic activities on the concentrations of heavy metals in honey bees and their products have also been observed by Ruschioni et al. (2013). For instance, honey bees and their products from areas with high density of vehicles exhibit greater concentrations of toxic substances compared to the ones located in the city outskirts (Conti and Botrè, 2001). Thus, honey bees as biomonitors can help to analyse the spatial and temporal variations of heavy metal pollutants across a geographical area.

Interestingly, the pattern of accumulation of heavy metals is similar in bee bread and bee tissues (Smith et al., 2019). Propolis is not a good choice to document differences in trace metal contamination across a spatial scale (Smith and Weis, 2020). Beeswax, on the contrary, has been found to be very efficient to monitor heavy metal pollution resulting from exposures to the sources for long duration.

This is because of the lipid base of beeswax that helps by acting as a reservoir of pollutants (Calatayud-Vernich et al., 2018). As far as the bees are concerned, there are experimental evidences of augmented expression of metallothionein encoding genes in forager bees due to increased exposure to heavy metals like Pb, Cr and Cd (Purać et al., 2019). Metallothioneins help to neutralize the effects of heavy metals shielding the honey bees from their toxic effects (Coyle et al., 2002). Thus, according to Purać et al. (2019), alteration in expression profiles of AmMT (metallothionein gene) in honey bees can reflect the states of Pb, Cu and Cd pollution in the environment. This gives a scope for application of molecular biology techniques in assessment of environmental quality.

Other Environmental Pollutants

In addition to the pesticides and heavy metals, the persistent organic pollutants (POPs) are also detrimental to environmental quality. They have the ability to travel long distances through aerial and aquatic environments because they are not easily degraded (Wania and MacKay, 1996). Therefore, these pollutants can bioaccumulate and biomagnify (Montory et al., 2020). Above all, they tend to be transported biologically by migratory species heightening their roles as environmental toxicants (Montory et al., 2020). Bee-based monitoring has been employed to study mainly two classes of POPs – the polycyclic aromatic hydrocarbons (PAHs) and polychlorinated biphenyls (PCBs) (Villalbla et al., 2020).

PAHs are naturally present in coal, gasoline and crude oil and are released to the atmosphere upon their combustion. Some common sources include burning of woods, cigarette smoke, fumes from asphalt roads and exhausts from vehicles (Cunningham et al., 2022). They are studied extensively because of their carcinogenic properties apart from being irritants to the eyes and lungs of humans (De La Torre-Roche et al., 2009). Honey bees have been reported to be very good indicators of PAH pollution (Lambert et al., 2012). Earlier, Perugini et al. (2009) reported different types of PAHs from worker bees and honey sampled at two different locations in Italy. Additionally, the concentrations of PAHs determined in honey bees or their products are influenced by the location of the bee hives (Lambert et al., 2012). The spatial variations of the amounts of PAHs present in honey could act as an indicator of anthropogenic activities like combustion of diesel or petrol (Al-Alam et al., 2019).

PCBs are synthetic organochlorine chemicals obtained mainly from discarded electrical equipment in landfills and burning of municipal wastes (Cunningham et al., 2022). Vapours from contaminated water bodies also contribute PCB to the environment (Cunningham et al., 2022). In humans, they have been reported to supress the immune system, cause cancer and increase the chances of cardiovascular diseases (Carpenter, 2006). Honey bees themselves have been found to be the best indicator for PCB contamination compared to their products (i.e. honey and pollen) (Sari et al., 2020). This is accentuated by the fact that PCBs show seasonal variation of contamination in honey bees with the highest proportion being recorded during the dry months with elevated temperatures (Sari et al., 2020). Higher rates of vapourization of PCBs from polluted soils cause greater exposure of honey bees to these contaminants during the hot and dry periods of the year (Sari et al., 2020). Poor water-solubility of PCBs does not make them easily available for uptake by the plants leading to their low amounts in pollen grains (Sari et al., 2020). Whatever data is available for PCB-contamination of pollen is primarily based on their surface contamination. Thus, bee-collected pollen is not an efficient indicator of PCB-contamination (Sari et al., 2020).

New/Emerging Pathogens

Plants can be affected by a wide variety of pathogens including fungi, bacteria and virus often causing huge agricultural losses. Many of these pathogens are transmitted by insect vectors (McArt et al., 2014). Honey bees, however, can transport these pathogens indirectly when they get adhered to their body while foraging (Pattemore et al., 2014). There are approximately 45 viruses that can be transmitted through pollen (Bhat and Rao, 2020). Some of these common viruses include *Ilavirus*, *Sobemovirus*, *Nepovirus* and *Tobamavirus* (Bhat and Rao, 2020). Currently, high throughput sequencing (HTS) technologies are emerging as affordable and efficient techniques to analyse complicated environmental samples for the presence of pathogens (Tremblay et al., 2019). The technologies have been found to be very useful to study bee-collected pollens for this purpose (Roberts et al., 2018).

Honey bee as biomonitors offer many advantages over regular monitoring techniques because the bees can travel long distances and can collect pollens from a wide variety of plants much faster than manual methods (Greenleaf et al., 2007). Additionally, they are very efficient gatherers of pollen making them excellent samplers (Roberts et al., 2018). Thus, they can be used to detect the prevalence of any pathogen(s) or their early detection much easily compared to isolation of pathogens directly from the plants (Tremblay et al., 2019). The surveillance of these pathogens are very important because they can be transmitted by insect pollinators (including honey bees) even to the self-fertilized crop plants (Roberts et al., 2018). The results can be used as warning signals for a disease outbreak and encourage undertaking necessary steps to prevent agriculture-related financial loses (Cunningham et al., 2022).

Bee-collected pollen is a better candidate for sampling pathogens compared to the honey bee themselves because pollens exhibit much more diverse species of plant and bee-related pathogens (Tremblay et al., 2019). Furthermore, the ease of collection of pollen from the bee hive makes such studies inexpensive and easier (Tremblay et al., 2019). Roberts et al. (2018) reported that bee-based surveillance coupled with HTS techniques are very useful for examining the emergence of exotic plant viruses since it is very difficult to differentiate infections caused by them from the ones caused by endemic viruses. This is because of the similarities of the symptoms they produce in plants (Roberts et al., 2018). They suggested the study of honey bee samples after being pooled together and using the information obtained therefrom in plant biosecurity projects.

Besides monitoring plant pathogens, honey bee based surveys can also be utilized to have an idea about the pathogens prevalent in the environment that affect different bee species (Di Prisco et al., 2016). This is important because many diseases have been reported to spread between managed and wild bees often causing disease spill over with dire consequences on ecosystem health (Grassl et al., 2018). For example, the presence of deformed wing virus (DWV) in bumble bee (*Bombus* spp.) population have often been associated with its presence in honey bee population as well (Manley et al., 2019). According to de Sousa Pereira et al. (2019) bee-collected pollen can also be analysed for the same purpose. Both the plant and bee pathogens can perturb the future food security which is a pertinent issue in the current scenario of climate change. This necessitates monitoring the environment for the presence of these pathogens to take precautionary steps in a timely manner (Tremblay et al., 2019).

THREATS TO THE POPULATION OF HONEY BEES

The major threats to the honey bee population have been enumerated below:

Source of Nutrition

The quantity and quality of pollen and nectar produced varies among different species of flowering plants. According to the observation of Alaux et al. (2017), honey bees are able to withstand harsh winters better if they have access to more diverse pollen supplies during the flowering season. This prompted Requier et al. (2015) to hypothesize that they are more concerned about the quantity of pollen resources than the nutritional quality. However, anthropogenic activities like monocultures and planting of exotic species (to which the local honey bee population is not well adapted) largely limit the pollen diversity with adverse effects on the health of honey bees (Vasiliev and Greenwood, 2021). Needless to say, the effect of climate change on loss of floral species is equally responsible for the declining population of honey bees (Reddy et al., 2012). This is even more destructive to the wild and solitary bee species which exhibit narrow dietary niche preferring only a few host plants to draw their nutrition (Wood et al., 2020).

Pesticides

Investigations on the effect of pesticides on honey bees have given contradictory results because despite their deleterious effects on individual honey bees, the colony somehow buffers against the adversities (Wood et al., 2020). Lucchetti et al. (2018) suggested that the honey bee workers possess capabilities to remove the toxins from the food materials prior to feeding their larvae thereby protecting them. However, many insecticides and fungicides, when present together, can exert synergistic toxic effects on them (Thompson and Wilkins, 2003). In general, wild bees (*Bombus* sp.) are less sensitive to pesticides compared to honey bees (*Apis* spp.) (Wood et al., 2020).

Habitat Loss

The last decade has witnessed an enormous growth of deforestation, industrialization, urbanization and agricultural intensification (Potts et al., 2010). These anthropogenic activities have led to simultaneous reduction of nesting sites and floral resources for the honey bees (Potts et al., 2010). For instance, use of herbicides due to agricultural intensification clears many weeds which form potential floral resource for the honey bees (Holzschuh et al., 2008). Other agrochemicals like insecticides or pollutants from industries may cause mortality of honey bees or impact their physiology negatively by exerting direct toxic effects (Cunningham et al., 2022). Land conversions and/or fragmentations also have negative impacts on their population by reducing the foraging sources and increasing the competition for food among them (Potts et al., 2010). This also happens when managed or feral species are introduced in an area for commercial pollination (Cuthbertson and Brown, 2009).

Introduction of Non-Native Plants and Pollinators

It has been observed that some grass species are so invasive that they may prevent the growth of flowering plants thereby restricting the food resources for the honey bees. Additionally, they may result in sudden onset of fires that can harm the honey bee populations of those areas (Cuthbertson and Brown, 2009). There are several other non-native plant species with ornate flowers which not only serve as alternative sources of nutrition for the honey bees but can also ensure a steady food supply in cases of resource shortage due to unfavourable environmental circumstances (Stout and Morales, 2009). However, these species disrupt the mutualistic associations between honey bees and native plants which may have detrimental ecological impacts in the long run (Potts et al., 2010). Introduction of non-native and/or managed pollinators for crop pollination in an area can not only increase their competition for food resources but can also cause genetic dilution through hybridizations (Vasiliev and Greenwood, 2021). This has negative impacts on pollinator biodiversity by wiping away the ecotypes (Potts et al., 2010). Furthermore, they may introduce parasites and pathogens in areas from where they were previously unrecorded with severe adverse effects on the local honey bee populations (Potts et al., 2010).

Pests and Diseases

The honey bee populations are susceptible to pests and diseases not only from the introduced pollinator populations but also from the worldwide trades of plant species. Exotic pests could be established by these plants which may be destructive to the native vegetation and thereby the nutritional sources of honey bees (Potts et al., 2010). Plant pests can also be introduced by the commercial feral bee species (Cunningham et al., 2022). Alternatively, plant parts may spread honey bee pests as well (e.g. the hive beetle, *Aethiva tumida* through imported fruits). Some common honey bee pathogens transmitted by managed bee species include *Melissococcus plutonius* [causes EFB (European Foulbrood)] and *Paenibacillus larvae* subsp. *larvae* [causes AFB (American foulbrood)] (Cuthbertson and Brown, 2009).

It is very difficult to discern the individual effects of the abovementioned factors on honey bees. Many of these exert synergistic effects which may be devastative for their populations. Above all, climate change has been proposed to be a major driving force behind the declining population of honey bees (Vasiliev and Greenwood, 2021). Recently, the problem of colony collapse disorder in honey bees has gained much research attention. Colony Collapse Disorder (CCD) is a phenomenon in which the number of worker bees decrease drastically in a honey bee colony which surprisingly exhibits an excessive brood population compared to the populations of adult bees. Eventually, the colony dies. Initially, no suitable reason for the colony collapse could be ascertained. Later on, a number of stress factors were identified. These include pathogens and parasites along with intensification of agriculture using excessive pesticides and herbicides (vanEngelsdorp et al., 2009). Lately, climate change has also been added to the list of reasons for the disorder (Flores et al., 2019). The impact of climate change on honey bees has been described in the following section.

CLIMATE CHANGE AND HONEY BEES

Climate change researchers have predicted temperature rises in the near future with changes in precipitation patterns and melting of icecaps (Le Conte and Navajas,

2008). Instead of sustained changes of weather conditions they have anticipated increasing frequency of brief climatic extremes. Many workers have warned about the negative impacts of climate change on pollinators (including honey bees) (Reddy et al., 2012; Abou-Shaara, 2016; Flores et al., 2019; Cunningham et al., 2022). Honey bees are ectothermic in nature and have a high surface-to-volume ratio and thereby tend to be affected by changes in climatic conditions (Vasiliev and Greenwood, 2021). Moreover, the existence of different ecotypes of honey bees all over the world prove the importance of climate in conditioning the honey bee population in a particular area. These ecotypes or races have developed over long periods of association with the dominant climates in the respective geographical areas together with the prevalent aboveground vegetation (Reddy et al., 2012). Therefore, climate change can affect honey bees in two ways – directly by impacting their physiology and life cycle and indirectly by affecting their food sources (i.e. the plants from which they draw nutrition) (Flores et al., 2019; Vasiliev and Greenwood, 2021). They have been elaborated below.

Direct Effects

Physiology and Foraging Behaviour

Honey bee being a colonial species enjoys certain advantages over solitary insect species in combating temperature extremes. For example, they can overwinter in the temperate regions by regulating the temperature of bee hives at conditions that favour their growth and survival (Cunningham et al., 2022). However, temperature rise beyond the threshold limit has been reported to hamper the homeostasis of honey bee colony with negative impacts on their thermoregulatory mechanisms (Cunningham et al., 2022). The outcomes vary with the thermal tolerance and plasticity of the insect pollinators (Reddy et al., 2012). Interestingly, different honey bee races have different thermal tolerance limits because they are adapted to the dominant geo-climatic conditions where they have evolved (Le Conte and Navajas, 2008; Abou-Shaara, 2016). For example, the workers of Carniolan honey bees (popular in Egypt) can tolerate temperatures as high as 49.1°C while those of the Arizonian feral honey bees can tolerate about 50.7°C (Atmowidjojo et al., 1997; Kafer et al., 2012). Additionally, Italian honey bees can tolerate temperatures higher than Carniolan honey bees (Kovac et al., 2014).

Honey bees are ectothermic and require warming for flight activities. The small sized honey bees have high surface-to-volume ratio enabling them to absorb heat quickly when the surrounding temperatures are high (Reddy et al., 2012). Therefore, the ambient temperature influences their foraging behaviour. In general, honey bees cannot tolerate high temperatures. Abou-Shaara (2016) reported that in vitro exposure of worker bees to about 75% relative humidity (RH) and 45°C temperature can cause their death within less than a day. Therefore, the bees need to modify their foraging strategies to avoid the negative impacts of increased temperatures (Abou-Shaara, 2016). Accordingly, the bees forage less frequently in hot summer months so that they can conserve their energies to regulate the temperature of their colonies. They do so by fanning, and/or evaporative cooling (Starks and Gilley, 1999). Le Conte and Navajas (2008) reported that the honey bees often collect water to cool their colonies under conditions of high ambient temperatures. Extremely high temperatures also impact their pollination services by affecting the pollen removal efficiency and its collection by the pollinators. Thus, temperature rises impair the honey harvesting capacity of the pollinators (Reddy et al., 2012). Furthermore, elevated temperatures may cause dehydration of the worker bees due to severe loss of water from their body (Atmowidjojo et al., 1997). This problem has been hypothesized to be aggravated in the coming years because increased environmental temperatures will cause rises in water scarcity (Le Conte and Navajas, 2008). Predictions show that precipitation will also increase in the future but the increased precipitation will probably not balance the thermal stress (Abou-Shaara, 2016). Rainfall patterns, though, will exert negative impacts on the foraging and pollinating behaviour of honey bees like elevated temperature does. This is because the honey bees usually do not leave their hives during the rains (Le Conte and Navajas, 2008).

Like high temperatures, low temperatures can also exert negative effects on the honey bees. For example, temperatures lower than 10°C can hamper the flight activity of honey bees discouraging them to leave their hives (Abou-Shaara, 2016). Research in this field shows that excessive high or low temperature leads to fertility reduction in honey bee queens (McAfee et al., 2020). McAfee et al. (2020) found that honey bees exposed to elevated temperatures have augmented expression of heat shock proteins (HSPs). Thus, variations in expression of HSPs in honey bees can be used as bioindicators of environmental warming (McAfee et al., 2020).

Genetic Effects

The honey bees may respond to climate change by three ways – adapting themselves to the new environment, moving to a more favourable area or by becoming extinct (Reddy et al., 2012). The researchers further suggested that climate change may promote the migration of honey bees towards the higher altitudes and polar regions. As such, the honey bees may explore newer habitats and territories in search for conducive environment and adequate food supplies. This may impact their interaction with other closely related bee species often leading to hybridizations (Le Conte and Navajas, 2008; Vasiliev and Greenwood, 2021). Danks (2006) proposed that the short generation times of insect pollinators gives them an inherent ability to undergo micro-evolution in order to adapt themselves with the changing environment. Moreover, stress factors like temperature and drought usually stimulate directional selection (Gienapp et al., 2008).

Pollinator species like honey bees are usually genetically diverse with races that are adapted to the local climatic regimes (Barghi et al., 2019). This arises due to the existence of genetic variations for a long period of time allowing the forces of natural selection to do their work (Hoffmann and Willy, 2008). Asynchronous range shifts of locally adapted pollinator species may give rise to introgressive hybridization (Hoffmann and Sgro, 2011). Additionally, it may also promote transgressive segregation ultimately giving an admixture of parental forms and other higher order genetic variants (Bergland et al., 2016). This eventually facilitates adaptation of pollinators in response to stress factors (Bergland et al., 2016). According to Salazar et al. (2005), hybridization can also lead to speciation. Based on these observations, it has been hypothesized that climate changes may drive micro-evolutionary trends in pollinators that may accelerate diversity at both intra and interspecies levels (Hill et al., 2011). The interactions may, thereby, prevent extinction of species (Skelly et al., 2007). However, the outcomes are not always simple. In some cases, adaptation to elevated temperatures can cause decline of chromosomal diversity instead of increased genetic variation (Rodrigez-Trelles and Rodrigez, 1998).

The directionality of range shifts in bees has been found to differ for different traits (like voltinism and habitat generalization) (Aguirre-Gutierrez et al., 2017). Several studies show that pollinator groups often perform range shifts in opposite direction causing disintegration of reproductive barriers leading to hybridization (Sanchez-Guillen et al., 2016). Observing the low levels of genetic disparities in range-edge populations the positive outcomes of hybridization seems to be an oversimplified expectation (Kagawa and Takimoto, 2018). This is because it always does not result in formation of stable hybrids or populations with high adaptability (Rhymer and Simberloff, 1996). For example, the hybridization may give rise to poorly adaptive offspring which may die causing a fall in their population (also called demographic swamping) (Todesco et al., 2016). Moreover, introgressive hybridization may cause extinction of rare species wiping away populations that are genetically distinct (also called genetic swamping) (Rhymer and Simberloff, 1996). Furthermore, hybridization can also cause species reversals resulting in erosion of local adaptations and merging of different evolutionary lineages (Allendorf et al., 2001). The net outcome is genetic homogenization (Todesco et al., 2016). The ease with which the losses of genetic diversity can easily be neglected due to lack of accompanying morphological changes further aggravates the loses of genetic diversity in pollinator population because they may remain unrecorded for a long time (Ellstrand and Reiseberg, 2016).

Day length, maximum daily temperature and number of days with mean temperature above a threshold value are the environmental factors that regulate the phenology of honey bees (Reddy et al., 2012). Lengthening of the breeding season of honey bees under altered climatic conditions may increase their chances of mating with closely related species that were previously reproductively isolated resulting in convergence of distinct evolutionary lineages and genetic homogenization (Taylor and Frisen, 2017). Thus, climate change can promote species extinctions (Seehausen et al., 2008). According to Lande (1993), genetic homogenization makes pollinator colonies susceptible to extinction under the pressures of environmental stress factors.

Indirect Effects

Availability of Food Resources

The study of impacts of climate change on pollinators remain incomplete if the effects on vegetation are not considered. Changing temperature and rainfall patterns have been hypothesized to affect the flowering periods of plants which directly impact the nutritional source of pollinators (Thuiller et al., 2005). Mismatches of alterations in plant phenology with the pollinator phenology may disrupt the mutualistic relationships between the plants and the pollinators with dire consequences on ecosystem health (Schweiger et al., 2010). Since this impacts the pollination services, the phenomenon has also been termed as 'pollination crisis' (Reddy et al., 2012).

The development of flowers and the production of pollen and nectar are governed by climate which, in turn, impacts the foraging behaviour of honey bees and their development (Abou-Shaara, 2016). According to Yoruk and Sahinler (2013) and Abou-Shaara (2016), elevated temperatures will result in an earlier flowering season. Harsh environmental conditions like high temperatures and low rainfall can also cause the flowers to wilt earlier in the season (Gordo and Sanz, 2010). This causes shortening of the flowering season that not only reduces the time of gathering pollen and nectar by the pollinators but also tends to increase the period of non-availability of floral resources until the next growing season (Flores et al., 2019). Consequently, the colony suffers from stress due to food scarcity and may succumb to malnourishment (Switanek et al., 2017).

Climate change, with respect to rainfall, occur in both ways – scarcity may lead to drought and excess may lead to flood. Though the exact effects of these events on flowering have not been studied extensively, existing data indicates that honey bee development will be affected adversely in both cases. Too much rain can wash the flowers (as in *Acacia* sp.) and dilute the nectar lowering their attractive influence on honey bees (Le Conte and Navajas, 2008). Dry weather conditions can hamper the production of nectars by flowers resulting in poor availability of resources for harvesting their honey. For example, lavender flowers growing in extremely dry climates are devoid of nectar. Thus, non-availability of nectar can cause mortality in honey bees (Le Conte and Navajas, 2008; Abou-Shaara, 2016).

Asynchronous Phenology of Plants and Honey Bees

Biological timing of the crop and its pollinators is important for pollination to be effective. This is especially important for crops like litchi and mangoes which bloom in masses for brief periods (Reddy et al., 2012). Efficient pollination during this period requires the presence of abundant pollinators. Climatic alterations may have an impact on their timings such that the flowers may not get their pollinators when required (Reddy et al., 2012). The effects may not only be temporal but also spatial where the location of the flowers and their pollinators may not match (due to climate-triggered range shifts). Moreover, the intensity of responses of the plants and pollinators may also vary to the same environmental cue (e.g. temperature) (Hegland et al., 2009). The differences in responses may also depend on the time of the year. Such mismatches are disastrous not only for the flowering plants but also the pollinators (Reddy et al., 2012).

Since the different races of honey bees are adapted to different dominant geo-climatic conditions (at a location), alterations in climate will affect their development (Le Conte and Navajas, 2008). The changes may be such that the honey bee colony may be severely affected because they are tuned to seasonal cycles to maintain themselves and their colony. In the temperate regions, the winters are very cold when the honey bees do not exhibit any activity and depend on the stored food for their survival (Le Conte and Navajas, 2008). With the arrival of spring, as the environment warms up, the queen bee begins laying eggs resulting in colony growth. The worker bees also increase in size. The honey bees forage to collect nectar and pollen from spring blooms. However, if the winter is too long (due to climatic alterations) the resources in the colony may get depleted and the honey bees may die of starvation (Le Conte and Navajas, 2008). Likewise, if the winter is shorter than usual, the honey bee population will be impacted negatively because the early emerging adults may not be able to adapt themselves to cold temperature conditions arising later (Vasiliev and Greenwood, 2021). Or, it may lead to starvation among the pollinators because the flowering season might not have begun by then (Vasiliev and Greenwood, 2021). Similar situations will arise due to occurrence of cold snaps in the spring season. It may be a stress for the honey bees, especially the ones with fast rates of development. They will face the maximal impact of food crisis. It has been observed that hybrid honey bee varieties have faster growth rates compared to the local ecotypes (Le Conte and Navajas, 2008). As such, the local ecotypes grow slowly and can avoid the negative effects of cold snaps. Hybrids are usually reared artificially to increase honey-harvest without considering the environmental fluctuations (Le Conte and Navajas, 2008). Therefore, under natural conditions the local ecotypes can withstand climatic fluctuations better than the hybrid varieties (Le Conte and Navajas, 2008).

Occurrence of unprecedented events of drought in the autumn season will lower the pollen production subsequently declining the pollen reserves for the overwintering honey bees. As such, the developing individuals will have a compromised immune system making them vulnerable to parasites and pathogens often causing mortality in honey bees (Le Conte and Navajas, 2008). Here, it needs to be mentioned that accumulation of too much pollen reserves during the flowering season does not offer an effective solution (Flores et al., 2019). This is because, pollen, when stored, gradually loses its nutritional qualities and is not much preferred by the honey bees as their diet but they are forced to utilize them during periods of food scarcity (Carroll et al., 2017). Additionally, old pollen may possess certain chemicals (e.g. acaricides) concentrated inside them which when consumed by the honey bees may be introduced into their systems. This may threaten their health and survivability (Calatayud-Vernich et al., 2018).

Range Shifts of Honey Bees

Non-availability of resources or climatic extremes may often force the honey bees to undertake migration in search of favourable conditions. Increasing environmental temperatures will make the earlier cold regions conducive for the honey bees encouraging the exploration of such areas (Chown et al., 2007). Diniz et al. (2003) and Pinto et al. (2005) reported that the geographical distribution of Africanised honey bees has now expanded to Argentina and further to the USA. Progression beyond this has not occurred probably because the temperatures are extremely cold to support the Africanised honey bees (Le Conte and Navajas, 2008). However, the migrations may not always be successful since the climate is also a key determinant of spatial distribution of different species of flowering plants from which the honey bees derive their nutrition (Thuiller et al., 2005). Any negative impacts on the diversity and distribution of floral species in a particular region may deprive the migrating honey bees of sufficient nutrition. Global warming has already been proposed to have profound effects on the aboveground vegetation in the form of homogeneity. Homogeneous landscapes offer little opportunity to the pollinators for range shifts to avoid detrimental weather conditions. This decreased micro-climate buffering will reduce the survivability of the pollinators under extreme weather events (Vasiliev and Greenwood, 2021).

The range shifts of honey bee species are further affected by changes in land use patterns (Fuchs et al., 2015). For example, presence of long stretches of monoculture croplands with intensive management can hamper the movement of pollinators because such croplands are unsuitable for them due to deleterious effects of pesticides, herbicides, etc. This has negative impacts on their range shifts (Yan and Roy, 2016). Similar effects may also be generated by infrastructural development of communication or any other land cover unfavourable for the pollinators (Bo and Florio, 2012). These restrictions in the movement of pollinators during range shifts often lead to local extinctions causing decreases in β-diversity (Devictor et al., 2012). The influence of climatic factors on β-diversity of pollinators has also been reported by Kaloveloni et al. (2018). According to Scheffer et al. (2012), loss of β-diversity may impair the resilience of local pollinator colonies making them vulnerable to the climatic extremes. Planting non-crop flowering plants as strip crops, cover plants or hedgerows may provide additional floral resources to honey bees and serve as micro-climate buffering, mitigating the problem to some extent (Reddy et al., 2012).

Exposure to and/or Emergence of New Pathogens and Parasites

The interactions between the hosts and parasites/pathogens can be influenced by climatic alterations in such a way that the parasite/pathogen may get established in the honey bee population. The same thing can also happen with the predators and pests when they expand their ranges in response to climate change and come in contact with the honey bees (Le Conte and Navajas, 2008). For example, the bird, bee-eater has expanded its range from its origin in the Mediterranean region. It feeds on hymenopteran insects (e.g. bees) but fortunately has not caused much trouble to the beekeeper. However, the climate change events in the near future can disrupt this balance with the birds causing destruction to the bee population (Le Conte and Navajas, 2008). Another example is that of the hive beetle (*Aethiva tumida*) having its origin in South Africa and later on being imported into the USA through citrus fruits. The hive beetle is an apiary pest and can also develop on citrus fruits (Le Conte and Navajas, 2008). It prefers hot and humid climate for its growth and survival and is discouraged by cold temperatures. However, climate change can lead to the development of warm and humid conditions even in the colder regions causing the pest to spread further and cause havoc to the honey bee population (Le Conte and Navajas, 2008).

Pathogens infecting honey bees exist as different haplotypes of diverse virulence. Some of them include *Nosema cerana* and *Nosema apis* (bacteria) infecting *Apis mellifera* as well as the mite, *Varroa destructor*, infecting *Apis mellifera* and *Apis cerana*. Climate change has been found to promote the transmission of several haplotypes of these pathogens in honey bee populations (Le Conte and Navajas, 2008). These pathogens are more or less found in all parts of the world. However, some pathogens are restricted to certain places having small ranges of distribution (e.g. *Tropilaelaps* sp. is limited to Asia only) (Sammataro et al., 2000). As the honey bee populations undertake migration in search of new and better resources, they may either introduce or become exposed to pathogens new to the area or they have not encountered earlier. This is exemplified by the Varroa mite (*Varroa destructor*) that has caused huge troubles to the migrating groups of *Apis mellifera* in different parts of the world (Solignac et al., 2005). Apart from migration, the honey bees can be distributed to new areas through bee exchanges among the beekeepers (Cox-Foster et al., 2007). The exposure of *Apis mellifera* population to the Israeli acute paralysis virus is considered to have been brought about by anthropogenic activities (Cox-Foster et al., 2007). Moreover, the development of certain pathogens can be encouraged by the altered climatic conditions (e.g. humid atmosphere favours the development of the fungus, *Ascosphera apis* which causes chalkbrood disease in honey bees) (Le Conte and Navajas, 2008). Additionally, the effect of climate change on honey bee life cycles may be such that the early emerging adults may not be immunologically competent enough to withstand infestations by pathogens and/or parasites (Le Conte and Navajas, 2008). Furthermore, alterations in the flowering season or negative impacts of climatic stress factors on pollen production and its nutritional quality may also lead to the development of weaker broods often leading to colony collapse (vanEngelsdorp et al., 2009).

FUTURE PERSPECTIVES

Honey bees are reared and exchanged by the beekeepers with the aim of commercial benefits. Therefore, they must be made aware of the long term impacts on environment their actions might cause. For example, selective breeding of the high temperature resistant honey bee races or hybridizing them with highly productive ones can cause extinction of certain ecotypes which are essential for pollinating native plant species (Vasiliev and Greenwood, 2021). This may have serious impacts on local plant diversity in the long run, the outcomes of which are still under investigation. Other way round, replacing native plant species with exotic ones can affect local bee populations adversely. Actually, the native plant species promote landscape heterogeneity. However, its importance in honey bee conservation is only being discussed recently. Under these circumstances, awareness programmes for the beekeepers as well as the local inhabitants are essential for honey bee conservation and maintenance of ecological balance. Taking suitable initiatives for community participation can accelerate biodiversity conservation drives.

Since the impacts of climatic stress factors on honey bee populations are complicated and integrated due to involvement of several factors, their mitigation strategies require a multidimensional and a multi-strategic approach. For instance, Abou-Shaara (2016) suggested the use of modified hives with artificial fans and humidifier as a good alternative to improve honey and pollen storage activities within the colony during hot climates. Development of these innovative techniques that do not involve genetic dilutions require further research for conservation of honey bee ecotypes.

CONCLUSION

The contribution of honey bees in the preservation of biodiversity and guaranteeing food security for humans is immense. These ecosystem services enable them to be used as environmental biomonitors as well. However, their population has declined rapidly since the last few years due to several reasons including climate change. Their genetic variability and plasticity has enabled them to get adapted to various geo-climatic conditions across the world in the form of ecotypes. These long-term adaptations together with their ectothermic nature make them vulnerable to climatic stress factors. Apart from direct effects on their physiology and life cycles, climate change can affect the flowering seasons which may impede the timely supply of nectar and pollen to the honey bees often resulting in colony collapses. The negative impacts of climatic stress factors are augmented by anthropogenic activities like agricultural intensification and landscape fragmentation. Mitigation strategies should thereby be taken in a multi-strategic manner following consultations with the environmentalists, policy makers, beekeepers and agricultural practitioners. The strategies should not only aim to conserve the honey bee ecotypes but also the landscape connectivity and heterogeneity for long-term solutions.

REFERENCES

Abou-Shaara, H. F. (2016). Expectations about the Potential Impacts of Climate Change on Honey Bee Colonies in Egypt. *Journal of Apiculture*, *31*(2), 157–164. doi:10.17519/apiculture.2016.06.31.2.157

Aguirre-Gutiérrez, J., Kissling, W. D., Biesmeijer, J. C., WallisDeVries, M. F., Reemer, M., & Carvalheiro, L. G. (2017). Historical changes in the importance of climate and land use as determinants of Dutch pollinator distributions. *Journal of Biogeography*, *44*(3), 696–707. doi:10.1111/jbi.12937

Al-Alam, J., Fajloun, Z., Chbani, A., & Millet, M. (2019). Determination of 16 PAHs and 22 PCBs in honey samples originated from different region of Lebanon and used as environmental biomonitors sentinel. *Journal of Environmental Sciences (China)*, *54*(1), 9–15. doi:10.1080/10934529.2018.1500782 PMID:30199354

Alaux, C., Allier, F., Decourtye, A., Odoux, J. F., Tamic, T., Chabirand, M., Delestra, E., Decugis, F., Le Conte, Y., & Henry, M. (2017). A 'Landscape physiology' approach for assessing bee health highlights the benefits of floral landscape enrichment and semi-natural habitats. *Scientific Reports*, *7*(1), 40568. doi:10.1038rep40568 PMID:28084452

Allendorf, F. W., Leary, R. F., Spruell, P., & Wenburg, J. K. (2001). The problems with hybrids: Setting conservation guidelines. *Trends in Ecology & Evolution*, *16*(11), 613–622. doi:10.1016/S0169-5347(01)02290-X

Álvarez-Ayuso, E., & Abad-Valle, P. (2017). Trace element levels in an area impacted by old mining operations and their relationship with beehive products. *The Science of the Total Environment*, *599-600*, 671–678. doi:10.1016/j.scitotenv.2017.05.030 PMID:28494292

Arias, M. C., & Sheppard, W. S. (1996). Molecular phylogenetics of honey bee subspecies (*Apis mellifera* L.) inferred from mitochondrial DNA sequence. *Molecular Phylogenetics and Evolution*, *5*(3), 557–566. doi:10.1006/mpev.1996.0050 PMID:8744768

Arias, M. C., & Sheppard, W. S. (2005). Phylogenetic relationships of honey bees (Hymenoptera:Apinae:Apini) inferred from nuclear and mitochondrial DNA sequence data. *Molecular Phylogenetics and Evolution*, *37*(1), 25–35. doi:10.1016/j.ympev.2005.02.017 PMID:16182149

Atmowidjojo, A. H., Wheeler, D. E., Erickson, E. H., & Cohen, A. C. (1997). Temperature tolerance and water balance in feral and domestic honey bees, *Apis mellifera* L. *Comparative Biochemistry and Physiology. Part A, Physiology*, *118*(4), 1399–1403. doi:10.1016/S0300-9629(97)00031-5

Barghi, N., Tobler, R., Nolte, V., Jakšić, A. M., Mallard, F., Otte, K. A., Dolezal, M., Taus, T., Kofler, R., & Schlötterer, C. (2019). Genetic redundancy fuels polygenic adaptation in *Drosophila*. *PLoS Biology*, *17*(2), e3000128. doi:10.1371/journal.pbio.3000128 PMID:30716062

Bergland, A. O., Tobler, R., González, J., Schmidt, P., & Petrov, D. (2016). Secondary contact and local adaptation contribute to genome-wide patterns of clinal variation in *Drosophila melanogaster*. *Molecular Ecology*, *25*(5), 1157–1174. doi:10.1111/mec.13455 PMID:26547394

Bhat, A. I., & Rao, G. P. (2020). Virus transmission through pollen. In A. I. Bhat & G. P. Rao (Eds.), *Characterization of Plant Viruses: Methods and Protocols* (pp. 61–64). Humana, Springer. doi:10.1007/978-1-0716-0334-5_9

Bishop, C. A., Woundneh, M. B., Maisonneuve, F., Common, J., Elliott, J. E., & Moran, A. J. (2020). Determination of neonicotinoids and butenolide residues in avian and insect pollinators and their ambient environment in Western Canada (2017, 2018). *The Science of the Total Environment*, *737*(1), 139386. doi:10.1016/j.scitotenv.2020.139386 PMID:32563110

Bo, C. F. D., & Florio, M. (2012). Infrastructure and growth in a spatial framework: Evidence from the EU regions. *European Planning Studies*, *20*(8), 1393–1414. doi:10.1080/09654313.2012.680587

Calatayud-Vernich, P., Calatayud, F., Simó, E., & Picó, Y. (2018). Pesticide residues in honey bees, pollen and beeswax: Assessing beehive exposure. *Environmental Pollution*, *241*, 106–114. doi:10.1016/j.envpol.2018.05.062 PMID:29803024

Carpenter, D. O. (2006). Polychlorinated biphenyls (PCBs): Routes of exposure and effects on human health. *Reviews on Environmental Health*, *21*(1), 1–23. doi:10.1515/REVEH.2006.21.1.1 PMID:16700427

Carroll, M. J., Brown, N., Goodall, C., Downs, A. M., Sheenan, T. H., & Anderson, K. E. (2017). Honey bees preferentially consume freshly-stored pollen. *PLoS One*, *12*(4), e0175933. doi:10.1371/journal.pone.0175933 PMID:28430801

Chown, S. L., Slabber, S., McGeouch, M., Janion, C., & Leinaas, H. P. (2007). Phenotypic plasticity mediates climate change responses among invasive and indigenous arthropods. *Proceedings. Biological Sciences*, *274*(1625), 2531–2537. doi:10.1098/rspb.2007.0772 PMID:17686728

Conti, M. E., & Botrè, F. (2001). Honeybees and their products as potential bioindicators of heavy metals contamination. *Environmental Monitoring and Assessment*, *69*(3), 267–282. doi:10.1023/A:1010719107006 PMID:11497382

Cox-Foster, D. L., Conlan, S., Holmes, E. C., Palacios, G., Evans, J. D., Moran, N. A., Quan, P. L., Briese, T., Hornig, M., Geiser, D. M., Martinson, V., van Engelsdorp, D., Kalkstein, A. L., Drysdale, A., Hui, J., Zhai, J. H., Cui, L. W., Hutchison, S. K., Simons, J. F., ... Lipkin, W. I. (2007). A metagenomic survey of microbes in honey bee colony collapse disorder. *Science*, *318*(5848), 283–287. doi:10.1126cience.1146498 PMID:17823314

Coyle, P., Philcox, J. C., Carey, L. C., & Rofe, A. M. (2002). Metallothionein: The multipurpose protein. *Cellular and Molecular Life Sciences*, *59*(4), 627–647. doi:10.100700018-002-8454-2 PMID:12022471

Cunningham, M. M., Tran, L., McKee, C. G., Polo, R. O., Newman, T., Lansing, L., Griffiths, J. S., Bilodeau, G. J., Rott, M., & Guarna, M. M. (2022). Honey bees as biomonitors of environmental contaminants, pathogens, and climate change. *Ecological Indicators*, *134*, 108457. doi:10.1016/j.ecolind.2021.108457

Cuthbertson, A. G. S., & Brown, M. A. (2009). Issues affecting British honey bee biodiversity and the need for conservation of this important ecological component. *International Journal of Environmental Science and Technology*, *6*(4), 695–699. doi:10.1007/BF03326110

Danks, H. V. (2006). Short life cycles in insects and mites. *Canadian Entomologist*, *138*(4), 407–463. doi:10.4039/n06-803

De La Torre-Roche, R. J., Lee, W.-Y., & Campos-Díaz, S. I. (2009). Soil-borne polycyclic aromatic hydrocarbons in El Paso, Texas: Analysis of a potential problem in the United States/Mexico border region. *Journal of Hazardous Materials*, *163*(2-3), 946–958. doi:10.1016/j.jhazmat.2008.07.089 PMID:18768257

de Oliveira, R. C., Queiroz, S., da Luz, C. F. P., Porto, R. S., & Rath, S. (2016). Bee pollen as a bioindicator of environmental pesticide contamination. *Chemosphere*, *163*, 525–534. doi:10.1016/j.chemosphere.2016.08.022 PMID:27567152

de Sousa Pereira, K., Meeus, I., & Smagghe, G. (2019). Honey bee-collected pollen is a potential source of *Ascosphaera apis* infection in managed bumble bees. *Scientific Reports*, *9*(1), 1–9. doi:10.103841598-019-40804-2 PMID:30862950

DesJardins, N. S., Fisher, A. II, Ozturk, C., Fewell, J. H., DeGrandi-Hoffman, G., Harrison, J. F., & Smith, B. H. (2021). A common fungicide, Pristine®, impairs olfactory associative learning performance in honey bees (*Apis mellifera*). *Environmental Pollution*, *288*, 117720. doi:10.1016/j.envpol.2021.117720 PMID:34252716

Devictor, V., van Swaay, C., Brereton, T., Brotons, L., Chamberlain, D., Heliölä, J., Herrando, S., Julliard, R., Kuussaari, M., Lindström, Å., Reif, J., Roy, D. B., Schweiger, O., Settele, J., Stefanescu, C., Strien, A. V., Turnhout, C. V., & Vermouzek, Z. (2012). Differences in the climatic debts of birds and butterflies at a continental scale. *Nature Climate Change*, *2*(2), 121–124. doi:10.1038/nclimate1347

Di Prisco, G., Annoscia, D., Margiotta, M., Ferrara, R., Varricchio, P., Zanni, V., Caprio, E., Nazzi, F., & Pennacchio, F. (2016). A mutualistic symbiosis between a parasitic mite and a pathogenic virus undermines honey bee immunity and health. *Proceedings of the National Academy of Sciences of the United States of America, 113*(12), 3203–3208. doi:10.1073/pnas.1523515113 PMID:26951652

Diniz, N. M., Soares, A. E. E., Sheppard, W. S., & Del Lama, M. A. (2003). Genetic structure of honeybee populations from southern Brazil and Uruguay. *Genetics and Molecular Biology, 26*(1), 47–52. doi:10.1590/S1415-47572003000100008

Dukku, U. H. (2016). Evaluation of morphometric characters of honeybee (*Apis mellifera* L.) populations in the Lake Chad Basin in Central Africa. *Advances in Entomology (Irvine, Calif.), 4*(02), 75–89. doi:10.4236/ae.2016.42009

Ellstrand, N. C., & Rieseberg, L. H. (2016). When gene flow really matters: Gene flow in applied evolutionary biology. *Evolutionary Applications, 9*(7), 833–836. doi:10.1111/eva.12402 PMID:27468301

Flores, J. M., Gil-Lebrero, S., Gámiz, V., Rodríguez, M. I., Ortiz, M. A., & Quiles, F. J. (2019). Effect of the climate change on honey bee colonies in a temperate Mediterranean zone assessed through remote hive weight monitoring system in conjunction with exhaustive colonies assessment. *The Science of the Total Environment, 653*, 1111–1119. doi:10.1016/j.scitotenv.2018.11.004 PMID:30759551

Franck, P., Garnery, L., Solignac, M., & Cornuet, J. M. (2000). Molecular confirmation of a fourth lineage in honeybees from the Near East. *Apidologie, 31*(2), 167–180. doi:10.1051/apido:2000114

Fuchs, R., Herold, M., Verburg, P. H., Clevers, J. G., & Eberle, J. (2015). Gross changes in reconstructions of historic land cover/use for Europe between 1900 and 2010. *Global Change Biology, 21*(1), 299–313. doi:10.1111/gcb.12714 PMID:25155867

Gienapp, P., Teplitsky, C., Alho, J. S., Mills, J. A., & Merilä, J. (2008). Climate change and evolution: Disentangling environmental and genetic responses. *Molecular Ecology, 17*(1), 167–178. doi:10.1111/j.1365-294X.2007.03413.x PMID:18173499

Gordo, O., & Sanz, J. J. (2010). Impact of climate change on plant phenology in Mediterranean ecosystems. *Global Change Biology, 16*(3), 1082–1106. doi:10.1111/j.1365-2486.2009.02084.x

Grassl, J., Holt, S., Cremen, N., Peso, M., Hahne, D., & Baer, B. (2018). Synergistic effects of pathogen and pesticide exposure on honey bee (*Apis mellifera*) survival and immunity. *Journal of Invertebrate Pathology, 159*, 78–86. doi:10.1016/j.jip.2018.10.005 PMID:30300630

Greenleaf, S. S., Williams, N. M., Winfree, R., & Kremen, C. (2007). Bee foraging ranges and their relationship to body size. *Oecologia*, *153*(3), 589–596. doi:10.100700442-007-0752-9 PMID:17483965

Hegland, S. J., Nielsen, A., Lázaro, A., Bjerknes, A. L., & Totland, Ø. (2009). How does climate warming affect plant-pollinator interactions? *Ecology Letters*, *12*(2), 184–195. doi:10.1111/j.1461-0248.2008.01269.x PMID:19049509

Hill, J. K., Griffiths, H. M., & Thomas, C. D. (2011). Climate change and evolutionary adaptations at species' range margins. *Annual Review of Entomology*, *56*(1), 143–159. doi:10.1146/annurev-ento-120709-144746 PMID:20809802

Hoffmann, A. A., & Sgro, C. M. (2011). Climate change and evolutionary adaptation. *Nature*, *470*(7335), 479–485. doi:10.1038/nature09670 PMID:21350480

Hoffmann, A. A., & Willy, Y. (2008). Detecting genetic responses to environmental change. *National Review*, *9*(6), 421–432. doi:10.1038/nrg2339 PMID:18463665

Holzschuh, A., Steffan-Dewenter, I., & Tscharntke, T. (2008). Agricultural landscapes with organic crops support higher pollinator diversity. *Oikos*, *117*(3), 354–361. doi:10.1111/j.2007.0030-1299.16303.x

Ilyasov, R. A., Lee, M., Takahashi, J.-i., Kwon, H. W., & Nikolenko, A. G. (2020). A revision of subspecies structure of western honey bee *Apis mellifera*. *Saudi Journal of Biological Sciences*, *27*(12), 3615–3621. doi:10.1016/j.sjbs.2020.08.001 PMID:33304172

Jyothi, N. R. (2020). Heavy metal sources and their effects on human health. In M. K. Nazal & H. Zhao (Eds.), *Heavy Metals – Their Environmental Impact and Mitigation*. IntechOpen. doi:10.5772/intechopen.95370

Kafer, H., Kovac, H., & Stabentheiner, A. (2012). Upper thermal limits of honeybee (*Apis mellifera*) and yellowjacket (*Vespula vulgaris*) foragers. *Mitteilungen der Deutschen Gesellschaft fur Allgemeine und Angewandte Entomologie*, *18*, 267–270.

Kagawa, K., & Takimoto, G. (2018). Hybridization can promote adaptive radiation by means of transgressive segregation. *Ecology Letters*, *21*(2), 264–274. doi:10.1111/ele.12891 PMID:29243294

Kaloveloni, A., Tscheulin, T., & Petanidou, T. (2018). Geography, climate, ecology: What is more important in determining bee diversity in the Aegean Archipelago? *Journal of Biogeography*, *45*(12), 2690–2700. doi:10.1111/jbi.13436

Klein, A. M., Vaissière, B. E., Cane, J. H., Steffan-Dewenter, I., Cunningham, S. A., Kremen, C., & Tscharntke, T. (2007). Importance of pollinators in changing landscapes for world crops. *Proceedings. Biological Sciences, 274*(1608), 303–313. doi:10.1098/rspb.2006.3721 PMID:17164193

Kovac, H., Kafer, H., Stabentheiner, A., & Costa, C. (2014). Metabolism and upper thermal limits of *Apis mellifera carnica* and *A. m. ligustica. Apidologie, 45*(6), 664–677. doi:10.100713592-014-0284-3 PMID:25378763

Lambert, O., Veyrand, B., Durand, S., Marchand, P., Le Bizec, B., Piroux, M., Puyo, S., Thorin, C., Delbac, F., & Pouliquen, H. (2012). Polycyclic aromatic hydrocarbons: Bees, honey and pollen as sentinels for environmental chemical contaminants. *Chemosphere, 86*(1), 98–104. doi:10.1016/j.chemosphere.2011.09.025 PMID:22051346

Lande, R. (1993). Risks of Population Extinction from Demographic and Environmental Stochasticity and Random Catastrophes. *American Naturalist, 142*(6), 911–927. doi:10.1086/285580 PMID:29519140

Le Conte, Y., & Navajas, M. (2008). Climate change: Impact on honey bee populations and diseases. *Revue Scientifique et Technique (International Office of Epizootics), 27*(2), 499–510. PMID:18819674

Lucchetti, M. A., Kilchenmann, V., Glauser, G., Praz, C., & Kast, P. (2018). Nursing protects honeybee larvae from secondary metabolites of pollen. *Proceedings. Biological Sciences, 285*(1875), 20172849. doi:10.1098/rspb.2017.2849 PMID:29563265

Manley, R., Temperton, B., Doyle, T., Gates, D., Hedges, S., Boots, M., & Wilfert, L. (2019). Knock-on community impacts of a novel vector: Spillover of emerging DWV-B from Varroa-infested honeybees to wild bumblebees. *Ecology Letters, 22*(8), 1306–1315. doi:10.1111/ele.13323 PMID:31190366

Martin, S. J., & Medina, L. M. (2004). Africanized honeybees have unique tolerance to Varroa mites. *Trends in Parasitology, 20*(3), 112–114. doi:10.1016/j.pt.2004.01.001 PMID:15036029

McAfee, A., Chapman, A., Higo, H., Underwood, R., Milone, J., Foster, L. J., Guarna, M. M., Tarpy, D. R., & Pettis, J. S. (2020). Vulnerability of honey bee queens to heat-induced loss of fertility. *Nature Sustainability, 3*(5), 367–376. doi:10.103841893-020-0493-x

McArt, S. H., Koch, H., Irwin, R. E., & Adler, L. S. (2014). Arranging the bouquet of disease: Floral traits and the transmission of plant and animal pathogens. *Ecology Letters*, *17*(5), 624–636. doi:10.1111/ele.12257 PMID:24528408

Michener, C. (2000). *The bees of the world*. Johns Hopkins University Press.

Montory, M., Habit, E., Fernandez, P., Grimalt, J. O., Kolok, A. S., Barra, R. O., & Ferrer, J. (2020). Biotransport of persistent organic pollutants in the southern Hemisphere by invasive Chinook salmon (*Oncorhynchus tshawytscha*) in the rivers of northern Chilean Patagonia, a UNESCO biosphere reserve. *Environment International*, *142*, 105803. doi:10.1016/j.envint.2020.105803 PMID:32563009

Niell, S., Jesús, F., P'erez, N., Pérez, C., Pareja, L., Abbate, S., Carrasco-Letelier, L., Díaz, S., Mendoza, Y., Cesio, V., & Heinzen, H. (2017). Neonicotinoids transference from the field to the hive by honey bees: Towards a pesticide residues biomonitor. *The Science of the Total Environment*, *581-582*, 25–31. doi:10.1016/j.scitotenv.2017.01.011 PMID:28073057

Pattemore, D. E., Goodwin, R. M., McBrydie, H. M., Hoyte, S. M., & Vanneste, J. L. (2014). Evidence of the role of honey bees (*Apis mellifera*) as vectors of the bacterial plant pathogen *Pseudomonas syringae*. *Australasian Plant Pathology*, *43*(5), 571–575. doi:10.100713313-014-0306-7

Perugini, M., Di Serafino, G., Giacomelli, A., Medrzycki, P., Sabatini, A. G., Persano Oddo, L., Marinelli, E., & Amorena, M. (2009). Monitoring of polycyclic aromatic hydrocarbons in bees (*Apis mellifera*) and honey in urban areas and wildlife reserves. *Journal of Agricultural and Food Chemistry*, *57*(16), 7440–7444. doi:10.1021/jf9011054 PMID:19627114

Pinto, M. A., Rubink, W. L., Patton, J. C., Coulson, R. N., & Johnston, J. S. (2005). Africanization in the United States: Replacement of feral European honeybees (*Apis mellifera* L.) by an African hybrid swarm. *Genetics*, *170*(4), 1653–1665. doi:10.1534/genetics.104.035030 PMID:15937139

Potts, S. G., Biesmeijer, J. C., Kremen, C., Neumann, P., Schweiger, O., & Kunin, W. E. (2010). Global pollinator declines: Trends, impacts and drivers. *Trends in Ecology & Evolution*, *25*(6), 345–353. doi:10.1016/j.tree.2010.01.007 PMID:20188434

Purać, J., Nikolić, T. V., Kojić, D., Ćelić, A. S., Plavša, J. J., Blagojević, D. P., & Petri, E. T. (2019). Identification of a metallothionein gene in honey bee *Apis mellifera* and its expression profile in response to Cd, Cu and Pb exposure. *Molecular Ecology*, *28*(4), 731–745. doi:10.1111/mec.14984 PMID:30575191

Reddy, P. V. R., Verghese, A., & Rajan, V. V. (2012). Potential impact of climate change on honeybees (*Apis* spp.) and their pollination services. *Pest Management in Horticultural Ecosystems*, *18*(2), 121–127.

Requier, F., Odoux, J. F., Tamic, T., Moreau, N., Henry, M., Decourtye, A., & Bretagnolle, V. (2015). Honey bee diet in intensive farmland habitats reveals an unexpectedly high flower richness and a major role of weeds. *Ecological Applications*, *25*(4), 881–890. doi:10.1890/14-1011.1 PMID:26465030

Rhymer, J. M., & Simberloff, D. (1996). Extinction by hybridization and introgression. *Annual Review of Ecology and Systematics*, *27*(1), 83–109. doi:10.1146/annurev.ecolsys.27.1.83

Roberts, J. M. K., Ireland, K. B., Tay, W. T., & Paini, D. (2018). Honey bee-assisted surveillance for early plant virus detection. *Annals of Applied Biology*, *173*(3), 285–293. doi:10.1111/aab.12461

Rodrigez-Trelles, F., & Rodrigez, M. A. (1998). Rapid micro-evolution and loss of chromosomal diversity in *Drosophila* in response to climate warming. *Evolutionary Ecology*, *12*(7), 829–838. doi:10.1023/A:1006546616462

Ruschioni, S., Riolo, P., Minuz, R. L., Stefano, M., Cannella, M., Porrini, C., & Isidoro, N. (2013). Biomonitoring with honeybees of heavy metals and pesticides in nature reserves of the Marche Region (Italy). *Biological Trace Element Research*, *154*(2), 226–233. doi:10.100712011-013-9732-6 PMID:23797576

Ruttner, F. (1988). *Biogeography and taxonomy of honeybees*. Springer. doi:10.1007/978-3-642-72649-1

Ruttner, F., Tassencourt, L., & Louveaux, J. (1978). Biometrical statistical analysis of the geographic variability of *Apis mellifera* L. *Apidologie*, *9*, 363–382. doi:10.1051/apido:19780408

Salazar, C. A., Jiggins, C. D., Arias, C. F., Tobler, A., Bermingham, E., & Linares, M. (2005). Hybrid incompatibility is consistent with a hybrid origin of *Heliconius heurippa* Hewitson from its close relatives, *Heliconius cydno* Doubleday and *Heliconius melpomene* Linnaeus. *Journal of Evolutionary Biology*, *18*(2), 247–256. doi:10.1111/j.1420-9101.2004.00839.x PMID:15715831

Sammataro, D., Gerson, U., & Needham, G. (2000). Parasitic mites of honey bees: Life history, implications, and impact. *Annual Review of Entomology*, *45*(1), 519–548. doi:10.1146/annurev.ento.45.1.519 PMID:10761588

Sánchez-Guillén, R. A., Córdoba-Aguilar, A., Hansson, B., Ott, J., & Wellenreuther, M. (2016). Evolutionary consequences of climate-induced range shifts in insects. *Biological Reviews of the Cambridge Philosophical Society*, *91*(4), 1050–1064. doi:10.1111/brv.12204 PMID:26150047

Sari, M. F., Ayyildiz, E. G., & Esen, F. (2020). Determination of polychlorinated biphenyls in honeybee, pollen, and honey samples from urban and semi-urban areas in Turkey. *Environmental Science and Pollution Research International*, *27*(4), 4414–4422. doi:10.100711356-019-07013-w PMID:31832954

Scheffer, M., Carpenter, S. R., Lenton, T. M., Bascompte, J., Brock, W., Dakos, V., van de Koppel, J., van de Leemput, I. A., Levin, S. A., van Nes, E. H., Pascual, M., & Vandermeer, J. (2012). Anticipating critical transitions. *Science*, *338*(6105), 344–348. doi:10.1126cience.1225244 PMID:23087241

Schweiger, O., Biesmeijer, J. C., Bommarco, R., Hickler, T., Hulme, P. E., Klotz, S., Kühn, I., Moora, M., Nielsen, A., Ohlemüller, R., Petanidou, T., Potts, S. G., Pyšek, P., Stout, J. C., Sykes, M. T., Tscheulin, T., Vilà, M., Walther, G. R., Westphal, C., ... Settele, J. (2010). Multiple stressors on biotic interactions: How climate change and alien species interact to affect pollination. *Biological Reviews of the Cambridge Philosophical Society*, *85*(4), 777–795. doi:10.1111/j.1469-185X.2010.00125.x PMID:20184567

Seehausen, O., Takimoto, G., Roy, D., & Jokela, J. (2008). Speciation reversal and biodiversity dynamics with hybridization in changing environments. *Molecular Ecology*, *17*(1), 30–44. doi:10.1111/j.1365-294X.2007.03529.x PMID:18034800

Skelly, D. K., Joseph, L. N., Possingham, H. P., Freidenburg, L. K., Farrugia, T. J., Kinnison, M. T., & Hendry, A. P. (2007). Evolutionary Responses to Climate Change. *Conservation Biology*, *21*(5), 1353–1355. doi:10.1111/j.1523-1739.2007.00764.x PMID:17883501

Smith, D. R. (1991). Mitochondrial DNA and honeybee biogeography. In D. R. Smith (Ed.), *Diversity in the genus Apis* (pp. 131–176). Westview Press.

Smith, K. E., & Weis, D. (2020). Evaluating spatiotemporal resolution of trace element concentrations and Pb isotopic compositions of honeybees and hive products as biomonitors for urban metal distribution. *GeoHealth, 4*(7). . doi:10.1029/2020GH000264

Smith, K. E., Weis, D., Amini, M., Shiel, A. E., Lai, V. W.-M., & Gordon, K. (2019). Honey as a biomonitor for a changing world. *Nature Sustainability*, *2*(3), 223–232. doi:10.103841893-019-0243-0

Solignac, M., Cornuet, J. M., Vautrin, D., Le Conte, Y., Anderson, D., Evans, J., Cros-Arteil, S., & Navajas, M. (2005). The invasive Korea and Japan types of *Varroa destructor*, ectoparasitic mites of the Western honeybee (*Apis mellifera*), are two partly isolated clones. *Proceedings. Biological Sciences, 272*(1561), 411–419. doi:10.1098/rspb.2004.2853 PMID:15734696

Starks, P. T., & Gilley, D. C. (1999). Heat shielding: A novel method of colonial thermoregulation in honey bees. *Naturwissenschaften, 86*(9), 438–440. doi:10.1007001140050648 PMID:10501692

Stout, J., & Morales, C. L. (2009). Ecological impacts of invasive alien species on bees. *Apidologie, 40*(3), 388–409. doi:10.1051/apido/2009023

Switanek, M., Crailsheim, K., Truhetz, H., & Brodschneider, R. (2017). Modelling seasonal effects of temperature and precipitation on honey bee winter mortality in a temperate climate. *The Science of the Total Environment, 579*, 1581–1587. doi:10.1016/j.scitotenv.2016.11.178 PMID:27916302

Taylor, R. S., & Frisen, V. L. (2017). The role of allochrony in speciation. *Molecular Ecology, 26*(13), 3330–3342. doi:10.1111/mec.14126 PMID:28370658

Thompson, H., & Wilkins, S. (2003). Assessment of the synergy and repellency of pyrethroid/fungicide mixtures. *Bulletin of Insectology, 56*, 131–134.

Thuiller, W., Lavorel, S., Araújo, M. B., Sykes, M. T., & Prentice, I. C. (2005). Climate change threats to plant diversity in Europe. *Proceedings of the National Academy of Sciences of the United States of America, 102*(23), 8245–8250. doi:10.1073/pnas.0409902102 PMID:15919825

Todesco, M., Pascual, M. A., Owens, G. L., Ostevik, K. L., Moyers, B. T., Hübner, S., Heredia, S. M., Hahn, M. A., Caseys, C., Bock, D. G., & Rieseberg, L. H. (2016). Hybridization and extinction. *Evolutionary Applications, 9*(7), 892–908. doi:10.1111/eva.12367 PMID:27468307

Tremblay, É. D., Duceppe, M.-O., Thurston, G. B., Gagnon, M.-C., Côté, M.-J., & Bilodeau, G. J. (2019). High-resolution biomonitoring of plant pathogens and plant species using metabarcoding of pollen pellet contents collected from a honey bee hive. *Environmental DNA, 1*(2), 155–175. doi:10.1002/edn3.17

vanEngelsdorp, D., Evans, J. D., Saegerman, C., Mullin, C., Haubruge, E., Nguyen, B. K., Frazier, M., Frazier, J., Cox-Foster, D., Chen, Y., Underwood, R., Tarpy, D. R., & Pettis, J. S. (2009). Colony collapse disorder: A descriptive study. *PLoS One, 4*(8), e6481. doi:10.1371/journal.pone.0006481 PMID:19649264

Vasiliev, D., & Greenwood, S. (2021). The role of climate change in pollinator decline across the Northern Hemisphere is underestimated. *The Science of the Total Environment, 775*, 145788. doi:10.1016/j.scitotenv.2021.145788 PMID:33618305

Villalba, A., Maggi, M., Ondarza, P. M., Szawarski, N., & Miglioranza, K. S. B. (2020). Influence of land use on chlorpyrifos and persistent organic pollutant levels in honey bees, bee bread and honey: Beehive exposure assessment. *The Science of the Total Environment, 713*, 136554. doi:10.1016/j.scitotenv.2020.136554 PMID:31955084

Wania, F., & MacKay, D. (1996). Tracking the Distribution of Persistent Organic Pollutants. *Environmental Science & Technology, 30*(9), 390A–396A. doi:10.1021/es962399q PMID:21649427

Wood, T. J., Michez, D., Paxton, R. J., Drossart, M., Neumann, P., Gérard, M., Vanderplanck, M., Barraud, A., Martinet, B., Leclercq, N., & Vereecken, N. J. (2020). Managed honey bees as a radar for wild bee decline? *Apidologie, 51*(6), 1100–1116. doi:10.100713592-020-00788-9

Wu, J. Y., Anelli, C. M., & Sheppard, W. S. (2011). Sub-lethal effects of pesticide residues in brood comb on worker honey bee (*Apis mellifera*) development and longevity. *PLoS One, 6*(2), e14720. doi:10.1371/journal.pone.0014720 PMID:21373182

Yan, L., & Roy, D. P. (2016). Conterminous United States field size quantification from multitemporal Landsat data. *Remote Sensing of Environment, 172*, 67–86. doi:10.1016/j.rse.2015.10.034

Yoruk, A., & Sahinler, N. (2013). Potential effects of global warming on the honey bee. *Uludag Bee Journal, 13*, 79–87.

Zarić, N. M., Ilijević, K., Stanisavljević, L., & Gržetić, I. (2017). Use of honeybees (*Apis mellifera* L.) as bioindicators for assessment and source appointment of metal pollution. *Environmental Science and Pollution Research International, 24*(33), 25828–25838. doi:10.100711356-017-0196-7 PMID:28936680

Compilation of References

Abarca, M., & Spahn, R. (2021). Direct and indirect effects of altered temperature regimes and phenological mismatches on insect populations. *Current Opinion in Insect Science*, *47*, 67–74. doi:10.1016/j.cois.2021.04.008 PMID:33989831

Abbassi, M. (2011). Acarien tétranyque menaces permanent sur les agrumes: Cas de l'acarien rouge oriental *Eutetranychus orientalis*. *Agriculture du Maghre, 50*.

Abou-Shaara, H. F. (2016). Expectations about the Potential Impacts of Climate Change on Honey Bee Colonies in Egypt. *Journal of Apiculture*, *31*(2), 157–164. doi:10.17519/apiculture.2016.06.31.2.157

Aboussaid, H., El-Aouame, L., El-Messoussi, S., & Oufdou, K. (2010). Biological Activity of *Bacillus Thuringiensis* (Berliner) Strains on Larvae and Adults of *Ceratitis Capitata* (Wiedemann) (Diptera: Tephritidae). *Journal of Environmental Protection*, *1*(4), 337–345. doi:10.4236/jep.2010.14040

Adler, L. S., Barber, N. A., Biller, O. M., & Irwin, R. E. (2020). Flowering plant composition shapes pathogen infection intensity and reproduction in bumble bee colonies. *Proceedings of the National Academy of Sciences of the United States of America*, *117*(21), 11559–11565. doi:10.1073/pnas.2000074117 PMID:32393622

Afshar, A. A., Parizi, M. H., Sharifi, I., Gorouhi, M. A., Sharifi, F., & Baafi, B. (2017). Evaluation of the ecological characteristics in the vector of anthroponotic cutaneous leishmaniasis in a new focus of Mohammad Abad, Kerman, southeast of Iran. *Asian Pacific Journal of Tropical Disease*, *7*(2), 84–87. doi:10.12980/apjtd.7.2017D6-354

Aghaei, A. A., Rassi, Y., Sharifi, I., Vatandoost, H., Mollaie, H. R., Oshaghi, M. A., Abai, M. R., & Rafizadeh, S. (2014). First report on natural Leishmania infection of Phlebotomus sergenti due Leishmania tropica by high resolution melting curve method in South-eastern Iran. *Asian Pacific Journal of Tropical Medicine*, *7*(2), 93–96. doi:10.1016/S1995-7645(14)60002-X PMID:24461520

Agriculture et alimentation. (n.d.). World Bank. Retrieved 28 November 2022, from https://www.banquemondiale.org/fr/topic/agriculture/overview

Aguirre-Gutiérrez, J., Kissling, W. D., Biesmeijer, J. C., WallisDeVries, M. F., Reemer, M., & Carvalheiro, L. G. (2017). Historical changes in the importance of climate and land use as determinants of Dutch pollinator distributions. *Journal of Biogeography*, *44*(3), 696–707. doi:10.1111/jbi.12937

Ajaoud, M., Es-sette, N., Hamdi, S., El-Idrissi, A. L., Riyad, M., & Lemrani, M. (2013). Detection and molecular typing of Leishmania tropica from Phlebotomus sergenti and lesions of cutaneous lesihmaniasis in an emerging focus of Morocco. *Parasites & Vectors*, *6*(1), 217. doi:10.1186/1756-3305-6-217 PMID:23890256

Alahyane, H. (2020). *Dynamique des populations de l'acarien rouge oriental, Eutetranychus orientalis (Klein) (Acari: Tetranychidae) ravageur des agrumes, et lutte biologique dans la région de Marrakech-Safi* [Thèse]. Cadi Ayyad University.

Alahyane, H., El Alaoui, A., Abousaid, H., Aimrane, A., Atibi, Y., Oufdou, K., & El Messoussi, S. (2019). Biological activity of some native *Bacillus thuringiensis* Berliner strains against *Eutetranychus orientalis* (Klein) (Acari: Tetranychidae). *Applied Ecology and Environmental Research*, *17*(2), 1967–1977. doi:10.15666/aeer/1702_19671977

Alahyane, H., Ouknin, M., Aboussaid, H., El Messoussi, S., Costa, J., & Majidi, L. (2022). Biological activities of essential oils from Moroccan plants against the honey bee ectoparasitic mite, *Varroa destructor. International Journal of Acarology*, *48*(1), 50–56. doi:10.1080/01647954.2021.2015436

Al-Alam, J., Fajloun, Z., Chbani, A., & Millet, M. (2019). Determination of 16 PAHs and 22 PCBs in honey samples originated from different region of Lebanon and used as environmental biomonitors sentinel. *Journal of Environmental Sciences (China)*, *54*(1), 9–15. doi:10.1080/10934529.2018.1500782 PMID:30199354

Alaux, C., Allier, F., Decourtye, A., Odoux, J. F., Tamic, T., Chabirand, M., Delestra, E., Decugis, F., Le Conte, Y., & Henry, M. (2017). A 'Landscape physiology' approach for assessing bee health highlights the benefits of floral landscape enrichment and semi-natural habitats. *Scientific Reports*, *7*(1), 40568. doi:10.1038rep40568 PMID:28084452

Ali, N., Ullah, A., Wahid, S., Khisroon, M., & Rasheed, S. B. (2016). Distribution, species composition and relative abundances of sandflies in North Waziristan Agency, Pakistan. *Medical and Veterinary Entomology*, *30*(1), 89–94. doi:10.1111/mve.12141 PMID:26582763

Ali, S., Liu, Y., Ishaq, M., Shah, T., Abdullah, Ilyas, A., & Din, I. (2017). Climate Change and Its Impact on the Yield of Major Food Crops: Evidence from Pakistan. *Foods*, *6*(6), 6. Advance online publication. doi:10.3390/foods6060039 PMID:28538704

Allendorf, F. W., Leary, R. F., Spruell, P., & Wenburg, J. K. (2001). The problems with hybrids: Setting conservation guidelines. *Trends in Ecology & Evolution*, *16*(11), 613–622. doi:10.1016/S0169-5347(01)02290-X

Aloni, B., Peet, M., Pharr, M., & Karni, L. (2001). The effect of high temperature and high atmospheric CO_2 on carbohydrate changes in bell pepper (*Capsicum annuum*) pollen in relation to its germination. *Physiologia Plantarum, 112*(4), 505–512. doi:10.1034/j.1399-3054.2001.1120407.x PMID:11473710

Alten, B., Maia, C., Afonso, M. O., Campino, L., Jiménez, M., González, E., Molina, R., Bañuls, A. L., Prudhomme, J., Vergnes, B., Toty, C., Cassan, C., Rahola, N., Thierry, M., Sereno, D., Bongiorno, G., Bianchi, R., Khoury, C., Tsirigotakis, N., ... Gradoni, L. (2016). Seasonal Dynamics of Phlebotomine Sand Fly Species Proven Vectors of Mediterranean Leishmaniasis Caused by Leishmania infantum. *PLoS Neglected Tropical Diseases, 10*(2), e0004458. doi:10.1371/journal.pntd.0004458 PMID:26900688

Álvarez-Ayuso, E., & Abad-Valle, P. (2017). Trace element levels in an area impacted by old mining operations and their relationship with beehive products. *The Science of the Total Environment, 599-600*, 671–678. doi:10.1016/j.scitotenv.2017.05.030 PMID:28494292

Amarendra, K. (2008). Effect of temperature on the development, fecundity, progeny sex ratio and life-table of Campoletis chlorideae, an endolarval parasitoid of the pod borer, Helicoverpa armigera. *BioControl, 53*(3), 461–471. doi:10.100710526-007-9083-3

Amiri, N., Lahlali, R., Amiri, S. E. L., Jarroudi, M., Khebiza, M. Y., & Messouli, M. (2021). Development of an Integrated Model to Assess the Impact of Agricultural Practices and Land Use on Agricultural Production in Morocco under Climate Stress over the Next Twenty Years. *Sustainability, 13*(21), 21. Advance online publication. doi:10.3390u132111943

Andresen, E. (2002). Dung beetles in a Central Amazonian rainforest and their ecological role as secondary seed dispersers. *Ecological Entomology, 27*(3), 257–270. doi:10.1046/j.1365-2311.2002.00408.x

Andrew, N. R., & Hill, S. J. (2017). Effect of climate change on insect pest management. In M. Coll & E. Wajnberg (Eds.), *Environmental Pest Management: Challenges for Agronomists, Ecologists, Economists and Policymakers* (1st ed., pp. 197–215). John Wiley and Sons Ltd. doi:10.1002/9781119255574.ch9

Ansaloni, T., Pascual-Ruiz, S., Hurtado, M. A., & Jacas, J. A. (2008). Can summer and fall vegetative growth regulate the incidence of *Tetranychus urticae* Koch on Clementine fruit? *Crop Protection (Guildford, Surrey), 27*(3-5), 459–464. doi:10.1016/j.cropro.2007.07.016

Aoki, C. F., Munro, H. L., & Gandhi, K. J. (2022). Responses and modeling of southern pine beetle and its host pines to climate change. In Bark Beetle Management, Ecology, and Climate Change (pp. 55-85). Academic Press. doi:10.1016/B978-0-12-822145-7.00015-5

Appert, J. (1982). Les acariens. In *Les ravageurs des cultures vivrières et maraîchères sous les tropiques* (p. 413). Edit.G.P. Maisonneuve et Larose.

Appiah, K., Du, J., & Poku, J. (2018). Causal relationship between agricultural production and carbon dioxide emissions in selected emerging economies. *Environmental Science and Pollution Research International, 25*(25), 24764–24777. doi:10.100711356-018-2523-z PMID:29926327

APRD. (2018). *Arthropod Pesticide Resistance Database (APRD).* Michigan State University. https://www.pesticideresistance.org/

Arbona, V., Flors, V., Jacas, J., García-Agustín, P., & Gómez-Cadenas, A. (2003). Enzymatic and Nonenzymatic Antioxidant Responses of Carrizo citrange, a Salt-Sensitive Citrus Rootstock, to Different Levels of Salinity. *Plant & Cell Physiology*, *44*(4), 388–394. doi:10.1093/pcp/pcg059 PMID:12721379

Arias, M. C., & Sheppard, W. S. (1996). Molecular phylogenetics of honey bee subspecies (*Apis mellifera* L.) inferred from mitochondrial DNA sequence. *Molecular Phylogenetics and Evolution*, *5*(3), 557–566. doi:10.1006/mpev.1996.0050 PMID:8744768

Arias, M. C., & Sheppard, W. S. (2005). Phylogenetic relationships of honey bees (Hymenoptera:Apinae:Apini) inferred from nuclear and mitochondrial DNA sequence data. *Molecular Phylogenetics and Evolution*, *37*(1), 25–35. doi:10.1016/j.ympev.2005.02.017 PMID:16182149

Asseng, S., Cao, W., Zhang, W., & Ludwig, F. (2009). Crop Physiology, Modelling and Climate Change: Impact and Adaptation Strategies. In V. Sadras & D. Calderini (Eds.), *Crop Physiology* (pp. 511–543). Academic Press. doi:10.1016/B978-0-12-374431-9.00020-7

Asseng, S., Zhu, Y., Basso, B., Wilson, T., & Cammarano, D. (2014). Simulation Modeling: Applications in Cropping Systems. In N. K. Van Alfen (Ed.), *Encyclopedia of Agriculture and Food Systems* (pp. 102–112). Academic Press. doi:10.1016/B978-0-444-52512-3.00233-3

Asseng, S., Zhu, Y., Wang, E., & Zhang, W. (2015). Crop modeling for climate change impact and adaptation. In V. O. Sadras & D. F. Calderini (Eds.), *Crop Physiology* (2nd ed., pp. 505–546). Academic Press. doi:10.1016/B978-0-12-417104-6.00020-0

Athanassiou, C. G., Kavallieratos, N. G., Benelli, G., Losic, D., Usha Rani, P., & Desneux, N. (2018). Nanoparticles for pest control: Current status and future perspectives. *Journal of Pest Science*, *91*(1), 1–15. doi:10.100710340-017-0898-0

Atmowidjojo, A. H., Wheeler, D. E., Erickson, E. H., & Cohen, A. C. (1997). Temperature tolerance and water balance in feral and domestic honey bees, *Apis mellifera* L. *Comparative Biochemistry and Physiology. Part A, Physiology*, *118*(4), 1399–1403. doi:10.1016/S0300-9629(97)00031-5

Aucejo-Romero, S., Gómez-Cadenas, A., & Jacas-Miret, J. A. (2004). Effects of NaCl-stressed citrus plants on life-history parameters of *Tetranychus urticae* (Acari: Tetranychidae). *Experimental & Applied Acarology*, *33*(1–2), 55–67. doi:10.1023/B:APPA.0000030026.77800.0c PMID:15285138

Augustina, I., Jabal, A. R., Permana, G. I., & Ratnasari, A. (2021). Distribution and ecology of mosquito larvae in Pahandut subdistrict, Palangkaraya city. *Journal of Physics: Conference Series*, *1918*(5), 052018. doi:10.1088/1742-6596/1918/5/052018

Awmack, C. S., Woodcock, C. M., & Harrington, R. (1997). Climate change may increase vulnerability of aphids to natural enemies. *Ecological Entomology*, *22*(3), 366–368. doi:10.1046/j.1365-2311.1997.00069.x

Ayt Ougougdal, H., Yacoubi Khebiza, M., Messouli, M., & Lachir, A. (2020). Assessment of Future Water Demand and Supply under IPCC Climate Change and Socio-Economic Scenarios, Using a Combination of Models in Ourika Watershed, High Atlas, Morocco. *Water (Basel)*, *12*(6), 6. Advance online publication. doi:10.3390/w12061751

Bailey, L., & Ball, B. V. (1991). *Index-Honey Bee Pathology* (2nd ed.). Academic press.

Bailey, L., & Fernando, E. F. W. (1972). Effects of sacbrood virus on adult honey-bees. *Annals of Applied Biology*, *72*(1), 27–35. doi:10.1111/j.1744-7348.1972.tb01268.x

Balakchowsky, A. S. (1966). *Entomologie appliquée à l'agriculture lépidoptères* (Vol. 2). Ed Masson et Cie.

Bale, J. S., & Hayward, S. A. L. (2010). Insect overwintering in a changing climate. *The Journal of Experimental Biology*, *213*(6), 980–994. doi:10.1242/jeb.037911 PMID:20190123

Bale, J., Masters, G. J., Hodkins, I. D., Awmack, C., Bezemer, T. M., Brown, V. K., Buterfield, J., Buse, A., Coulson, J. C., Farrar, J., Good, J. E. G., Harrigton, R., Hartley, S., Jones, T. H., Lindroth, R. L., Press, M. C., Symrnioudis, I., Watt, A. D., & Whittaker, J. B. (2002). Herbivory in global climate change research: Direct effects of rising temperature on insect herbivores. *Global Change Biology*, *8*(1), 1–16. doi:10.1046/j.1365-2486.2002.00451.x

Bardenhagen, C. J., Howard, P. H., & Gray, S. A. (2020). Farmer Mental Models of Biological Pest Control: Associations With Adoption of Conservation Practices in Blueberry and Cherry Orchards. *Frontiers in Sustainable Food Systems*, *4*, 54. doi:10.3389/fsufs.2020.00054

Barghi, N., Tobler, R., Nolte, V., Jakšić, A. M., Mallard, F., Otte, K. A., Dolezal, M., Taus, T., Kofler, R., & Schlötterer, C. (2019). Genetic redundancy fuels polygenic adaptation in *Drosophila. PLoS Biology*, *17*(2), e3000128. doi:10.1371/journal.pbio.3000128 PMID:30716062

Bartholomew, A., Hafezi, S. A., & Karimi, S. (2016). Effects of habitat complexity on the abundance, species richness and size of darkling beetles (Tenebrionidae) in artificial vegetation. *Journal of Arid Environments*, *129*, 35–41. doi:10.1016/j.jaridenv.2016.02.004

Barton, M. G., Clusella-Trullas, S., & Terblanche, J. S. (2019). Spatial scale, topography and thermoregulatory behaviour interact when modelling species' thermal niches. *Ecography*, *42*(2), 376–389. doi:10.1111/ecog.03655

Bassanezi, R. B., Lopes, S. A., de Miranda, M. P., Wulff, N. A., Volpe, H. X. L., & Ayres, A. J. (2020). Overview of citrus huanglongbing spread and management strategies in Brazil. *Tropical Plant Pathology*, *45*(3), 251–264. doi:10.100740858-020-00343-y

Battisti, A., Stastny, M., Buffo, E., & Larsson, S. (2006). A rapid altitudinal range expansion in the pine processionary moth produced by the 2003 climatic anomaly. *Global Change Biology*, *12*(4), 662–671. doi:10.1111/j.1365-2486.2006.01124.x

Bebber, D. P. (2015). Range-Expanding Pests and Pathogens in a Warming World. *Annual Review of Phytopathology*, *53*(1), 335–356. doi:10.1146/annurev-phyto-080614-120207 PMID:26047565

Bebber, D. P., Ramotowski, M. A. T., & Gurr, S. J. (2013). Crop pests and pathogens move polewards in a warming world. *Nature Climate Change*, *3*(11), 11. Advance online publication. doi:10.1038/nclimate1990

Belen, A., & Alten, B. (2011). Seasonal dynamics and altitudinal distributions of sand fly (Diptera: Psychodidae) populations in a cutaneous leishmaniasis endemic area of the Cukurova region of Turkey. *Journal of Vector Ecology*, *36*(s1), S87–S94. Advance online publication. doi:10.1111/j.1948-7134.2011.00116.x PMID:21366785

Bennas, N., & Sàinz-Cantero, C. E. (2006). Coléoptères aquatiques Adephaga de la chaîne rifaine marocaine (Coleoptera Gyrinidae, Haliplidae, Noteridae, Hygrobiidae, Dytiscidae). *Memorie della Societa Entomologica Italiana*, *85*(1), 31–73. doi:10.4081/memorieSEI.2006.31

Benziane, T., Abbassi, M., & Jahaj, A. (2001). Comparaison de deux méthodes de lutte intégrée contre les ravageurs en vergers d'agrumes dans la région du gharb. Symp.Prot.Int.Cul.Rég.Mé., 257–269.

Benziane, T., Abbassi, T., & Bihi, M. (2003). Evaluation de deux méthodes de lutte intégrée contre les ravageurs en vergers d'agrumes. *Journal of Applied Entomology*, *127*(1), 51–63. doi:10.1046/j.1439-0418.2003.00713.x

Bergland, A. O., Tobler, R., González, J., Schmidt, P., & Petrov, D. (2016). Secondary contact and local adaptation contribute to genome-wide patterns of clinal variation in *Drosophila melanogaster. Molecular Ecology*, *25*(5), 1157–1174. doi:10.1111/mec.13455 PMID:26547394

Bergstrom, B. C., Maki, L. R., & Werner, B. A. (1976). Small dung beetles as biological control agents: Laboratory studies of beetle action on trichostrongylid eggs in sheep and cattle feces. *Proceedings of the Helminthological Society of Washington*, *43*, 171–174.

Bernal, J. S., & Medina, R. F. (2018). Agriculture sows pests: How crop domestication, host shifts, and agricultural intensification can create insect pests from herbivores. *Current Opinion in Insect Science*, *26*, 76–81. doi:10.1016/j.cois.2018.01.008 PMID:29764664

Bertrand, J. (2001). *Agriculture et biodiversité: un partenariat à valoriser*. Educagri éditions.

Bezděk, J., & Baselga, A. (2015). Revision of western Palaearctic species of the *Oulema melanopus* group, with description of two new species from Europe (Coleoptera: Chrysomelidae: Criocerinae). *Acta Entomologica Musei Nationalis Pragae*, *55*(1), 273–304. https://www.biotaxa.org/AEMNP/article/view/12440

Bhat, A. I., & Rao, G. P. (2020). Virus transmission through pollen. In A. I. Bhat & G. P. Rao (Eds.), *Characterization of Plant Viruses: Methods and Protocols* (pp. 61–64). Humana, Springer. doi:10.1007/978-1-0716-0334-5_9

Biddinger, D. J., Weber, D. C., & Hull, L. A. (2009). Coccinellidae as predators of mites: *Stethorini* in biological control. *Biological Control*, *51*(2), 268–283. doi:10.1016/j.biocontrol.2009.05.014

Birthal, P. S., & Hazrana, J. (2019). Crop diversification and resilience of agriculture to climatic shocks: Evidence from India. *Agricultural Systems*, *173*, 345–354. doi:10.1016/j.agsy.2019.03.005

Birthal, P. S., Hazrana, J., Negi, D. S., & Bhan, S. C. (2021). Climate change and land-use in Indian agriculture. *Land Use Policy*, *109*, 105652. doi:10.1016/j.landusepol.2021.105652

Birthal, P. S., Negi, D. S., Khan, M., & Agarwal, S. (2015a). Is Indian agriculture becoming resilient to droughts? Evidence from rice production systems. *Food Policy*, *56*, 1–12. doi:10.1016/j.foodpol.2015.07.005

BirthalP. S.NegiD. S.KumarS.AggarwalS.SureshA.KhanT. (2014). How Sensitive Is Indian Agriculture to Climate Change? https://papers.ssrn.com/abstract=3867263

Birthal, P. S., Roy, D., & Negi, D. S. (2015b). Assessing the Impact of Crop Diversification on Farm Poverty in India. *World Development*, *72*, 70–92. doi:10.1016/j.worlddev.2015.02.015

Bishop, C. A., Woundneh, M. B., Maisonneuve, F., Common, J., Elliott, J. E., & Moran, A. J. (2020). Determination of neonicotinoids and butenolide residues in avian and insect pollinators and their ambient environment in Western Canada (2017, 2018). *The Science of the Total Environment*, *737*(1), 139386. doi:10.1016/j.scitotenv.2020.139386 PMID:32563110

Biswal, D. (2018). *Contemporary Health Issues and Environmental Impact.* In M. Bose, S. A. Bandyopadyay, & S. Poddar (Eds.), *Impact of climate change on parasitic diseases: Do we need to bother?* (pp. 32–40). Lincoln University College. doi:10.31674/books.2018.chiei.ch05

Biswal, D. (2021). Recent Advancements in Eco-Friendly Mosquito Control Strategies. In D. Ghosh & S. Saha (Eds.), *Recent Trends in Environmental Science and Applied Ecology* (pp. 124–129). Bhumi Publishing.

Blake, S., Foster, G. N., Fisher, G. E., & Ligertwood, G. L. (1996). Effects of management practices on the carabid faunas of newly established wildflower meadows in southern Scotland. *Annales Zoologici Fennici*, *33*, 139–147. https://www.jstor.org/stable/23735413

Blank, R. H., & Gill, G. S. C. (1997). Citrus thrips in New Zealand. *Proceedings of International Society of Citriculture. South Africa*, *1*, 515–518.

Bo, C. F. D., & Florio, M. (2012). Infrastructure and growth in a spatial framework: Evidence from the EU regions. *European Planning Studies*, *20*(8), 1393–1414. doi:10.1080/09654313.2012.680587

Böhme, F., Bischoff, G., Zebitz, C. P., Rosenkranz, P., & Wallner, K. (2018). Pesticide residue survey of pollen loads collected by honeybees (*Apis mellifera*) in daily intervals at three agricultural sites in South Germany. *PLoS One*, *13*(7), e0199995. doi:10.1371/journal.pone.0199995 PMID:29979756

Boller, E. F., Avilla, J., Joerg, E., Malavolta, C., Wijnands, F. G., & Esbjerg, P. (2004). *Integrated Production: Principles and Technical Guidelines*. IOBC/WPRS. Available online: https://www.iobcwprs.org/ip_ipm/01_IOBC_Principles_and_Tech_Guidelines_2004.pdf

Boubidi, S., Benallal, K., Boudrissa, A., Bouiba, L., Bouchareb, B., Garni, R., Bouratbine, A., Ravel, C., Dvorak, V., Votypka, J., Volf, P., & Harrat, Z. (2011). Phlebotomus sergenti (Parrot, 1917) identified as Leishmania killicki host in Ghardaia, south Algeria. *Microbes and Infection, 13*(7), 691–696. doi:10.1016/j.micinf.2011.02.008 PMID:21382502

Bouchard, P., Bousquet, Y., Aalbu, R. L., Alonso-Zarazaga, M. A., Merkl, O., & Davies, A. E. (2021). Review of genus-group names in the family Tenebrionidae (Insecta, Coleoptera*). ZooKeys, 1050*, 1–633. doi:10.3897/zookeys.1050.64217 PMID:34385881

Boussaade, L. (2003). *Etude comparative des principaux ravageurs des agrumes et leurs ennemis naturels en vergers bio, sous protection intégrée et conventionnelle* [Thesis]. Mediterranean Agronomic Institute Bari (IAMB).

Bouvet, J. P. R., Urbaneja, A., Pérez-Hedo, M., & Monzó, C. (2019). Contribution of predation to the biological control of a key herbivorous pest in citrus agroecosystems. *Journal of Animal Ecology, 88*(6), 915–926. doi:10.1111/1365-2656.12982 PMID:30895609

Branco, S., Alves-Pires, C., Maia, C., Cortes, S., Cristovão, J. M. S., Gonçalves, L., Campino, L., & Afonso, M. O. (2013). Entomological and ecological studies in a new potential zoonotic leishmaniasis focus in Torres Novas municipality, Central Region, Portugal. *Acta Tropica, 125*(3), 339–348. doi:10.1016/j.actatropica.2012.12.008 PMID:23262215

Brander, L. (2013). *Guidance manual on value transfer methods for ecosystem services. United Nations Environment Programme*. UNEP.

Brown, M. J., Dicks, L. V., Paxton, R. J., Baldock, K. C., Barron, A. B., Chauzat, M. P., Freitas, B. M., Goulson, D., Jepsen, S., Kremen, C., Li, J., Neumann, P., Pattemore, D. E., Potts, S. G., Schweiger, O., Seymour, C. L., & Stout, J. C. (2016). A horizon scan of future threats and opportunities for pollinators and pollination. *PeerJ, 4*, e2249. doi:10.7717/peerj.2249 PMID:27602260

Buckley, L. B. (2022). Temperature-sensitive development shapes insect phenological responses to climate change. *Current Opinion in Insect Science, 52*, 100897. doi:10.1016/j.cois.2022.100897 PMID:35257968

Bush, A., Mokany, K., Catullo, R., Hoffmann, A., Kellermann, V., Sgrò, C., McEvey, S., & Ferrier, S. (2016). Incorporating evolutionary adaptation in species distribution modelling reduces projected vulnerability to climate change. *Ecology Letters, 19*, 1468–1478.

Busse, M., Zoll, F., Siebert, R., Bartels, A., Bokelmann, A., & Scharschmidt, P. (2021). How farmers think about insects: Perceptions of biodiversity, biodiversity loss and attitudes towards insect-friendly farming practices. *Biodiversity and Conservation, 30*(11), 3045–3066. doi:10.100710531-021-02235-2

CABI. (2021). Bactrocera tau. CABI Compendium. *CABI Compendium, 8741*. Advance online publication. doi:10.1079/cabicompendium.8741

Cajaiba, R. L., Périco, E., Dalzochio, M. S., da Silva, W. B., Bastos, R., Cabral, J. A., & Santos, M. (2017). Does the composition of Scarabaeidae (Coleoptera) communities reflect the extent of land use changes in the Brazilian Amazon? *Ecological Indicators*, *74*, 285–294. doi:10.1016/j.ecolind.2016.11.018

Calatayud-Vernich, P., Calatayud, F., Simó, E., & Picó, Y. (2018). Pesticide residues in honey bees, pollen and beeswax: Assessing beehive exposure. *Environmental Pollution*, *241*, 106–114. doi:10.1016/j.envpol.2018.05.062 PMID:29803024

Cammell, M. E., & Knight, J. D. (1991). Effects of climate change on the population dynamics of Crop pests. *Advances in Ecological Research*, *22*, 117–162. doi:10.1016/S0065-2504(08)60135-X

Cardoso, P., Barton, P. S., Birkhofer, K., Chichorro, F., Deacon, C., Fartmann, T., Fukushima, C. S., Gaigher, R., Habel, J. C., Hallmann, C. A., Hill, M. J., Hochkirch, A., Kwak, M. L., Mammola, S., Ari Noriega, J., Orfinger, A. B., Pedraza, F., Pryke, J. S., Roque, F. O., ... Samways, M. J. (2020). Scientists' warning to humanity on insect extinctions. *Biological Conservation*, *242*, 108426. doi:10.1016/j.biocon.2020.108426

Carpenter, D. O. (2006). Polychlorinated biphenyls (PCBs): Routes of exposure and effects on human health. *Reviews on Environmental Health*, *21*(1), 1–23. doi:10.1515/REVEH.2006.21.1.1 PMID:16700427

Carroll, A. L., Taylor, S. W., Regniere, J., & Safranyik, D. L. (2004). Effects of climate change on range expansion by the mountain pine beetle in British Columbia. In *Mountain Pine Beetle Symposium: Challenges and Solutions*. Natural Resources Canada, Canadian Forest Service, Pacific Forestry Centre.

Carroll, M. J., Brown, N., Goodall, C., Downs, A. M., Sheenan, T. H., & Anderson, K. E. (2017). Honey bees preferentially consume freshly-stored pollen. *PLoS One*, *12*(4), e0175933. doi:10.1371/journal.pone.0175933 PMID:28430801

Cavalcanti, S. C., Niculau, E. S., Blank, A. F., Câmara, C. A., Araújo, I. N., & Alves, P. B. (2010). Composition and acaricidal activity of *Lippia sidoides* essential oil against two-spotted spider mite (*Tetranychus urticae* Koch). *Bioresource Technology*, *101*(2), 829–832. doi:10.1016/j.biortech.2009.08.053 PMID:19758799

Cayol, J. P., Causse, R., Louis, C., & Barthes, J. (1994). Medfly Ceratitis capitata Wiedemann (Dipt., Trypetidae) as a rot vector in laboratory conditions. *Journal of Applied Entomology*, *117*(1–5), 338–343. doi:10.1111/j.1439-0418.1994.tb00744.x

Chandra, K., Gupta, D., Uniyal, V. P., Bharadwaj, M., & Sanyal, A. K. (2012). Studies on scarabaeid beetles (Coleoptera) of Govind wildlife sanctuary, Garhwal, Uttarakhand, India. *Biological Forum: An International Journal*, *4*(1), 48–54.

Chantawannakul, P., Ramsey, S., Khongphinitbunjong, K., & Phokasem, P. (2018). Tropilaelaps mite: An emerging threat to European honey bee. *Current Opinion in Insect Science*, *26*, 69–75. doi:10.1016/j.cois.2018.01.012 PMID:29764663

Chapman, R. F. (1998). *The Insects-Structure and Function* (4th ed.). Cambridge University Press. doi:10.1017/CBO9780511818202

Cheli, G. H., Bosco, T., & Flores, G. E. (2022). The role of *Nyctelia circumundata* (Coleoptera: Tenebrionidae) on litter fragmentation processes and soil fertility in northeastern arid patagonia. *Geoderma, 415*, 115770. doi:10.1016/j.geoderma.2022.115770

Chen, D., Guo, R., Xiong, C., Zheng, Y., Hou, C., & Fu, Z. (2018). Morphological and molecular identification of chalkbrood disease pathogen *Ascosphaera apis* in *Apis cerana cerana. Journal of Apicultural Research, 57*(4), 516–521. doi:10.1080/00218839.2018.1475943

Cheng, Z., Wang, D., Han, S., Zuo, C., & He, Y. (2022). Transcriptome analysis in the thiamethoxam resistance of seven-spot ladybird beetle, *Coccinella septempunctata* (Coleoptera: Coccinellidae). *Ecotoxicology and Environmental Safety, 230*, 113144. doi:10.1016/j.ecoenv.2021.113144 PMID:34998260

Chen, X. Y. (2013). Harm and control of chalkbrood disease and Aspergillus flavus disease. *Mifeng Zazhi, 33*, 24.

Chen, X., Wang, W., Chen, J., Zhu, X., Shen, M., Gan, L., & Cao, X. (2020). Does any phenological event defined by remote sensing deserve particular attention? An examination of spring phenology of winter wheat in Northern China. *Ecological Indicators, 116*, 106456. doi:10.1016/j.ecolind.2020.106456

Chen, Y. H. (2016). Crop domestication, global human-mediated migration, and the unresolved role of geography in pest control. *Elementa, 4*, 000106. doi:10.12952/journal.elementa.000106

Chen, Y. H., Gols, R., & Benrey, B. (2015). Crop Domestication and Its Impact on Naturally Selected Trophic Interactions. *Annual Review of Entomology, 60*(1), 35–58. doi:10.1146/annurev-ento-010814-020601 PMID:25341108

Chen, Y. P., Evans, J., & Feldlaufer, M. (2006). Horizontal and vertical transmission of viruses in the honeybee *Apis mellifera. Journal of Invertebrate Pathology, 92*(3), 152–159. doi:10.1016/j.jip.2006.03.010 PMID:16793058

Chen, Y. P., Pettis, J. S., Corona, M., Chen, W. P., Li, C. J., Spivak, M., Visscher, P. K., DeGrandi-Hoffman, G., Boncristiani, H., Zhao, Y., Vanengelsdorp, D., Delaplane, K., Solter, L., Drummond, F., Kramer, M., Lipkin, W. I., Palacios, G., Hamilton, M. C., Smith, B., ... Evans, J. D. (2014). Israeli acute paralysis virus: Epidemiology, pathogenesis and implications for honey bee health. *PLoS Pathogens, 10*(7), e1004261. doi:10.1371/journal.ppat.1004261 PMID:25079600

Choi, J. B., Choi, E. Y., Kwon, T., Jeong, C. H., Oh, S. N., Han, G. Y., Kim, M. C., & Park, J. K. (2022). A checklist of the family Carabidae (Coleoptera: Adephaga: Caraboidea) from Cambodia. *Journal of Asia-Pacific Biodiversity, 15*(2), 180–195. doi:10.1016/j.japb.2021.12.007

Chown, S. L., Slabber, S., McGeouch, M., Janion, C., & Leinaas, H. P. (2007). Phenotypic plasticity mediates climate change responses among invasive and indigenous arthropods. *Proceedings. Biological Sciences, 274*(1625), 2531–2537. doi:10.1098/rspb.2007.0772 PMID:17686728

Cleland, E. E., Chuine, I., Menzel, A., Mooney, H. A., & Schwartz, M. D. (2007). Shifting plant phenology in response to global change. *Trends in Ecology & Evolution*, *22*(7), 357–365. doi:10.1016/j.tree.2007.04.003 PMID:17478009

Colares, C., Roza, A. S., Mermudes, J. R., Silveira, L. F., Khattar, G., Mayhew, P. J., Monteiro, R. F., Nunes, M. F., & Macedo, M. V. (2021). Elevational specialization and the monitoring of the effects of climate change in insects: Beetles in a Brazilian rainforest mountain. *Ecological Indicators*, *120*, 106888. doi:10.1016/j.ecolind.2020.106888

Coll, M., & Hughes, L. (2008). Effects of elevated CO2 on an insect omnivore: A test for nutritional effects mediated by host plants and prey. *Agriculture, Ecosystems & Environment*, *123*(4), 271–279. doi:10.1016/j.agee.2007.06.003

Conti, F., Tumminelli, R., Amico, C., Fisicaro, R., Frittitta, C., Perrotta, G., & Marullo, R. (2001). Monitoring of *Pezothrips kellyanus* on citrus in eastern Sicily. In *Proceedings of the 7th International Symposium on Thysanoptera*. University of Reggio Calabria.

Conti, M. E., & Botrè, F. (2001). Honeybees and their products as potential bioindicators of heavy metals contamination. *Environmental Monitoring and Assessment*, *69*(3), 267–282. doi:10.1023/A:1010719107006 PMID:11497382

Costanza, R., d'Arge, R., de Groot, R., Farber, S., Grasso, M., Hannon, B., Limburg, K., Naeem, S., O'Neill, R. V., Paruelo, J., Raskin, R. G., Sutton, P., & van den Belt, M. (1987). The Value of the World's Ecosystem Services and Natural Capital. *Nature*, *387*(6630), 253–260. doi:10.1038/387253a0

Cotton, J. M., Cerling, T. E., Hoppe, K. A., Mosier, T. M., & Still, C. J. (2016). Climate, CO2, and the history of North American grasses since the Last Glacial Maximum. *Science Advances*, *2*(3), e1501346. doi:10.1126ciadv.1501346 PMID:27051865

Coviella, C. E., & Trumble, J. T. (1999). Effects of elevated atmospheric carbon dioxide on insect-plant interactions. *Conservation Biology*, *13*(4), 700–712. doi:10.1046/j.1523-1739.1999.98267.x

Cox-Foster, D. L., Conlan, S., Holmes, E. C., Palacios, G., Evans, J. D., Moran, N. A., Quan, P. L., Briese, T., Hornig, M., Geiser, D. M., Martinson, V., van Engelsdorp, D., Kalkstein, A. L., Drysdale, A., Hui, J., Zhai, J. H., Cui, L. W., Hutchison, S. K., Simons, J. F., ... Lipkin, W. I. (2007). A metagenomic survey of microbes in honey bee colony collapse disorder. *Science*, *318*(5848), 283–287. doi:10.1126cience.1146498 PMID:17823314

Coyle, P., Philcox, J. C., Carey, L. C., & Rofe, A. M. (2002). Metallothionein: The multipurpose protein. *Cellular and Molecular Life Sciences*, *59*(4), 627–647. doi:10.100700018-002-8454-2 PMID:12022471

Cresswell, J. E., Desneux, N., & VanEngelsdorp, D. (2012). Dietary traces of neonicotinoid pesticides as a cause of population declines in honey bees: An evaluation by Hill's epidemiological criteria. *Pest Management Science*, *68*(6), 819–827. doi:10.1002/ps.3290 PMID:22488890

Crossley, M. S., Meier, A. R., Baldwin, E. M., Berry, L. L., Crenshaw, L. C., Hartman, G. L., Lagos-Kutz, D., Nichols, D. H., Patel, K., Varriano, S., Snyder, W. E., & Moran, M. D. (2020). No net insect abundance and diversity declines across US Long Term Ecological Research sites. *Nature Ecology & Evolution*, *4*(10), 10. Advance online publication. doi:10.103841559-020-1269-4 PMID:32778751

Crovetti, A., Conti, B., & Delrio, G. (1986). *Effect of abiotic factors on Ceratitis capitata (Wied.) (Diptera: Tephritidae). 2. Pupal development under constant temperatures.* Working Group on Fruit Flies of Economic Importance; CEC/IOBC ad-hoc meeting, Hamburg (Germany, F.R.), 23 Aug 1984. https://scholar.google.com/scholar_lookup?title=Effect+of+abiotic+factors+on+Ceratitis+capitata+%28Wied.%29+%28Diptera%3A+Tephritidae%29.+2.+Pupal+development+under+constant+temperatures&author=Crovetti%2C+A.+%28Pisa+Univ.+%28Italy%29.+Istituto+di+Entomologia+Agraria%29&publication_year=1986

Cunningham, M. M., Tran, L., McKee, C. G., Polo, R. O., Newman, T., Lansing, L., Griffiths, J. S., Bilodeau, G. J., Rott, M., & Guarna, M. M. (2022). Honey bees as biomonitors of environmental contaminants, pathogens, and climate change. *Ecological Indicators*, *134*, 108457. doi:10.1016/j.ecolind.2021.108457

Cuthbertson, A. G. S., & Brown, M. A. (2009). Issues affecting British honey bee biodiversity and the need for conservation of this important ecological component. *International Journal of Environmental Science and Technology*, *6*(4), 695–699. doi:10.1007/BF03326110

Dainese, M., Martin, E. A., Aizen, M. A., Albrecht, M., Bartomeus, I., Bommarco, R., Carvalheiro, L. G., Chaplin-Kramer, R., Gagic, V., Garibaldi, L. A., Ghazoul, J., Grab, H., Jonsson, M., Karp, D. S., Kennedy, C. M., Kleijn, D., Kremen, C., Landis, D. A., Letourneau, D. K., ... Steffan-Dewenter, I. (2019). A global synthesis reveals biodiversity-mediated benefits for crop production. *Science Advances*, *5*(10), eaax0121. Advance online publication. doi:10.1126ciadv.aax0121 PMID:31663019

Danks, H. V. (2006). Short life cycles in insects and mites. *Canadian Entomologist*, *138*(4), 407–463. doi:10.4039/n06-803

Davis, R. S., Peterson, R. K., & Macedo, P. A. (2007). An ecological risk assessment for insecticides used in adult mosquito management. *Integrated Environmental Assessment and Management*, *3*(3), 373–382. doi:10.1002/ieam.5630030308 PMID:17695110

De La Torre-Roche, R. J., Lee, W.-Y., & Campos-Díaz, S. I. (2009). Soil-borne polycyclic aromatic hydrocarbons in El Paso, Texas: Analysis of a potential problem in the United States/Mexico border region. *Journal of Hazardous Materials*, *163*(2-3), 946–958. doi:10.1016/j.jhazmat.2008.07.089 PMID:18768257

de Oliveira, R. C., Queiroz, S., da Luz, C. F. P., Porto, R. S., & Rath, S. (2016). Bee pollen as a bioindicator of environmental pesticide contamination. *Chemosphere*, *163*, 525–534. doi:10.1016/j.chemosphere.2016.08.022 PMID:27567152 de Sousa Pereira, K., Meeus, I., & Smagghe, G. (2019). Honey bee-collected pollen is a potential source of *Ascosphaera apis* infection in managed bumble bees. *Scientific Reports*, *9*(1), 1–9. doi:10.103841598-019-40804-2 PMID:30862950

de Vries, H. H. (1994). Size of habitat and presence of ground beetle species. In K. Desender, M. Dufrêne, & J. P. Maelfait (Eds.), *Carabid Beetles. Ecology and Evolution* (pp. 253–259). Kluwer Academic Publishers. doi:10.1007/978-94-017-0968-2_39

Delabie, J. H. C., Serrao, J. E., Mariano, C. S. F., & Matile-Ferrero, D. (2008). Comportements de communication de la cochenille neotropicale *Neochavesia caldasiae* (Balachowsky 1957) (*Pseudococcidae*: Rhizoecinae) avec sa fourmi symbiote *Acropyga fuhrmanni* (Forel 1914) (Formicidae: Formicinae). *Annales de la Société Entomologique de France*, *44*(4), 471–475. do i:10.1080/00379271.2008.10697582

Delaplane, K. S., & Mayer, D. F. (2000). *Crop pollination by bees*. CAB. doi:10.1079/9780851994482.0000

Dell, M., Jones, B. F., & Olken, B. A. (2012). Temperature Shocks and Economic Growth: Evidence from the Last Half Century. *American Economic Journal. Macroeconomics*, *4*(3), 66–95. doi:10.1257/mac.4.3.66

DeLucia, E. H., Casteel, C. L., Nabity, P. D., & O'Neill, B. F. (2008). Insects take a bigger bite out of plants in a warmer, higher carbon dioxide world. *Proceedings of the National Academy of Sciences of the United States of America*, *105*(6), 1781–1782. doi:10.1073/pnas.0712056105 PMID:18256185

Demir, S., & Karakuş, M. (2015). Natural Leishmania infection of Phlebotomus sergenti (Diptera: Phlebotominae) in an endemic focus of cutaneous leishmaniasis in Şanlıurfa, Turkey. *Acta Tropica*, *149*, 45–48. doi:10.1016/j.actatropica.2015.05.010 PMID:25997884

DesJardins, N. S., Fisher, A. II, Ozturk, C., Fewell, J. H., DeGrandi-Hoffman, G., Harrison, J. F., & Smith, B. H. (2021). A common fungicide, Pristine®, impairs olfactory associative learning performance in honey bees (*Apis mellifera*). *Environmental Pollution*, *288*, 117720. doi:10.1016/j.envpol.2021.117720 PMID:34252716

Deutsch, C. A., Tewksbury, J. J., Tigchelaar, M., Battisti, D. S., Merrill, S. C., Huey, R. B., & Naylor, R. L. (2018). Increase in crop losses to insect pests in a warming climate. *Science*, *361*(6405), 916–919. doi:10.1126cience.aat3466 PMID:30166490

Devictor, V., van Swaay, C., Brereton, T., Brotons, L., Chamberlain, D., Heliölä, J., Herrando, S., Julliard, R., Kuussaari, M., Lindström, Å., Reif, J., Roy, D. B., Schweiger, O., Settele, J., Stefanescu, C., Strien, A. V., Turnhout, C. V., & Vermouzek, Z. (2012). Differences in the climatic debts of birds and butterflies at a continental scale. *Nature Climate Change*, *2*(2), 121–124. doi:10.1038/nclimate1347

Di Prisco, G., Annoscia, D., Margiotta, M., Ferrara, R., Varricchio, P., Zanni, V., Caprio, E., Nazzi, F., & Pennacchio, F. (2016). A mutualistic symbiosis between a parasitic mite and a pathogenic virus undermines honey bee immunity and health. *Proceedings of the National Academy of Sciences of the United States of America, 113*(12), 3203–3208. doi:10.1073/pnas.1523515113 PMID:26951652

Diagne, C., Leroy, B., Vaissière, A.-C., Gozlan, R. E., Roiz, D., Jarić, I., Salles, J.-M., Bradshaw, C. J. A., & Courchamp, F. (2021). High and rising economic costs of biological invasions worldwide. *Nature, 592*(7855), 7855. Advance online publication. doi:10.103841586-021-03405-6 PMID:33790468

Diffenbaugh, N. S., Krupke, C. H., White, M. A., & Alexander, C. E. (2008). Global warming presents new challenges for maize pest management. *Environmental Research Letters, 3*(4), 1–9. doi:10.1088/1748-9326/3/4/044007

Diffenbaugh, N. S., Singh, D., Mankin, J. S., Horton, D. E., Swain, D. L., Touma, D., Charland, A., Liu, Y., Haugen, M., Tsiang, M., & Rajaratnam, B. (2017). Quantifying the influence of global warming on unprecedented extreme climate events. *Proceedings of the National Academy of Sciences of the United States of America, 114*(19), 4881–4886. doi:10.1073/pnas.1618082114 PMID:28439005

Diniz, N. M., Soares, A. E. E., Sheppard, W. S., & Del Lama, M. A. (2003). Genetic structure of honeybee populations from southern Brazil and Uruguay. *Genetics and Molecular Biology, 26*(1), 47–52. doi:10.1590/S1415-47572003000100008

Doha, S. A., & Samy, A. M. (2010). Bionomics of phlebotomine sand flies (Diptera: Psychodidae) in the province of Al-Baha, Saudi Arabia. *Memorias do Instituto Oswaldo Cruz, 105*(7), 850–856. doi:10.1590/S0074-02762010000700002 PMID:21120352

Donatelli, M., Magarey, R. D., Bregaglio, S., Willocquet, L., Whish, J. P. M., & Savary, S. (2017). Modelling the impacts of pests and diseases on agricultural systems. *Agricultural Systems, 155*, 213–224. doi:10.1016/j.agsy.2017.01.019 PMID:28701814

Do, Y., & Choi, M. B. (2022). Effects of urbanization on carabid beetles in an urban riparian area. *Entomological Research, 52*(6), 281–294. doi:10.1111/1748-5967.12601

Drossart, M., Rasmont, P., Vanormelingen, P., Dufrêne, M., Folschweiller, M., Pauly, A., Vereecken N. J., Vray, S., Zambra, E., D'Haeseleer, J., & Michez, D. (2019). Belgian red list of bees. *Multidisciplinary assessment of BELgian wild BEE decline to adapt mitigation management policy*, 3–140.

Du Plessis, H., Schlemmer, M.-L., & Van den Berg, J. (2020). The Effect of Temperature on the Development of Spodoptera frugiperda (Lepidoptera: Noctuidae). *Insects, 11*(4), 4. Advance online publication. doi:10.3390/insects11040228 PMID:32272548

Dukku, U. H. (2016). Evaluation of morphometric characters of honeybee (*Apis mellifera* L.) populations in the Lake Chad Basin in Central Africa. *Advances in Entomology (Irvine, Calif.), 4*(02), 75–89. doi:10.4236/ae.2016.42009

Duyck, P. F., & Quilici, S. (2002). Survival and development of different life stages of three Ceratitis spp. (Diptera: Tephritidae) reared at five constant temperatures. *Bulletin of Entomological Research*, *92*(6), 461–469. doi:10.1079/BER2002188 PMID:17598297

Dyer, L. A., Richards, L. A., Short, S. A., & Dodson, C. D. (2013). Effects of CO2 and Temperature on Tritrophic Interactions. *PLoS One*, *8*(4), e62528. doi:10.1371/journal.pone.0062528 PMID:23638105

Early, R., González-Moreno, P., Murphy, S. T., & Day, R. (2018). *Forecasting the global extent of invasion of the cereal pest Spodoptera frugiperda, the fall armyworm* (p. 391847). bioRxiv. doi:10.1101/391847

Ecosostenibile. (2021, April 7). Ceratitis capitata: Sistematica, Habitat, Ciclo biologico, Lotta. Un Mondo Ecosostenibile. https://antropocene.it/2021/04/07/ceratitis-capitata/

Eddoughri, F., Lkammarte, F. Z., El Jarroudi, M., Lahlali, R., Karmaoui, A., Yacoubi Khebiza, M., & Messouli, M. (2022). Analysis of the Vulnerability of Agriculture to Climate and Anthropogenic Impacts in the Beni Mellal-Khénifra Region, Morocco. *Sustainability*, *14*(20), 20. Advance online publication. doi:10.3390u142013166

El Messoussi, S. E., Hafid, H., Lahrouni, A., & Afif, M. (2007). Simulation of Temperature Effect on the Population Dynamic of the Mediterranean Fruit Fly Ceratitis capitata (Diptera; Tephritidae). *Journal of Agronomy*, *6*(2), 374–377. doi:10.3923/ja.2007.374.377

El Messoussi, S., Lahrouni, A., & Afif, M. (2003). Simulation of the population dynamics of insect pests about biotic and abiotic factors. *Proceedings of the Journée Scientifique Sur la Lutte Biologique*.

Ellstrand, N. C., & Rieseberg, L. H. (2016). When gene flow really matters: Gene flow in applied evolutionary biology. *Evolutionary Applications*, *9*(7), 833–836. doi:10.1111/eva.12402 PMID:27468301

European Commission Agriculture & Rural Development. (2020). *Honey Market Presentation*. Available online: https://ec.europa.eu/info/sites/info/files/food-farming fish eries/animals_and_animal_products/documents/market-presentat ion honey_autumn2020_en.pdf

European Commission. (n.d.). *Honey Bees*. Available online: https://ec.europa.eu/food/animals/live_animals/bees_en

Eyre, M. D., & Luff, M. L. (1990). A preliminary classification of European grassland habitats using carabid beetles. In N. E. Stork (Ed.), *The Role of Ground Beetles in Ecological and Environmental Studies* (pp. 227–236). Intercept.

Ezcurra, E., Rapoport, E. H., & Marino, C. R. (1978). The Geographical Distribution of Insect Pests. *Journal of Biogeography*, *5*(2), 149–157. doi:10.2307/3038169

Fall armyworm, Spodoptera frugiperda (J.E. Smith). (n.d.). Retrieved 23 October 2022, from https://entnemdept.ufl.edu/creatures/field/fall_armyworm.htm

Fand, B. B., Kamble, A. L., & Kumar, M. (2012). Will climate change pose serious threat to crop pest management: A critical review. *International Journal of Scientific Research, 2*, 1–14.

FAO. (2008). *Climate Related Transboundary Pests and Diseases.* https://www.fao.org/3/a-ai785e.pdf

Farré-Armengol, G., Filella, I., Llusià, J., Niinemets, Ü., & Peñuelas, J. (2014). Changes in floral bouquets from compound-specific responses to increasing temperatures. *Global Change Biology, 20*(12), 3660–3669. doi:10.1111/gcb.12628 PMID:24817412

Farré-Armengol, G., Filella, I., Llusia, J., & Peñuelas, J. (2013). Floral volatile organic compounds: Between attraction and deterrence of visitors under global change. *Perspectives in Plant Ecology, Evolution and Systematics, 15*(1), 56–67. doi:10.1016/j.ppees.2012.12.002

Faye, E., Herrera, M., Bellomo, L., Silvain, J.-F., & Dangles, O. (2014). Strong Discrepancies between Local Temperature Mapping and Interpolated Climatic Grids in Tropical Mountainous Agricultural Landscapes. *PLoS One, 9*(8), e105541. https://doi.org/10.1371/journal.pone.0105541

Faye, E., Rebaudo, F., Carpio, C., Herrera, M., & Dangles, O. (2017). Does heterogeneity in crop canopy microclimates matter for pests? Evidence from aerial high-resolution thermography. *Agriculture, Ecosystems & Environment, 246*, 124–133. https://doi.org/10.1016/j.agee.2017.05.027

Ferguson, L. V., Heinrichs, D. E., & Sinclair, B. J. (2016). Paradoxical acclimation responses in the thermal performance of insect immunity. *Oecologia, 181*(1), 77–85. doi:10.100700442-015-3529-6 PMID:26846428

Firon, N., Shaked, R., Peet, M. M., Pharr, D. M., Zamski, E., Rosenfeld, K., Althan, L., & Pressman, E. (2006). Pollen grains of heat tolerant tomato cultivars retain higher carbohydrate concentration under heat stress conditions. *Scientia Horticulturae, 109*(3), 212–217. doi:10.1016/j.scienta.2006.03.007

First Detection of Fall Armyworm in China. (n.d.). International Plant Protection Convention. Retrieved 23 October 2022, from https://www.ippc.int/fr/news/first-detection-of-fall-armyworm-in-china/

Fleming, R. A., & Volney, W. J. (1995). Effects of climate change on insect defoliator population processes in Canada's boreal forests: Some plausible scenarios. *Water, Air, and Soil Pollution, 82*(1-2), 445–454. doi:10.1007/BF01182854

Flores, J. M., Gil-Lebrero, S., Gámiz, V., Rodríguez, M. I., Ortiz, M. A., & Quiles, F. J. (2019). Effect of the climate change on honey bee colonies in a temperate Mediterranean zone assessed through remote hive weight monitoring system in conjunction with exhaustive colonies assessment. *The Science of the Total Environment, 653*, 1111–1119. doi:10.1016/j.scitotenv.2018.11.004 PMID:30759551

Forsgren, E., Lundhagen, A. C., Imdorf, A., & Fries, I. (2005). Distribution of *Melissococcus plutonius* in honeybee colonies with and without symptoms of European foulbrood. *Microbial Ecology, 50*(3), 369–374. doi:10.100700248-004-0188-2 PMID:16254762

Franchimont, J., & Saadaoui, E. M. (2001). *National Study on Biodiversity Synthesis Report. Kingdom of Morocco Ministery of Territorial Planing, Water and Environment*. Department of Environment.

Franck, P., Garnery, L., Solignac, M., & Cornuet, J. M. (2000). Molecular confirmation of a fourth lineage in honeybees from the Near East. *Apidologie*, *31*(2), 167–180. doi:10.1051/apido:2000114

Free, J. B. (1993). *Insect Pollination of Crops*. Academic Press.

Fries, I. (1991). Treatment of sealed honey bee brood with formic acid for control of Varroa jacobsoni. *American Bee Journal*.

Fuchs, R., Herold, M., Verburg, P. H., Clevers, J. G., & Eberle, J. (2015). Gross changes in reconstructions of historic land cover/use for Europe between 1900 and 2010. *Global Change Biology*, *21*(1), 299–313. doi:10.1111/gcb.12714 PMID:25155867

Gallai, N., Salles, J. M., & Vaissiere, B. E. (2009). Évaluation de la contribution économique du service de la pollinisation à l'agriculture européenne. *Bulletin Technique Apicole*, *36*(2), 110–116.

Ganuza, C., Redlich, S., Uhler, J., Tobisch, C., Rojas-Botero, S., Peters, M. K., Zhang, J., Benjamin, C. S., Englmeier, J., Ewald, J., Fricke, U., Haensel, M., Kollmann, J., Riebl, R., Uphus, L., Müller, J., & Steffan-Dewenter, I. (2022). Interactive effects of climate and land use on pollinator diversity differ among taxa and scales. *Science Advances*, *8*(18), eabm9359. Advance online publication. doi:10.1126ciadv.abm9359 PMID:35544641

Garibaldi, L. A., Steffan-Dewenter, I., Winfree, R., Aizen, M. A., Bommarco, R., Cunningham, S. A., Kremen, C., Carvalheiro, L. G., Harder, L. D., Afik, O., Bartomeus, I., Benjamin, F., Boreux, V., Cariveau, D., Chacoff, N. P., Dudenhoffer, J. H., Freitas, B. M., Ghazoul, J., Greenleaf, S., ... Klein, A. M. (2013). Wild pollinators enhance fruit set of crops regardless of honey bee abundance. *Science*, *339*(6127), 1608–1611. doi:10.1126cience.1230200 PMID:23449997

Genersch, E. (2010). American Foulbrood in honeybees and its causative agent, *Paenibacillus larvae*. *Journal of Invertebrate Pathology*, *103*, S10–S19. doi:10.1016/j.jip.2009.06.015 PMID:19909971

Gerstmeier, R. (2002). *Trichodes audouini-dubreuili* Reymond—Rediscovery and Redescription of an Obscure Species (Coleoptera: Cleridae). *Coleopterists Bulletin*, *56*(2), 273–278. doi:10.1649/0010-065X(2002)056[0273:TADRRA]2.0.CO;2

Ghalbane, I., Alahyane, H., Aboussaid, H., Chouikh, N. E., Costa, J., Romane, A., & El Messoussi, S. (2022). Chemical composition and insecticidal properties of Moroccan *Lavandula dentata* and *Lavandula stoechas* essential oils against Mediterranean Fruit Fly, *Ceratitis capitata*. *Neotropical Entomology*, *51*(4), 628–636. doi:10.100713744-022-00975-6 PMID:35737225

Gienapp, P., Teplitsky, C., Alho, J. S., Mills, J. A., & Merilä, J. (2008). Climate change and evolution: Disentangling environmental and genetic responses. *Molecular Ecology*, *17*(1), 167–178. doi:10.1111/j.1365-294X.2007.03413.x PMID:18173499

Gill, H. K., Goyal, G., & Chahil, G. (2017). Insect diapause: A review. *Journal of Agricultural Science and Technology A, 7*(7), 454–473. doi:10.17265/2161-6256/2017.07.002

Gilliam, M. (1990). Chalkbrood disease of honey bees, *Apis mellifera*, caused by the fungus, *Ascosphaera apis*: A review of past and current research. *Proceedings and abstracts, Vth International Colloquium on Invertebrate Pathology and Microbial Control*, 398-402.

Gilman, S. E., Urban, M. C., Tewksbury, J., Gilchrist, G. W., & Holt, R. D. (2010). A framework for community interactions under climate change. *Trends in Ecology & Evolution, 25*(6), 325–331. https://doi.org/10.1016/j.tree.2010.03.002

Goergen, G., Kumar, P. L., Sankung, S. B., Togola, A., & Tamò, M. (2016). First Report of Outbreaks of the Fall Armyworm Spodoptera frugiperda (J E Smith) (Lepidoptera, Noctuidae), a New Alien Invasive Pest in West and Central Africa. *PLoS One, 11*(10), e0165632. https://doi.org/10.1371/journal.pone.0165632

Gomy, Y. (2007). Description d'une nouvelle espèce de *Teretrius* Erichson, 1834 du Maroc (Coleoptera, Histeridae). *Bulletin Mensuel de la Societe Linneenne de Lyon, 76*(5), 113–119. doi:10.3406/linly.2007.13656

Gomy, Y., de Castro-Arrazola, I., Sanchez-Piñero, F., & Labrique, H. (2016). Contribution à la connaissance des Histeridae du Maroc (quatrième note)(Coleoptera). *Entomologiste, 72*(1), 7–32.

González-Varo, J. P., Biesmeijer, J. C., Bommarco, R., Potts, S. G., Schweiger, O., Smith, H. G., Steffan-Dewenter, I., Szentgyorgyi, H., Woyciechowski, M., & Vila, M. (2013). Combined effects of global change pressures on animal-mediated pollination. *Trends in Ecology & Evolution, 28*(9), 524–530. doi:10.1016/j.tree.2013.05.008 PMID:23746938

Gordo, O., & Sanz, J. J. (2005). Phenology and climate change: A long-term study in a Mediterranean locality. *Oecologia, 146*(3), 484–495. doi:10.100700442-005-0240-z PMID:16170564

Gordo, O., & Sanz, J. J. (2010). Impact of climate change on plant phenology in Mediterranean ecosystems. *Global Change Biology, 16*(3), 1082–1106. doi:10.1111/j.1365-2486.2009.02084.x

Goulson, D., Lye, G. C., & Darvill, B. (2008). Decline and conservation of bumble bees. *Annual Review of Entomology, 53*(1), 191–208. doi:10.1146/annurev.ento.53.103106.093454 PMID:17803456

Grassl, J., Holt, S., Cremen, N., Peso, M., Hahne, D., & Baer, B. (2018). Synergistic effects of pathogen and pesticide exposure on honey bee (*Apis mellifera*) survival and immunity. *Journal of Invertebrate Pathology, 159*, 78–86. doi:10.1016/j.jip.2018.10.005 PMID:30300630

Greenleaf, S. S., & Kremen, C. (2006). Wild bees enhance honey bees' pollination of hybrid sunflower. *Proceedings of the National Academy of Sciences of the United States of America, 103*(37), 13890–13895. doi:10.1073/pnas.0600929103 PMID:16940358

Greenleaf, S. S., Williams, N. M., Winfree, R., & Kremen, C. (2007). Bee foraging ranges and their relationship to body size. *Oecologia, 153*(3), 589–596. doi:10.100700442-007-0752-9 PMID:17483965

Gregory, P. J., Johnson, S. N., Newton, A. C., & Ingram, J. S. I. (2009). Integrating pests and pathogens into the climate change/food security debate. *Journal of Experimental Botany*, *60*(10), 2827–2838. doi:10.1093/jxb/erp080 PMID:19380424

Grüter, C., & Ratnieks, F. L. (2011). Flower constancy in insect pollinators: Adaptive foraging behaviour or cognitive limitation. *Communicative & Integrative Biology*, *4*(6), 633–636. doi:10.4161/cib.16972 PMID:22446521

Guerroudj, F. Z., & Berchi, S. (2019). Composition of the Necrophagous Fauna on a Rabbit Corpse in Constantine, Northeastern Algeria. *Energy Procedia*, *157*, 1083–1088. doi:10.1016/j.egypro.2018.11.275

Hagstrum, D. W., & Hagstrum, W. R. (1970). A Simple Device for Producing Fluctuating Temperatures, with an Evaluation of the Ecological Significance of Fluctuating Temperatures1. *Annals of the Entomological Society of America*, *63*(5), 1385–1389. https://doi.org/10.1093/aesa/63.5.1385

Hajek, A. E. (2004). Introduction to biological control. In A. E. Hajek Book (Ed.), *Natural Enemies: An Introduction to Biological Control* (pp. 19–38). Cambridge University Press. doi:10.1017/CBO9780511811838.003

Halsch, C. A., Shapiro, A. M., Fordyce, J. A., Nice, C. C., Thorne, J. H., Waetjen, D. P., & Forister, M. L. (2021). Insects and recent climate change. *Proceedings of the National Academy of Sciences of the United States of America*, *118*(2), e2002543117. doi:10.1073/pnas.2002543117 PMID:33431560

Hamim. (2005). Photosynthesis of C3 and C4 species in response to increased CO_2 concentration and drought stress. *HAYATI Journal of Biosciences, 12*(4), 131-138. doi:10.1016/S1978-3019(16)30340-0

Hance, T., van Baaren, J., Vernon, P., & Boivin, G. (2007). Impact of Extreme Temperature on Parasitoids in a Climate Change Perspective. *Annual Review of Entomology*, *52*, 107–126. https://doi.org/10.1146/annurev.ento.52.110405.091333

Haokip, S. W., Shankar, K., & Lalrinngheta, J. (2020). Climate change and its impact on fruit crops. *Journal of Pharmacognosy and Phytochemistry*, *9*(1), 435–438.

Harrington, R., Clark, S. J., Weltham, S. J., Virrier, P. J., Denhol, C. H., Hullé, M., Maurice, D., Roun-sevell, M. D., & Cocu, N. (2007). Environmental change and the phenology of European aphids. *Global Change Biology*, *13*(8), 1556–1565. doi:10.1111/j.1365-2486.2007.01394.x

Harrington, R., Woiwod, I., & Sparks, T. (1999). Climate change and trophic interactions. *Trends in Ecology & Evolution*, *14*(4), 146–150. https://doi.org/10.1016/S0169-5347(99)01604-3

Háva, J. (2004). Notes on Dermestidae (Coleoptera) with new synonyms and description of three new species. *Annales Historico-Naturales Musei Nationalis Hungarici*, *96*, 81–95.

HCEFLCD. (2007). *Etude d'aménagement du Parc national d'Iriqui.* Rapport de synthèse. HCEFLCD-DREF SO. 2007.

Heeb, L., Jenner, E., & Cock, M. J. (2019). Climate-smart pest management: Building resilience of farms and landscapes to changing pest threats. *Journal of Pest Science*, *92*(3), 951–969. doi:10.100710340-019-01083-y

Hegland, S. J., Nielsen, A., Lázaro, A., Bjerknes, A. L., & Totland, Ø. (2009). How does climate warming affect plant-pollinator interactions. *Ecology Letters*, *12*(2), 184–195. doi:10.1111/j.1461-0248.2008.01269.x PMID:19049509

Hellmann, J. J., Byers, J. E., Bierwagen, B. G., & Dukes, J. S. (2008). Five potential consequences of climate change for invasive species. *Conservation Biology*, *22*(3), 534–543. doi:10.1111/j.1523-1739.2008.00951.x PMID:18577082

Hill, J. K., Griffiths, H. M., & Thomas, C. D. (2011). Climate change and evolutionary adaptations at species' range margins. *Annual Review of Entomology*, *56*(1), 143–159. doi:10.1146/annurev-ento-120709-144746 PMID:20809802

Hoffmann, A. A., & Sgro, C. M. (2011). Climate change and evolutionary adaptation. *Nature*, *470*(7335), 479–485. doi:10.1038/nature09670 PMID:21350480

Hoffmann, A. A., & Willy, Y. (2008). Detecting genetic responses to environmental change. *National Review*, *9*(6), 421–432. doi:10.1038/nrg2339 PMID:18463665

Holt, J. G., Krieg, R. N., Sneath, A. P. H., Staley, J. T., & Williams, S. T. (2000). In W. R. Hensyl (Ed.), *Bergey's Manual of Determinative Bacteriology* (9th ed., pp. 559–562). William and Willkins.

Holzschuh, A., Steffan-Dewenter, I., & Tscharntke, T. (2008). Agricultural landscapes with organic crops support higher pollinator diversity. *Oikos*, *117*(3), 354–361. doi:10.1111/j.2007.0030-1299.16303.x

Hong, E., Kim, Y., Jeong, J. C., Kang, S. H., Jung, J. K., & Suk, S. W. (2017). Community structure and distribution of ground beetles (Coleoptera: Carabidae) in Sobaeksan National Park, Korea. *Journal of Ecology and Environment*, *41*(1), 1–7. doi:10.118641610-017-0036-1

Hoover, S. E., Ladley, J. J., Shchepetkina, A. A., Tisch, M., Gieseg, S. P., & Tylianakis, J. M. (2012). Warming, CO_2, and nitrogen deposition interactively affect a plant-pollinator mutualism. *Ecology Letters*, *15*(3), 227–234. doi:10.1111/j.1461-0248.2011.01729.x PMID:22221802

Horn, A. (2020, February 12). A very hungry caterpillar that decimated crops around the world has arrived in Australia. *ABC News*. https://www.abc.net.au/news/2020-02-12/worm-moth-fall-armyworm-detected-found-australia-torres-strait/11957838

Howden, S. M., Soussana, J. F., Tubiello, F. N., Chhetri, N., Dunlop, M., & Meinke, H. (2007). Adapting agriculture to climate change. *Proceedings of the National Academy of Sciences of the United States of America*, *104*(50), 19691–19696. doi:10.1073/pnas.0701890104 PMID:18077402

Huang, Y., Gu, X., Peng, X., Tao, M., Chen, G., & Zhang, X. (2020). Effect of short-term high-temperatures on the growth, development and reproduction in the fruit fly, Bactrocera tau (Diptera: Tephritidae). *Scientific Reports*, *10*, 6418. https://doi.org/10.1038/s41598-020-63502-w

Hull, M., D'Acier, A. C., Bankhead-Dronnet, S., & Harrington, R. (2010). Aphids in the face of global changes. *Comptes Rendus Biologies*, *333*(6-7), 497–503. doi:10.1016/j.crvi.2010.03.005 PMID:20541161

Hussain, M., Munir, S., Jamal, M. A., Ayaz, S., Akhoundi, M., & Mohamed, K. (2017). Epidemic outbreak of anthroponotic cutaneous leishmaniasis in Kohat District, Khyber Pakhtunkhwa, Pakistan. *Acta Tropica*, *172*, 147–155. doi:10.1016/j.actatropica.2017.04.035 PMID:28476600

Ilyasov, R. A., Lee, M., Takahashi, J.-i., Kwon, H. W., & Nikolenko, A. G. (2020). A revision of subspecies structure of western honey bee *Apis mellifera*. *Saudi Journal of Biological Sciences*, *27*(12), 3615–3621. doi:10.1016/j.sjbs.2020.08.001 PMID:33304172

IMD. (2010). *Annual Climate Summary. India Meteorological Department*. Government of India, Ministry of Earth Sciences.

IPBES Secretariat. (2019). *Summary for Policymakers of the Global Assessment Report on Biodiversity and Ecosystem Services of the Intergovernmental Science-Policy Platform on Biodiversity and Ecosystem Services*. Author.

IPCC. (2007). Climate Change-Impacts, Adaptation and Vulnerability. *Cambridge University Press*.

IPCC. (2014). *AR5 Synthesis Report: Climate Change 2014*. Retrieved 26 December 2022, from https://www.ipcc.ch/report/ar5/syr/

IPCC. (2022). *AR6 Synthesis Report: Climate Change 2022*. Retrieved 28 November 2022, from https://www.ipcc.ch/report/sixth-assessment-report-cycle/

Irvin, N. A., Pierce, C., & Hoddle, M. S. (2021). Evaluating the potential of flowering plants for enhancing predatory hoverflies (Syrphidae) for biological control of *Diaphorina citri* (Liviidae) in California. *Biological Control*, *157*, 104574. doi:10.1016/j.biocontrol.2021.104574

Isaacs, R., Tuell, J., Fiedler, A., Gardiner, M., & Landis, D. (2009). Maximizing arthropod-mediated ecosystem services in agricultural landscapes: The role of native plants. *Frontiers in Ecology and the Environment*, *7*(4), 196–203. https://doi.org/10.1890/080035

Isenring, E., Fehr, J., Gültekin, N., & Schlagenhauf, P. (2018). Infectious disease profiles of Syrian and Eritrean migrants presenting in Europe: A systematic review. *Travel Medicine and Infectious Disease*, *25*, 65–76. doi:10.1016/j.tmaid.2018.04.014 PMID:29702253

Islam, Y., Güncan, A., Fan, Y., Zhou, X., Naeem, A., & Shah, F. M. (2022). Age-stage, two-sex life table and predation parameters of *Harmonia axyridis* Pallas (Coleoptera: Coccinellidae), reared on *Acyrthosiphon pisum* (Harris)(Hemiptera: Aphididae), at four different temperatures. *Crop Protection (Guildford, Surrey)*, *106029*. Advance online publication. doi:10.1016/j.cropro.2022.106029

Jactel, H., Koricheva, J., & Castagneyrol, B. (2019). Responses of forest insect pests to climate change: Not so simple. *Current Opinion in Insect Science*, *35*, 103–108. doi:10.1016/j.cois.2019.07.010 PMID:31454625

Jakobsen, H. B., & Kritjánsson, K. (1994). Influence of temperature and floret age on nectar secretion in *Trifolium repens* L. *Annals of Botany*, *74*(4), 327–334. doi:10.1006/anbo.1994.1125

Jaleel, W., Lu, L., & He, Y. (2018). Biology, taxonomy, and IPM strategies of Bactrocera tau Walker and complex species (Diptera; Tephritidae) in Asia: A comprehensive review. *Environmental Science and Pollution Research International*, *25*(20), 19346–19361. https://doi.org/10.1007/s11356-018-2306-6

Janzen, D. H., & Hallwachs, W. (2021). To us insectometers, it is clear that insect decline in our Costa Rican tropics is real, so let's be kind to the survivors. *Proceedings of the National Academy of Sciences of the United States of America*, *118*(2), e2002546117. https://doi.org/10.1073/pnas.2002546117

Jaouadi, K., Depaquit, J., Haouas, N., Chaara, D., Gorcii, M., Chargui, N., Dedet, J.-P., Pratlong, F., Boubabous, R., & Babba, H. (2012). Twenty-four new human cases of cutaneous leishmaniasis due to Leishmania killicki in Metlaoui, southwestern Tunisia. Probable role of Phlebotomus sergenti in the transmission. *Acta Tropica*, *122*(3), 276–283. doi:10.1016/j.actatropica.2012.01.014 PMID:22306359

Jaouad, M., Moinina, A., Ezrari, S., & Lahlali, R. (2020). Key pests and diseases of citrus trees with emphasis on root rot diseases: An overview. *Moroccan Journal of Agricultural Sciences*, *1*(3).

Jaskula, R., Rewicz, T., & Kwiatkowski, K. (2015). Tiger beetle fauna (Coleoptera: Carabidae, Cicindelinae) of Morocco: distribution, phenology and list of taxa. *Entomologica Fennica*, *26*(3), 132–155. doi:10.33338/ef.52825

Jasrotia, P., Yadav, J., Lal Kashyap, P., Kumar Bhardwaj, A., Kumar, S., & Singh, G. P. (2021). Impact of climate change on insect pests of rice–wheat cropping system: Recent trends and mitigation strategies. In S. Sareen, P. Sharma, C. Singh, P. Jasrotia, G. Pratap Singh, & A. K. Sarial (Eds.), Improving Cereal Productivity Through Climate Smart Practices (pp. 225–239). Woodhead Publishing. https://doi.org/10.1016/B978-0-12-821316-2.00013-3

Jeffery, S., Gardi, C., Jones, A., Montanarella, L., Marmo, L., Miko, L., Ritz, K., Peres, G., Römbke, J., & van der Putten, W. H. (Eds.). (2010). *European Atlas of Soil Biodiversity*. Publications Office of the European Union.

Jia, Z. C., Lu, C., Zhou, R. J., & Jiang, L. (2020). Comparative morphology between the white grubs *Pseudosymmachia tumidifrons* and *Brahmina faldermanni* (Coleoptera: Scarabaeidae: Melolonthinae) using scanning electron microscopy. *Zoologischer Anzeiger*, *289*, 8–17. doi:10.1016/j.jcz.2020.08.008

Jyothi, N. R. (2020). Heavy metal sources and their effects on human health. In M. K. Nazal & H. Zhao (Eds.), *Heavy Metals – Their Environmental Impact and Mitigation*. IntechOpen. doi:10.5772/intechopen.95370

Kafer, H., Kovac, H., & Stabentheiner, A. (2012). Upper thermal limits of honeybee (*Apis mellifera*) and yellowjacket (*Vespula vulgaris*) foragers. *Mitteilungen der Deutschen Gesellschaft fur Allgemeine und Angewandte Entomologie*, *18*, 267–270.

Kagawa, K., & Takimoto, G. (2018). Hybridization can promote adaptive radiation by means of transgressive segregation. *Ecology Letters*, *21*(2), 264–274. doi:10.1111/ele.12891 PMID:29243294

Kaloveloni, A., Tscheulin, T., & Petanidou, T. (2018). Geography, climate, ecology: What is more important in determining bee diversity in the Aegean Archipelago? *Journal of Biogeography*, *45*(12), 2690–2700. doi:10.1111/jbi.13436

Kannan, R., & James, D. A. (2009). Effects of climate change on global diversity: A review of key literature. *Tropical Ecology*, *50*, 31–39.

Karmaoui, A. (2019). Drought and desertification in Moroccan Pre-Sahara, Draa valleys: Exploring from the perspective of young people. *Geoenvironmental Disasters*, *6*(1), 2. doi:10.1186/s40677-019-0118-8

Karmaoui, A. (2020). Seasonal Distribution of Phlebotomus papatasi, Vector of Zoonotic Cutaneous Leishmaniasis. *Acta Parasit.* doi:10.2478/s11686-020-00201-6

Karmaoui, A., El Jaafari, S., Chaachouay, H., & Hajji, L. (2021). The socio-ecological system of the pre-Sahara zone of Morocco: A conceptual framework to analyse the impact of drought and desertification. *GeoJournal*, 1–14. doi:10.100710708-021-10546-8

Karmaoui, A., Sereno, D., Maia, C., Campino, L., El Jaafari, S., Taybi, A., & Hajj, L. (2022). A conceptual model for understanding the zoonotic cutaneous leishmaniasis transmission risk in the Moroccan pre-Saharan area. *Parasite Epidemiology and Control*, *00243*, e00243. Advance online publication. doi:10.1016/j.parepi.2022.e00243 PMID:35243048

Kasap, Ö. E., Belen, A., Kaynas, S., Simsek, F. M., Biler, L., Ata, N., & Alten, B. (2009). *Activity patterns of sand fly (Diptera: Psychodidae) species and comparative performance of different traps in an endemic cutaneous leishmaniasis focus in Cukurova Plain.* Southern Anatolia. doi:10.2754/avb200978020327

Kayedi, M. H., Abdali, N., & Rostami, R. (2016). Seasonal activity of sand flies (Diptera Psychodidae) in Khorramabad County, Lorestan Province, Iran. *Nature and Science*, *14*(1).

Kerr, J. T., Pindar, A., Galpern, P., Packer, L., Potts, S. G., Roberts, S. M., Rasmont, P., Schweiger, O., Colla, S. R., Richardson, L. L., Wagner, D. L., Gall, L. F., Sikes, D. S., & Pantoja, A. (2015). Climate change impacts on bumblebees converge across continents. *Science*, *349*(6244), 177–180. doi:10.1126cience.aaa7031 PMID:26160945

Kevan, P. G. (2001). Pollination: a plinth, pedestal, and pillar for terrestrial productivity. The why, how, and where of pollination protection, conservation, and promotion. Bees and Crop Pollination: Crisis, Crossroads, Conservation, 7–68.

Khan, Z. H., Khan, A. A., Ahmad, S. B., Dar, S., & Khan, Z. H. (2011). *Biopesticides, its prospects and limitations: An Overview Shahnawaz a biodiversity and management of aphid in temperate horticultural ecosystem of Kashmir View project Mass Production and Supply of bio-agent for the management of horticultural Pests View project.* Division of Entomology, Sher-eKashmir University of Agricultural Sciences and Technology of Kashmir, Shalimar Campus.

Kleijn, D., & Raemakers, I. (2008). A retrospective analysis of pollenhost plant use by stable and declining bumble bee species. *Ecology*, *89*(7), 1811–1823. doi:10.1890/07-1275.1 PMID:18705369

Klein, A., Vaissiere, B. E., Cane, J. H., Steffan, D., Cunningham, S. A., Kremen, C., & Tscharntke, T. (2007). Importance of pollinators in changing landscapes for world crops. *Proceeding of the Royal Society B: Biological Sciences, 274*, 303–313.

Klein, A. M., Boreux, V., Fornoff, F., Mupepele, A. C., & Pufal, G. (2018). Relevance of wild and managed bees for human well-being. *Current Opinion in Insect Science*, *26*, 82–88. doi:10.1016/j.cois.2018.02.011 PMID:29764666

Klein, A. M., Vaissière, B. E., Cane, J. H., Steffan-Dewenter, I., Cunningham, S. A., Kremen, C., & Tscharntke, T. (2007). Importance of pollinators in changing landscapes for world crops. *Proceedings. Biological Sciences*, *274*(1608), 303–313. doi:10.1098/rspb.2006.3721 PMID:17164193

Kluser, S., & Peduzzi, P. (2007). *Global pollinator decline: A literature review*. Academic Press.

Kocmánková, E., Trnka, M., Juroch, J., Dubrovský, M., Semerádová, D., Možný, M., & Žalud, Z. (2010). Impact of climate change on the occurrence and activity of harmful organisms. *Plant Protection Science*, *45*(Special Issue), S48–S52. https://doi.org/10.17221/2835-PPS

Kolimenakis, A., Heinz, S., Wilson, M. L., Winkler, V., Yakob, L., Michaelakis, A., Papachristos, D., Richardson, C., & Horstick, O. (2021). The role of urbanisation in the spread of *Aedes* mosquitoes and the diseases they transmit—A systematic review. *PLoS Neglected Tropical Diseases*, *15*(9), e0009631. doi:10.1371/journal.pntd.0009631 PMID:34499653

Kollberg, I., Bylund, H., Jonsson, T., Schmidt, A., Gershenzon, J., & Björkman, C. (2015). Temperature affects insect outbreak risk through tritrophic interactions mediated by plant secondary compounds. *Ecosphere*, *6*(6), art102. https://doi.org/10.1890/ES15-000021.1

Kovac, H., Kafer, H., Stabentheiner, A., & Costa, C. (2014). Metabolism and upper thermal limits of *Apis mellifera carnica* and *A. m. ligustica*. *Apidologie*, *45*(6), 664–677. doi:10.100713592-014-0284-3 PMID:25378763

Kremen, C., Williams, N. M., Aizen, M. A., Gemmill-Herren, B., LeBuhn, G., Minckley, R., & Ricketts, T. H. (2007). Pollination and other ecosystem services produced by mobile organisms: A conceptual framework for the effects of land-use change. *Ecology Letters*, *10*(4), 299–314. doi:10.1111/j.1461-0248.2007.01018.x PMID:17355569

Kremen, C., Williams, N. M., & Thorp, R. W. (2002). Crop pollination from native bees at risk from agricultural intensification. *Proceedings of the National Academy of Sciences of the United States of America*, *99*(26), 16812–16816. doi:10.1073/pnas.262413599 PMID:12486221

Kumar, S., & Archana, S. (2015). Biopesticides: Present status and the future prospects. *Journal of Fertilizer Pesticide*, *6*(2), 12. doi:10.4172/2471-2728.1000e129

Kumral, N. A., Çobanoglu, S., & Yalçın, C. (2010). Acaricidal, repellent and oviposition deterrent activities of *Datura stramonium* L. against adult *Tetranychus urticae* (Koch). *Journal of Pest Science*, *83*(2), 173–180. doi:10.100710340-009-0284-7

Lambert, O., Veyrand, B., Durand, S., Marchand, P., Le Bizec, B., Piroux, M., Puyo, S., Thorin, C., Delbac, F., & Pouliquen, H. (2012). Polycyclic aromatic hydrocarbons: Bees, honey and pollen as sentinels for environmental chemical contaminants. *Chemosphere*, *86*(1), 98–104. doi:10.1016/j.chemosphere.2011.09.025 PMID:22051346

Lande, R. (1993). Risks of Population Extinction from Demographic and Environmental Stochasticity and Random Catastrophes. *American Naturalist*, *142*(6), 911–927. doi:10.1086/285580 PMID:29519140

Latest Pest Reports. (n.d.). International Plant Protection Convention. Retrieved 23 October 2022, from https://www.ippc.int/en/countries/thailand/pestreports/2018/12/first-detection-of-fall-army-worm-on-the-border-of-thaila
nd/

Laughton, A. M., O'Connor, C. O., & Knell, R. J. (2017). Responses to a warming world: Integrating life history, immune investment, and pathogen resistance in a model insect species. *Ecology and Evolution*, *7*(22), 9699–9710. doi:10.1002/ece3.3506 PMID:29188001

Lautenbach, S., Seppelt, R., Liebscher, J., & Dormann, C. F. (2012). Spatial and temporal trends of global pollination benefit. *PLoS One*, *7*(4), e35954. doi:10.1371/journal.pone.0035954 PMID:22563427

Lawton, D., Huseth, A. S., Kennedy, G. G., Morey, A. C., Hutchison, W. D., Reisig, D. D., Dorman, S. J., Dillard, D., Venette, R. C., Groves, R. L., Adamczyk, J. J., Barbosa Dos Santos, I., Baute, T., Brown, S., Burkness, E., Dean, A., Dively, G. P., Doughty, H. B., Fleischer, S. J., ... Zuefle, M. (2022). Pest population dynamics are related to a continental overwintering gradient. *Proceedings of the National Academy of Sciences of the United States of America*, *119*(37), e2203230119. doi:10.1073/pnas.2203230119 PMID:36067290

Le Conte, Y., & Navajas, M. (2008). Climate change: Impact on honey bee populations and diseases. *Revue Scientifique et Technique (International Office of Epizootics)*, *27*(2), 499–510. PMID:18819674

Le Gall, M., & Behmer, S. T. (2014). Effects of Protein and Carbohydrate on an Insect Herbivore: The Vista from a Fitness Landscape. *Integrative and Comparative Biology*, *54*(5), 942–954. https://doi.org/10.1093/icb/icu102

Legrand, A., Gaucherel, C., Baudry, J., & Meynard, J. M. (2011). Long-term effects of organic, conventional, and integrated crop systems on Carabids. *Agronomy for Sustainable Development*, *31*(3), 515–524. doi:10.100713593-011-0007-3

Lehner, F., Coats, S., Stocker, T. F., Pendergrass, A. G., Sanderson, B. M., Raible, C. C., & Smerdon, J. E. (2017). Projected drought risk in 1.5°C and 2°C warmer climates. *Geophysical Research Letters*, *44*(14), 7419–7428. https://doi.org/10.1002/2017GL074117

Lewis, T. (1997). Flight and dispersal. In T. Lewis (Ed.), *Thrips as crop pests* (pp. 175–196). CAB International. doi:10.1079/9780851991788.0000

Li, J. L., Cornman, R. S., Evans, J. D., Pettis, J. S., Zhao, Y., Murphy, C., Peng, W. J., Wu, J., Hamilton, M., Boncristiani, H. F. Jr, Zhou, L., Hammond, J., & Chen, Y. P. (2014). Systemic spread and propagation of a plant-pathogenic virus in European honeybees, *Apis mellifera. mBio, 5*(1), e00898–e13. doi:10.1128/mBio.00898-13 PMID:24449751

Liquido, N. J., Cunningham, R. T., & Nakagawa, S. (1990). Host Plants of Mediterranean Fruit Fly (Diptera: Tephritidae) on the Island of Hawaii (1949-1985 Survey). *Journal of Economic Entomology, 83*(5), 1863–1878. https://doi.org/10.1093/jee/83.5.1863

Lisi, O., D'Urso, V., Vaccalluzzo, V., Bongiorno, G., Khoury, C., Severini, F., Di Muccio, T., Gramiccia, M., Gradoni, L., & Maroli, M. (2014). Persistence of phlebotomine Leishmania vectors in urban sites of Catania (Sicily, Italy). *Parasites & Vectors, 7*(1), 560. doi:10.118613071-014-0560-0 PMID:25487039

Lister, B. C., & Garcia, A. (2018). Climate-driven declines in arthropod abundance restructure a rainforest food web. *Proceedings of the National Academy of Sciences of the United States of America, 115*(44), E10397–E10406. https://doi.org/10.1073/pnas.1722477115

Liu, H., Wang, X., Chen, Z., & Lu, Y. (2022). Characterization of Cold and Heat Tolerance of Bactrocera tau (Walker). *Insects, 13*(4), 4. https://doi.org/10.3390/insects13040329

Liu, J., Zhao, Z., Yan, Y., Ali, A., Ahmed, Z., He, D., Yu, M., Hang, J., Perven, M., Nazir, T., Tayyib, M., & Zhao, Z. (2021). Effect of alfalfa habitat change on dispersal behavior of *Harmonia axyridis Pallas* and *Hippodamia variegata Goeze* (Coleoptera: Coccinellidae). *Journal of Asia-Pacific Entomology, 24*(4), 997–1003. doi:10.1016/j.aspen.2021.09.002

Li, Y., Hou, R., & Tao, F. (2020). Interactive effects of different warming levels and tillage managements on winter wheat growth, physiological processes, grain yield and quality in the North China Plain. *Agriculture, Ecosystems & Environment, 295*, 106923. https://doi.org/10.1016/j.agee.2020.106923

Lobell, D. B., Sibley, A., & Ivan Ortiz-Monasterio, J. (2012). Extreme heat effects on wheat senescence in India. *Nature Climate Change, 2*(3), 3. https://doi.org/10.1038/nclimate1356

Logan, J. A., Regniere, J., & Powell, J. A. (2003). Assessing the impacts of global warming on forest pest dynamics. *Frontiers in Ecology and the Environment, 1*(3), 130–137. doi:10.1890/1540-9295(2003)001[0130:ATIOGW]2.0.CO;2

Losey, J. E., & Vaughan, M. (2006). The economic value of ecological services provided by insects. *Bioscience, 56*(4), 311–323. doi:10.1641/0006-3568(2006)56[311:TEVOES]2.0.CO;2

Loussert, R. (1987). Les agrumes, Vol 2, production, technique agricole méditerranéenne. Academic Press.

Louveaux, J., Albisetti, M., Delangue, M., & Theurkauff, J. (1966). Les modalités de l'adaptation des abeilles (*Apis mellifica* L.) au milieu naturel. *Anné Abeille*, *9*(4), 323–350. doi:10.1051/apido:19660402

Lucchetti, M. A., Kilchenmann, V., Glauser, G., Praz, C., & Kast, P. (2018). Nursing protects honeybee larvae from secondary metabolites of pollen. *Proceedings. Biological Sciences*, *285*(1875), 20172849. doi:10.1098/rspb.2017.2849 PMID:29563265

Lumaret, J. P., Kadiri, N., & Martinez Morales, I. (2022). The Global Decline of Dung Beetles. Imperiled: The Encyclopedia of Conservation, 553-562. doi:10.1016/B978-0-12-821139-7.00018-0

Lutz, E. K., Lahondère, C., Vinauger, C., & Riffell, J. A. (2017). Olfactory learning and chemical ecology of olfaction in disease vector mosquitoes: A life history perspective. *Current Opinion in Insect Science*, *20*, 75–83. doi:10.1016/j.cois.2017.03.002 PMID:28602240

Ma, C.-S., Ma, G., & Pincebourde, S. (2021). Survive a Warming Climate: Insect Responses to Extreme High Temperatures. *Annual Review of Entomology*, *66*(1), 163–184. https://doi.org/10.1146/annurev-ento-041520-074454

Macfadyen, S., & Bohan, D. A. (2010). Crop domestication and the disruption of species interactions. *Basic and Applied Ecology*, *11*(2), 116–125. https://doi.org/10.1016/j.baae.2009.11.008

Madkouri, M. (1973). Etude Préliminaire sur la Bio écologie de *Parlatoria blanchardi* Targ (Hom. Diaspididae) dans la Vallée du Draa. *Al-Awamia*, *48*, 39–72.

Ma, F.-Z., Lü, Z.-C., Wang, R., & Wan, F.-H. (2014). Heritability and Evolutionary Potential in Thermal Tolerance Traits in the Invasive Mediterranean Cryptic Species of Bemisia tabaci (Hemiptera: Aleyrodidae). *PLoS One*, *9*(7), e103279. https://doi.org/10.1371/journal.pone.0103279

Ma, G., & Ma, C.-S. (2022). Potential distribution of invasive crop pests under climate change: Incorporating mitigation responses of insects into prediction models. *Current Opinion in Insect Science*, *49*, 15–21. https://doi.org/10.1016/j.cois.2021.10.006

Malavasi, A., Zucchi, R. A., & Sugayama, R. L. (2000). Biogeografia. In A. Malavasi & R. A. Zucchi (Eds.), *Moscas-das-frutas de importênciaeconômica no Brasil - conhecimento básico e aplicado* (pp. 41–48). Ribeião Preto. https://agris.fao.org/agris-search/search.do?recordID=US201300072841

Manley, R., Temperton, B., Doyle, T., Gates, D., Hedges, S., Boots, M., & Wilfert, L. (2019). Knock-on community impacts of a novel vector: Spillover of emerging DWV-B from Varroa-infested honeybees to wild bumblebees. *Ecology Letters*, *22*(8), 1306–1315. doi:10.1111/ele.13323 PMID:31190366

MAPM. (2015). *Investor's guide in the agricultral sector in Morocco.pdf*. Retrieved 22 December 2022, from https://www.agriculture.gov.ma/sites/default/files/investors_guide_in_the_agricultural_sector_in_morocco.pdf

MAPMDREF. (2020). *Statistique du ministère de l'agriculture et de la pêche maritimes, du développement rural et des eaux et forêts*. MAPMDREF.

Marchiori, C. H. (2021). *Study of the ecology of the Tephritidae family (Insecta: Diptera)*. Academic Press.

Martin, S. J., & Medina, L. M. (2004). Africanized honeybees have unique tolerance to Varroa mites. *Trends in Parasitology, 20*(3), 112–114. doi:10.1016/j.pt.2004.01.001 PMID:15036029

Mazih, A. (2008). Current situation of Citrus pests and the control methods in use in Morocco. *IOBC/WPRS Bulletin, 38*, 10–16.

Mazih, A. (2011). Citrus IPM in Morocco: Current status. In Proceedings of Meeting of Working Group "Integrated Control in Citrus Fruit Crops" (Agadir, Morocco). *Bulletin, IOBC/WPRS, 62*, 266.

Mazih, A., & Chitoukli, M. S. (2001). Studies on annual population dynamics and damage of citrus leafminer (*Phyllocnistis citrella*) in sous area (Morocco). *Proceedings du symposium sur protection intégrée des cultures dans la région méditerranéenne*, 159-171.

Mazih, A. (2015). Status of citrus IPM in the southern Mediterranean basin Morocco, North Africa. In *ISHS Acta Horticulturae 1065: XII International Citrus Congress*. International Society of Citriculture.

McAfee, A., Chapman, A., Higo, H., Underwood, R., Milone, J., Foster, L. J., Guarna, M. M., Tarpy, D. R., & Pettis, J. S. (2020). Vulnerability of honey bee queens to heat-induced loss of fertility. *Nature Sustainability, 3*(5), 367–376. doi:10.103841893-020-0493-x

McArt, S. H., Koch, H., Irwin, R. E., & Adler, L. S. (2014). Arranging the bouquet of disease: Floral traits and the transmission of plant and animal pathogens. *Ecology Letters, 17*(5), 624–636. doi:10.1111/ele.12257 PMID:24528408

McGregor, S. E. (1976). Insect pollination of cultivated crop plants (No. 496). *Agricultural Research Service, US Department of Agriculture., 496*, 93–98.

McKee, F. R., Windmuller-Campione, M. A., Althoff, E. R., Reinikainen, M. R., Dubuque, P. A., & Aukema, B. H. (2022). Eastern larch beetle, a changing climate, and impacts to northern tamarack forests. In Bark Beetle Management, Ecology, and Climate Change (pp. 261-300). Academic Press. doi:10.1016/B978-0-12-822145-7.00001-5

MEA (Millennium Ecosystem Assessment). (2005). *Ecosystems and Human Well-being: Synthesis*. Island Press.

Meats, A., & Smallridge, C. J. (2007). Short- and long-range dispersal of medfly, Ceratitis capitata (Dipt., Tephritidae), and its invasive potential. *Journal of Applied Entomology, 131*(8), 518–523. https://doi.org/10.1111/j.1439-0418.2007.01168.x

Meeus, I., Vercruysse, V., & Smagghe, G. (2012). Molecular detection of *Spiroplasma apis* and *Spiroplasma melliferum* in bees. *Journal of Invertebrate Pathology, 109*(1), 172–174. doi:10.1016/j.jip.2011.11.006 PMID:22138255

Memmott, J., Carvell, C., Pywell, R. F., & Craze, P. G. (2010). The potential impact of global warming on the efficacy of field margins sown for the conservation of bumble-bees. *Philosophical Transactions of the Royal Society of London. Series B, Biological Sciences*, *365*(1549), 2071–2079. doi:10.1098/rstb.2010.0015 PMID:20513715

Menéndez, R. (2007). How are insects responding to global warming? *Tijdschrift voor Entomologie*, *150*, 355–365.

Mesas, A., Jaramillo, A., & Castañeda, L. E. (2021). Experimental evolution on heat tolerance and thermal performance curves under contrasting thermal selection in Drosophila subobscura. *Journal of Evolutionary Biology*, *34*(5), 767–778. https://doi.org/10.1111/jeb.13777

Metcalf, R. L. (1996). Applied Entomology in the Twenty-First Century. *American Entomologist (Lanham, Md.)*, *42*(4), 216–227. https://doi.org/10.1093/ae/42.4.216

Mevi-Schutz, J., & Erhardt, A. (2005). Amino acids in nectar enhance butterfly fecundity: A long-awaited link. *American Naturalist*, *165*(4), 411–419. doi:10.1086/429150 PMID:15791533

Meynard, C. N., Migeon, A., & Navajas, M. (2013). Uncertainties in Predicting Species Distributions under Climate Change: A Case Study Using Tetranychus evansi (Acari: Tetranychidae), a Widespread Agricultural Pest. *PLoS One*, *8*(6), e66445. https://doi.org/10.1371/journal.pone.0066445

Michener, C. (2000). *The bees of the world*. Johns Hopkins University Press.

Milićević, V., Radojičić, S., Kureljušić, J., Šekler, M., Nešić, K., Veljović, L., Maksimović Zorić, J., & Radosavljević, V. (2018). Molecular detection of black queen cell virus and Kashmir bee virus in honey. *AMB Express*, *8*(1), 1–5. doi:10.118613568-018-0655-7 PMID:30088183

Mohanasundaram, A., Monobrullah, M. D., Sharma, K. K., Anees, K., Singh, R. K., Meena, S. C., & Verma, S. (2014). Climate change: Effect of weather parameters on production of summer season crop of rangeeni strain of Indian lac insect, Kerria lacca (Kerr) at Ranchi, Jharkhand. *Journal of Agrometeorology*, *16*(1), 108–113.

Mohtasebi, S., Teimouri, A., Afshar, M. J. A., Mobedi, I., Abbasian, H., Totonchian, N., & Mowlavi, G. (2021). First report of *Spirocerca lupi* larva in dung beetles (*Scarabaeus armeniacus*) in the central region of Iran: A morphological and molecular identification. *Comparative Immunology, Microbiology and Infectious Diseases*, *77*, 101671. doi:10.1016/j.cimid.2021.101671 PMID:34049036

Montory, M., Habit, E., Fernandez, P., Grimalt, J. O., Kolok, A. S., Barra, R. O., & Ferrer, J. (2020). Biotransport of persistent organic pollutants in the southern Hemisphere by invasive Chinook salmon (*Oncorhynchus tshawytscha*) in the rivers of northern Chilean Patagonia, a UNESCO biosphere reserve. *Environment International*, *142*, 105803. doi:10.1016/j.envint.2020.105803 PMID:32563009

Morales, P., Cermeli, M., Godoy, F., & Salas, B. (2004). A list of Mediterranean fruit fly Ceratitis capitata Wiedemann (Diptera: Tephritidae) host plants based on the records of INIA-CENIAP Museum of Insects of Agricultural Interest. *Entomotrópica*, *19*(1), 51–54.

Moritz, R. F., De Miranda, J., Fries, I., Le Conte, Y., Neumann, P., & Paxton, R. J. (2010). Research strategies to improve honeybee health in Europe. *Apidologie*, *41*(3), 227–242. doi:10.1051/apido/2010010

Morris, M. G., & Schön, K. (2010). *Corimalia martini* (Brisout, 1883)(Insecta, Coleoptera, Nanophyidae), a weevil new to the Canary Islands from Fuerteventura. *Museum Scientiarum Naturalium Nivariense*, *38*(Vieraea 38), 83–90. doi:10.31939/vieraea.2010.38.06

Mouches, C., Bové, J. M., Tully, J. G., Rose, D. L., McCoy, R. E., Carle-Junca, P., Garnier, M., & Saillard, C. (1983). Spiroplasma apis, a new species from the honey-bee Apis mellifera. In Annales de l'Institut Pasteur/Microbiologie (Vol. 134, No. 3, pp. 383-397). Elsevier Masson.

Moumane, A., El Ghazali, F. E., Al Karkouri, J., Delorme, J., Batchi, M., Chafiki, D., & Karmaoui, A. (2021). Monitoring spatiotemporal variation of groundwater level and salinity under land use change using integrated field measurements, GIS, geostatistical, and remote-sensing approach: Case study of the Feija aquifer, Middle Draa watershed, Moroccan Sahara. *Environmental Monitoring and Assessment*, *193*(12), 1–21. doi:10.100710661-021-09581-2 PMID:34735624

Mound, L. A., & Jackman, D. J. (1998). Thrips in the economy and ecology of Australia. *Proceedings of the Sixth Australian Applied Entomological Research Conference*, 472–478.

Mozaffari, G. A. (2022). Climate Change and Its Consequences in Agriculture. *The Nature, Causes, Effects and Mitigation of Climate Change on the Environment*, 83.

Mukezangango, J., & Page, S. (2017). *Statistical overview of the Canadian honey and bee industry and the economic contribution of honey bee pollination*. Horticulture and Cross Sectoral Division. Agriculture and Agri-Food Canada.

Muñoz-Valencia, V., González-Obando, R., & Díaz, F. (2016). Rapid responses to a strong experimental selection for heat hardening in the invasive whitefly Bemisia tabaci MEAM 1. *Entomologia Experimentalis et Applicata*, *160*(2), 147–155. https://doi.org/10.1111/eea.12466

Mutters, R. G., Ferreira, L. G. R., & Hall, A. E. (1989). Proline content of the anthers and pollen of heat-tolerant and heat-sensitive cowpea subjected to different temperatures. *Crop Science*, *29*(6), 1497–1500. doi:10.2135/cropsci1989.0011183X002900060036x

Nath, R., & Sikha, D. (2019). Insect pests of citrus and their management. *International Journal of Plant Production*, *12*(2), 188–196. doi:10.15740/HAS/IJPP/12.2/188-196

Navarro, C. C., Aguilar, A., & Marí, F. G. (2008). *Pezothrips Kellyanus, Thrips causante* de daños en frutos de cítricos. *Levante agrícola: Revista internacional de cítricos, troisième trimestre*, *299*, 298–303.

Navarro, C. C., Aguilar, A., & Marí, F. G. (2011). Population trend and fruit damage of *Pezothrips kellyanus* in Citrus orchards in Valencia (Spain). Integrated Control in Citrus Fruit Crops. *IOBC/WPRS Bulletin, 38*, 204–220.

Nawaz, R., Abbasi, N. A., Hafiz, I. A., Khan, M. F., & Khalid, A. (2021). Environmental variables influence the developmental stages of the citrus leafminer, infestation level and mined leaves physiological response of Kinnow mandarin. *Scientific Reports, 11*(1), 1–20. doi:10.103841598-021-87160-8 PMID:33833311

Neelaboina, B. K., Khan, G. A., Kumar, S., Gani, M., Ahmad, M. N., & Ghosh, M. K. (2018). Impact of climate change on agriculture and sericulture. *Journal of Entomology and Zoology Studies, 6*(5), 426–429.

Newbold, T., Hudson, L. N., Hill, S. L. L., Contu, S., Lysenko, I., Senior, R. A., Börger, L., Bennett, D. J., Choimes, A., Collen, B., Day, J., De Palma, A., Díaz, S., Echeverria-Londoño, S., Edgar, M. J., Feldman, A., Garon, M., Harrison, M. L. K., Alhusseini, T., … Purvis, A. (2015). Global effects of land use on local terrestrial biodiversity. *Nature, 520*(7545). doi:10.1038/nature14324

Newton, A. C., Begg, G., & Swanston, J. S. (2009). Deployment of diversity for enhanced crop function. *Annals of Applied Biology, 154*(3), 309–322. doi:10.1111/j.1744-7348.2008.00303.x

Nichols, E., Alarcón, V., Forgie, S., Gomez-Puerta, L. A., & Jones, M. S. (2017). Coprophagous insects and the ecology of infectious diseases of wildlife. *ILAR Journal, 58*(3), 336–342. doi:10.1093/ilar/ilx022 PMID:29036417

Nichols, E., Spector, S., Louzada, J., Larsen, T., Amezquita, S., Favila, M. E., & Network, T. S. R. (2008). Ecological functions and ecosystem services provided by Scarabaeinae dung beetles. *Biological Conservation, 141*(6), 1461–1474. doi:10.1016/j.biocon.2008.04.011

Niell, S., Jesús, F., P'erez, N., Pérez, C., Pareja, L., Abbate, S., Carrasco-Letelier, L., Díaz, S., Mendoza, Y., Cesio, V., & Heinzen, H. (2017). Neonicotinoids transference from the field to the hive by honey bees: Towards a pesticide residues biomonitor. *The Science of the Total Environment, 581-582*, 25–31. doi:10.1016/j.scitotenv.2017.01.011 PMID:28073057

Noriega, J. A., Hortal, J., Azcárate, F. M., Berg, M. P., Bonada, N., Briones, M. J. I., Toro, I. D., Goulson, D., Ibanez, S., Landis, D. A., Moretti, M., Potts, S. G., Slade, E. M., Stout, J. C., Ulyshen, M. D., Wackers, F. L., Woodcock, B. A., & Santos, A. M. C. (2018). Research trends in ecosystem services provided by insects. *Basic and Applied Ecology, 26*, 8–23. doi:10.1016/j.baae.2017.09.006

Obrycki, J. J., Harwood, J. D., Kring, T. J., & O'Neil, R. J. (2009). Aphidophagy by Coccinellidae: Application of biological control in agroecosystems. *Biological Control, 51*(2), 244–254. doi:10.1016/j.biocontrol.2009.05.009

Oerke, E.-C. (2006). Crop losses to pests. *The Journal of Agricultural Science, 144*(1), 31–43. https://doi.org/10.1017/S0021859605005708

Ogden, N. H. (2017). Climate change and vector-borne diseases of public health significance. *FEMS Microbiology Letters*, *364*(19). Advance online publication. doi:10.1093/femsle/fnx186 PMID:28957457

Oliveira, C. M., Auad, A. M., Mendes, S. M., & Frizzas, M. R. (2014). Crop losses and the economic impact of insect pests on Brazilian agriculture. *Crop Protection (Guildford, Surrey)*, *56*, 50–54. https://doi.org/10.1016/j.cropro.2013.10.022

Oliver, T. H., Marshall, H. H., Morecroft, M. D., Brereton, T., Prudhomme, C., & Huntingford, C. (2015). Interacting effects of climate change and habitat fragmentation on drought-sensitive butterflies. *Nature Climate Change*, *5*(10), 941–945. doi:10.1038/nclimate2746

Oliver, T. H., & Morecroft, M. D. (2014). Interactions between climate change and land use change on biodiversity: Attribution problems, risks, and opportunities. *Wiley Interdisciplinary Reviews: Climate Change*, *5*(3), 317–335. https://doi.org/10.1002/wcc.271

Olivier, V., Blanchard, P., Chaouch, S., Lallemand, P., Schurr, F., Celle, O., Dubois, E., Tordo, N., Thiéry, R., Houlgatte, R., & Ribiere, M. (2008). Molecular characterisation and phylogenetic analysis of chronic bee paralysis virus, a honey bee virus. *Virus Research*, *132*(1-2), 59–68. doi:10.1016/j.virusres.2007.10.014 PMID:18079012

ONSSA. (2022). *La santé des colonnies d'abeilles au Maroc*. http://www.onssa.gov.ma

Ouknin, M., Alahyane, H., Aboussaid, H., Oufdou, K., Messoussi, S. E., Costa, J., & Majidi, L. (2021). Acaricidal properties of essential oils from Moroccan Thyme against Oriental Red mite, *Eutetranychus orientalis* (Klein) (Acari: Tetranychidae). *Journal of Essential Oil-Bearing Plants*, *24*(2), 329–341. doi:10.1080/0972060X.2021.1927854

Outhwaite, C. L., Gregory, R. D., Chandler, R. E., Collen, B., & Isaac, N. J. B. (2020). Complex long-term biodiversity change among invertebrates, bryophytes and lichens. *Nature Ecology & Evolution*, *4*(3), 3. https://doi.org/10.1038/s41559-020-1111-z

Outhwaite, C. L., McCann, P., & Newbold, T. (2022). Agriculture and climate change are reshaping insect biodiversity worldwide. *Nature*, *605*(7908), 7908. https://doi.org/10.1038/s41586-022-04644-x

Paini, D. R., Sheppard, A. W., Cook, D. C., De Barro, P. J., Worner, S. P., & Thomas, M. B. (2016). Global threat to agriculture from invasive species. *Proceedings of the National Academy of Sciences of the United States of America*, *113*(27), 7575–7579.

Papadopoulos, N. T., Katsoyannos, B. I., Carey, J. R., & Kouloussis, N. A. (2001). Seasonal and Annual Occurrence of the Mediterranean Fruit Fly (Diptera: Tephritidae) in Northern Greece. *Annals of the Entomological Society of America*, *94*(1), 41–50. https://doi.org/10.1603/0013-8746(2001)094[0041:SAAOOT]2.0.CO;2

Papa, G., Maier, R., Durazzo, A., Lucarini, M., Karabagias, I. K., Plutino, M., Bianchetto, E., Aromolo, R., Pignatti, G., Ambrogio, A., Pellecchia, M., & Negri, I. (2022). The Honey Bee *Apis mellifera*: An Insect at the Interface between Human and Ecosystem Health. *Biology (Basel), 11*(2), 233. doi:10.3390/biology11020233 PMID:35205099

Parachnowitsch, A. L., & Manson, J. S. (2015). The chemical ecology of plant-pollinator interactions: Recent advances and future directions. *Current Opinion in Insect Science, 8*, 41–46. doi:10.1016/j.cois.2015.02.005 PMID:32846674

Parisi, V., Menta, C., Gardi, C., Jacomini, C., & Mozzanica, E. (2005). Microarthropod communities as a tool to assess soil quality and biodiversity: A new approach in Italy. *Agriculture, Ecosystems & Environment, 105*(1-2), 323–333. doi:10.1016/j.agee.2004.02.002

Park, C., Kang, H. S., Jeong, J., Kang, I., Choi, K., Yoo, M. S., & Chae, C. (2016). In-situ hybridization for the detection of sacbrood virus in infected larvae of the honey bee (*Apis cerana*). *Journal of Comparative Pathology, 154*(2-3), 258–262. doi:10.1016/j.jcpa.2015.12.003 PMID:26852344

Park, S., Lee, S., & Hong, K. J. (2015). Review of the family Bostrichidae (Coleoptera) of Korea. *Journal of Asia-Pacific Biodiversity, 8*(4), 298–304. doi:10.1016/j.japb.2015.10.015

Parmesan, C. (2007). Influences of species, latitudes and methodologies on estimates of phenological response to global warming. *Global Change Biology, 13*(9), 1860–1872. doi:10.1111/j.1365-2486.2007.01404.x

Parmesan, C., Ryrholm, N., Stefanescu, C., Hill, J. K., Thomas, C. D., Descimon, H., Huntley, B., Kaila, L., Kullberg, J., Tammaru, T., Tennent, W. J., Thomas, J. A., & Warren, M. (1999). Poleward shifts in geographical ranges of butterfly species associated with regional warming. *Nature, 399*(6736), 6736. https://doi.org/10.1038/21181

Parmesan, C., & Yohe, G. (2003). A globally coherent fingerprint of climate change impacts across natural systems. *Nature, 421*(6918), 37–42. doi:10.1038/nature01286 PMID:12511946

Partel, V., Nunes, L., Stansly, P., & Ampatzidis, Y. (2019). Automated vision-based system for monitoring Asian citrus psyllid in orchards utilizing artificial intelligence. *Computers and Electronics in Agriculture, 162*, 328–336. doi:10.1016/j.compag.2019.04.022

Pathak, H., Aggarwal, P. K., & Singh, S. D. (2012). *Climate Change Impact, Adaptation and Mitigation in agriculture: Methodology for Assessment and Application Division of Environmental Sciences*. Indian Agricultural Research Institute New Delhi.

Pattemore, D. E., Goodwin, R. M., McBrydie, H. M., Hoyte, S. M., & Vanneste, J. L. (2014). Evidence of the role of honey bees (*Apis mellifera*) as vectors of the bacterial plant pathogen *Pseudomonas syringae*. *Australasian Plant Pathology, 43*(5), 571–575. doi:10.100713313-014-0306-7

Patterson, D. T., Westbrook, J. K., Joyce, R. J. V., Lingren, P. D., & Rogasik, J. (1999). Weeds, insects and diseases. *Climatic Change, 43*(4), 711–727. doi:10.1023/A:1005549400875

Pérez-Hedo, M., Riahi, C., & Urbaneja, A. (2021). Use of zoophytophagous mirid bugs in horticultural crops: Current challenges and future perspectives. *Pest Management Science*, *77*(1), 33–42. doi:10.1002/ps.6043 PMID:32776672

Peris, D. (2020). Coleoptera in amber from Cretaceous resiniferous forests. *Cretaceous Research*, *113*, 104484. doi:10.1016/j.cretres.2020.104484

Perugini, M., Di Serafino, G., Giacomelli, A., Medrzycki, P., Sabatini, A. G., Persano Oddo, L., Marinelli, E., & Amorena, M. (2009). Monitoring of polycyclic aromatic hydrocarbons in bees (*Apis mellifera*) and honey in urban areas and wildlife reserves. *Journal of Agricultural and Food Chemistry*, *57*(16), 7440–7444. doi:10.1021/jf9011054 PMID:19627114

Petzoldt, C., & Seamann, A. (2010). Climate Change Effects on Insects and Pathogens. *Climate change and Agriculture: Promoting Practical and Profitable responses*, 6-16. Accessed online at http://www.climateandfarming.org/pdfs/ FactSheets/III.2Insects.Pathogens. pdf

Pike, V. L., Lythgoe, K. A., & King, K. C. (2019). On the diverse and opposing effects of nutrition on pathogen virulence. *Proceedings of the Royal Society B*, *286*(1906), 20191220. 10.1098/rspb.2019.1220

Pimental, D. (1976). World Food Crisis: Energy and Pests. *Bulletin of the Entomological Society of America*, *22*(1), 20–26. https://doi.org/10.1093/besa/22.1.20

Pimentel, D. (1986). *Agroecology and economics*. https://scholar.google.com/scholar_lookup?title=Agroecology+and+economics.&author=Pimentel+D.&publication_year=1986

Pinto, M. A., Rubink, W. L., Patton, J. C., Coulson, R. N., & Johnston, J. S. (2005). Africanization in the United States: Replacement of feral European honeybees (*Apis mellifera* L.) by an African hybrid swarm. *Genetics*, *170*(4), 1653–1665. doi:10.1534/genetics.104.035030 PMID:15937139

Piperno, D. R. (2006). Quaternary Environmental History and Agricultural Impact on Vegetation in Central America. *Annals of the Missouri Botanical Garden*, *93*(2), 274–296.

Piperno, D. R., Moreno, J. E., Iriarte, J., Holst, I., Lachniet, M., Jones, J. G., Ranere, A. J., & Castanzo, R. (2007). Late Pleistocene and Holocene environmental history of the Iguala Valley, Central Balsas Watershed of Mexico. *Proceedings of the National Academy of Sciences of the United States of America*, *104*(29), 11874–11881. https://doi.org/10.1073/pnas.0703442104

Platia, G. (2008). Descriptions of new species of click-beetles from the Palearctic region and Taiwan with chorological and synonymical notes. *Quaderno di Studi e Notizie di Storia Naturale della Romagna*, *27*, 187–215.

Porhajašová, J. I., & Babošová, M. (2022). Impact of arable farming management on the biodiversity of Carabidae (Coleoptera). *Saudi Journal of Biological Sciences*, *29*(9), 103371. doi:10.1016/j.sjbs.2022.103371 PMID:35865322

Porter, J. H., Parry, M. L., & Carter, T. R. (1991). The potential effects of climatic change on agricultural insect pests. *Agricultural and Forest Meteorology, 57*(1), 221–240. https://doi.org/10.1016/0168-1923(91)90088-8

Potts, S. G., Ngo, H. T., Biesmeijer, J. C., Breeze, T. D., Dicks, L. V., Garibaldi, L. A., Hill, R., Settele, J., & Vanbergen, A. (2016a). *The assessment report of the Intergovernmental Science-Policy Platform on Biodiversity and Ecosystem Services on pollinators, pollination and food production.* Bonn, Germany, Secretariat of the Intergovernmental Science-Policy Platform on Biodiversity and Ecosystem Services. Available on https://nora.nerc.ac.uk/id/eprint/519227/

Potts, S. G., Biesmeijer, J. C., Kremen, C., Neumann, P., Schweiger, O., & Kunin, W. E. (2010). Global pollinator declines: Trends, impacts and drivers. *Trends in Ecology & Evolution, 25*(6), 345–353. doi:10.1016/j.tree.2010.01.007 PMID:20188434

Potts, S. G., Imperatriz-Fonseca, V., Ngo, H. T., Aizen, M. A., Biesmeijer, J. C., Breeze, T. D., Dicks, L. V., Garibaldi, L. A., Hill, R., Settele, J., & Vanbergen, A. J. (2016b). Safeguarding pollinators and their values to human well-being. *Nature, 540*(7632), 220–229. doi:10.1038/nature20588 PMID:27894123

Potts, S. G., Vulliamy, B., Dafni, A., Ne'eman, G., & Willmer, P. (2003). Linking bees and flowers: How do floral communities structure pollinator communities. *Ecology, 84*(10), 2628–2642. doi:10.1890/02-0136

Prakash, A., Rao, J., Mukherjee, A. K., Jeyaveeran, B., Pokhare, S., Adak, T., Munda, S., & R., S. (2014). *Climate Change: Impact on Crop Pests.* Academic Press.

Purać, J., Nikolić, T. V., Kojić, D., Ćelić, A. S., Plavša, J. J., Blagojević, D. P., & Petri, E. T. (2019). Identification of a metallothionein gene in honey bee *Apis mellifera* and its expression profile in response to Cd, Cu and Pb exposure. *Molecular Ecology, 28*(4), 731–745. doi:10.1111/mec.14984 PMID:30575191

Rader, R., Bartomeus, I., Garibaldi, L. A., Garratt, M. P., Howlett, B. G., Winfree, R., ... Woyciechowski, M. (2016). Non-bee insects are important contributors to global crop pollination. *Proceedings of the National Academy of Sciences of the United States of America, 113*(1), 146–151. doi:10.1073/pnas.1517092112 PMID:26621730

Rainio, J., & Niemelä, J. (2003). Ground beetles (Coleoptera: Carabidae) as bioindicators. *Biodiversity and Conservation, 12*(3), 487–506. doi:10.1023/A:1022412617568

Rao, G. G. S. N., Rao, A. V. M. S., & Rao, V. U. M. (2009). Trends in rainfall and temperature in rainfed India in the previous century. In P. K. Aggarwal (Ed.), *Global climate change and Indian Agriculture case studies from ICAR network project* (pp. 71–73). ICAR Publication.

Rasmont, P., Franzén, M., Lecocq, T., Harpke, A., Roberts, S. P., Biesmeijer, J. C., & Schweiger, O. (2015). *Climatic risk and distribution atlas of European bumblebees* (Vol. 10). Pensoft Publishers.

Rasmont, P., & Iserbyt, S. (2012). The Bumblebees Scarcity Syndrome: Are heat waves leading to local extinctions of bumblebees (Hymenoptera: Apidae: Bombus)? *Annales de la Société Entomologique de France*, *48*(3–4), 275–280. doi:10.1080/00379271.2012.10697776

Ratcliffe, B. C., & Jameson, M. L. (2004). The Revised classification for Scarabaeoidea: What the Hell is going on? *Papers in Entomology*, 25. http://www-museum.unl.edu/research/entomology/Scarabs-Newsletter.htm

Raza, A., Razzaq, A., Mehmood, S. S., Zou, X., Zhang, X., Lv, Y., & Xu, J. (2019). Impact of climate change on crops adaptation and strategies to tackle its outcome: A review. *Plants*, *8*(2), 34. doi:10.3390/plants8020034 PMID:30704089

Reddy, P. V. R., Verghese, A., & Rajan, V. V. (2012). Potential impact of climate change on honeybees (*Apis* spp.) and their pollination services. *Pest Management in Horticultural Ecosystems*, *18*(2), 121–127.

Requier, F., Odoux, J. F., Tamic, T., Moreau, N., Henry, M., Decourtye, A., & Bretagnolle, V. (2015). Honey bee diet in intensive farmland habitats reveals an unexpectedly high flower richness and a major role of weeds. *Ecological Applications*, *25*(4), 881–890. doi:10.1890/14-1011.1 PMID:26465030

Rhymer, J. M., & Simberloff, D. (1996). Extinction by hybridization and introgression. *Annual Review of Ecology and Systematics*, *27*(1), 83–109. doi:10.1146/annurev.ecolsys.27.1.83

Ritchie, S. A., Zborowski, P., Banks, D., Walsh, I., & Davis, J. (2008). Efficacy of novel updraft traps for collection of mosquitoes in Cairns, Australia. *Journal of the American Mosquito Control Association*, *24*(4), 520–527. doi:10.2987/5698.1 PMID:19181059

Roberts, J. M. K., Ireland, K. B., Tay, W. T., & Paini, D. (2018). Honey bee-assisted surveillance for early plant virus detection. *Annals of Applied Biology*, *173*(3), 285–293. doi:10.1111/aab.12461

Rochlin, I., White, G., Reissen, N., Martheswaran, T., & Faraji, A. (2022). Effects of aerial adulticiding for mosquito management on nontarget insects: A Bayesian and community ecology approach. *Ecosphere*, *13*(1), e3896. doi:10.1002/ecs2.3896

Rodrigez-Trelles, F., & Rodrigez, M. A. (1998). Rapid micro-evolution and loss of chromosomal diversity in *Drosophila* in response to climate warming. *Evolutionary Ecology*, *12*(7), 829–838. doi:10.1023/A:1006546616462

Rondeau, S., Giovenazzo, P., & Fournier, V. (2018). Risk assessment and predation potential of Stratiolaelaps scimitus (Acari: Laelapidae) to control *Varroa destructor* (Acari: Varroidae) in honey bees. *PLoS One*, *13*(12), e0208812. doi:10.1371/journal.pone.0208812 PMID:30532269

Root, T. L., Price, J. T., Hall, K. R., Schneider, S. H., Rosenzweig, C., & Pounds, J. A. (2003). Fingerprints of global warming on wild animals and plants. *Nature*, *421*(6918), 57–60. doi:10.1038/nature01333 PMID:12511952

Roy, D. B., & Sparks, T. H. (2000). Phenology of British butterflies and climate change. *Global Change Biology*, *6*(4), 407–416. doi:10.1046/j.1365-2486.2000.00322.x

Ruschioni, S., Riolo, P., Minuz, R. L., Stefano, M., Cannella, M., Porrini, C., & Isidoro, N. (2013). Biomonitoring with honeybees of heavy metals and pesticides in nature reserves of the Marche Region (Italy). *Biological Trace Element Research, 154*(2), 226–233. doi:10.100712011-013-9732-6 PMID:23797576

Russell, T. L., Kay, B. H., & Skilleter, G. A. (2009). Environmental effects of mosquito insecticides on saltmarsh invertebrate fauna. *Aquatic Biology*, *6*, 77–90. doi:10.3354/ab00156

Ruttner, F. (1988). *Biogeography and taxonomy of honeybees*. Springler. doi:10.1007/978-3-642-72649-1

Ruttner, F., Tassencourt, L., & Louveaux, J. (1978). Biometrical statistical analysis of the geographic variability of *Apis mellifera* L. *Apidologie*, *9*, 363–382. doi:10.1051/apido:19780408

Sahraoui, H. (2012). *Influence des pratiques agroécologiques et de la protection phytosanitaire sur les communautés d'acariens Phytoseiidae (Acari : mesostigmata) dans les vergers agrumicoles tunisiens. Thésis of doctorat*. Institut National Agronomique.

Sak, B., Sakova, K., & Ditrich, O. (2004). Effects of a novel anti-exospore monoclonal antibody on microsporidial development in vitro. *Parasitology Research*, *92*, 74–80. doi:10.100700436-003-0988-1 PMID:14610668

Salazar, C. A., Jiggins, C. D., Arias, C. F., Tobler, A., Bermingham, E., & Linares, M. (2005). Hybrid incompatibility is consistent with a hybrid origin of *Heliconius heurippa* Hewitson from its close relatives, *Heliconius cydno* Doubleday and *Heliconius melpomene* Linnaeus. *Journal of Evolutionary Biology*, *18*(2), 247–256. doi:10.1111/j.1420-9101.2004.00839.x PMID:15715831

Sammataro, D., Gerson, U., & Needham, G. (2000). Parasitic mites of honey bees: Life history, implications, and impact. *Annual Review of Entomology*, *45*(1), 519–548. doi:10.1146/annurev.ento.45.1.519 PMID:10761588

Sánchez-Guillén, R. A., Córdoba-Aguilar, A., Hansson, B., Ott, J., & Wellenreuther, M. (2016). Evolutionary consequences of climate-induced range shifts in insects. *Biological Reviews of the Cambridge Philosophical Society*, *91*(4), 1050–1064. doi:10.1111/brv.12204 PMID:26150047

Sari, M. F., Ayyildiz, E. G., & Esen, F. (2020). Determination of polychlorinated biphenyls in honeybee, pollen, and honey samples from urban and semi-urban areas in Turkey. *Environmental Science and Pollution Research International*, *27*(4), 4414–4422. doi:10.100711356-019-07013-w PMID:31832954

Saxena, R., Joshi, A., Joshi, S., Borkotoky, S., Singh, K., Rai, P. K., Mueed, Z., & Sharma, R. (2023). The role of artificial intelligence strategies to mitigate abiotic stress and climate change in crop production. In A. Srivastav, A. Dubey, A. Kumar, S. Kumar Narang, & M. Ali Khan (Eds.), Visualization Techniques for Climate Change with Machine Learning and Artificial Intelligence (pp. 273–293). Elsevier. https://doi.org/10.1016/B978-0-323-99714-0.00006-6.

Scheffer, M., Carpenter, S. R., Lenton, T. M., Bascompte, J., Brock, W., Dakos, V., van de Koppel, J., van de Leemput, I. A., Levin, S. A., van Nes, E. H., Pascual, M., & Vandermeer, J. (2012). Anticipating critical transitions. *Science*, *338*(6105), 344–348. doi:10.1126cience.1225244 PMID:23087241

Schelhaas, M. J., Nabuurs, G. J., & Schuck, A. (2003). Natural disturbances in the European forests in the 19th and 20th centuries. *Global Change Biology*, *9*(11), 1620–1633. doi:10.1046/j.1365-2486.2003.00684.x

Schneider, L., Rebetez, M., & Rasmann, S. (2022). The effect of climate change on invasive crop pests across biomes. *Current Opinion in Insect Science*, *50*, 100895. doi:10.1016/j.cois.2022.100895 PMID:35240333

Schweiger, O., Biesmeijer, J. C., Bommarco, R., Hickler, T., Hulme, P. E., Klotz, S., & Settele, J. (2010). Multiple stressors on biotic interactions: How climate change and alien species interact to affect pollination. *Biological Reviews of the Cambridge Philosophical Society*, *85*(4), 777–795. doi:10.1111/j.1469-185X.2010.00125.x PMID:20184567

Seehausen, O., Takimoto, G., Roy, D., & Jokela, J. (2008). Speciation reversal and biodiversity dynamics with hybridization in changing environments. *Molecular Ecology*, *17*(1), 30–44. doi:10.1111/j.1365-294X.2007.03529.x PMID:18034800

Sekkat, A. (2008). *Les pucerons des agrumes au Maroc*. Symposium Méditerranéen sur la Protection Phytosanitaire des agrumes, Rabat, Maroc.

Settele, J., Bishop, J., & Potts, S. G. (2016). Climate change impacts on pollination. *Nature Plants*, *2*(7), 1–3. doi:10.1038/nplants.2016.92 PMID:27364126

Settele, J., Kudrna, O., Harpke, A., Kühn, I., Van Swaay, C., Verovnik, R., & Schweiger, O. (2008). Climatic risk atlas of European butterflies (Vol. 1). Sofia: Pensoft. *BioRisk*, *1*, 1–712. doi:10.3897/biorisk.1

Sharanabasappa, K. C. M., Asokan, R., Swamy, H. M. M., Maruthi, M. S., Pavithra, H. B., Hegbe, K., Navi, S., Prabhu, S. T., & Goergen, G. E. (2018). *First report of the fall armyworm, Spodoptera frugiperda (J E Smith) (Lepidoptera: Noctuidae), an alien invasive pest on maize in India*. https://biblio1.iita.org/handle/20.500.12478/5181

Sharma, H. C. (2002). Host plant resistance to insects: principles and practices. In Resources Management in Plant Protection, Vol. 1. Plant Protection Association of India. Rajendranagar.

Sharma, H. C. (2010). *Effect of climate change on IPM in grain legumes*. In Fifth International Food Legumes Research Conference (IFLRC V), and the Seventh European Conference on Grain Legumes (AEP VII), Anatalaya, Turkey.

Sharma, H. C. (2016). *Climate Change vis-à-vis Pest Management*. In Conference on National Priorities in Plant Health Management (Organised by Plant Protection Association of India, NBPGR, Hyderabad), Tirupati.

Shoukry, A., & Hafez, M. (1979). Studies on the Biology of the Mediterranean Fruit Fly Ceratitis Capitata. *Entomologia Experimentalis et Applicata, 26*(1), 33–39. https://doi.org/10.1111/j.1570-7458.1979.tb02894.x

Sidhu, A. K., & Mehta, H. S. (2008). Role of butterflies in the natural ecosystem with special reference to high altitude (Pangi Valley, Himachal Pradesh). In *Proceedings of International Conference on Climate Change, Biodiversity and Food Security in the South Asian Region.* Punjab State Council for Science and Technology, Chandigarh and United Nations Educational, Scientific and Cultural Organization, New Delhi.

Silva, R., Alves, R. C., & Zingaretti, S. M. (2020). Increased [CO2] Causes Changes in Physiological and Genetic Responses in C4 Crops: A Brief Review. *Plants, 9*(11), 1567. doi:10.3390/plants9111567 PMID:33202833

Skelly, D. K., Joseph, L. N., Possingham, H. P., Freidenburg, L. K., Farrugia, T. J., Kinnison, M. T., & Hendry, A. P. (2007). Evolutionary Responses to Climate Change. *Conservation Biology, 21*(5), 1353–1355. doi:10.1111/j.1523-1739.2007.00764.x PMID:17883501

Skendžić, S., Zovko, M., Živković, I. P., Lešić, V., & Lemić, D. (2021). The Impact of Climate Change on Agricultural Insect Pests. *Insects, 12*(5), 440. doi:10.3390/insects12050440 PMID:34066138

Smaili, C., Affelah, M., & Aarab, A. (2001). *Biologie, écologie et lutte biologique de la mineuse des agrumes Phyllocnistis citrella Lep. (Gracillariidae) sur clémentinier dans la région du Gharb.1ère colloque de la région du Gharb Kénitra.* Academic Press.

Smaili, M. C., & Wadjinny, J. (2008). Evaluation de la kaolinite comme insecticide naturels contre la cératite des agrumes *Ceraitis capitata* (Dip, Tephritidae) dans la région du Nord du Gharb. *Symposium méditerranéen sur la protection phytosanitaire des agrumes.*

Smaili, M. C., Abbassi, M., Boutaleb, J. A., & Blenzar, A. (2013). Richesse spécifique des ennemis naturels associés aux vergers d'agrumes au Maroc: Intérêt et implication pour la lutte biologique. *Bulletin OEPP. EPPO Bulletin. European and Mediterranean Plant Protection Organisation, 43*(1), 155–166. doi:10.1111/epp.12013

Smith, K. E., & Weis, D. (2020). Evaluating spatiotemporal resolution of trace element concentrations and Pb isotopic compositions of honeybees and hive products as biomonitors for urban metal distribution. *GeoHealth, 4*(7). . doi:10.1029/2020GH000264

Smith, D. R. (1991). Mitochondrial DNA and honeybee biogeography. In D. R. Smith (Ed.), *Diversity in the genus Apis* (pp. 131–176). Westview Press.

Smith, K. E., Weis, D., Amini, M., Shiel, A. E., Lai, V. W.-M., & Gordon, K. (2019). Honey as a biomonitor for a changing world. *Nature Sustainability, 2*(3), 223–232. doi:10.103841893-019-0243-0

Solignac, M., Cornuet, J. M., Vautrin, D., Le Conte, Y., Anderson, D., Evans, J., & Navajas, M. (2005). The invasive Korea and Japan types of *Varroa destructor*, ectoparasitic mites of the Western honeybee (*Apis mellifera*), are two partly isolated clones. *Proceedings. Biological Sciences*, *272*(1561), 411–419. doi:10.1098/rspb.2004.2853 PMID:15734696

Somwang, P., Khositharattanakool, P., Pathawong, N., Pongsiri, A., Davidson, S. A., & Ponlawat, A. (2021). Field evaluation of four commercial light traps, trap placement, and effect of carbon dioxide on phlebotomine sand fly collection in northern Thailand. *Acta Tropica*, *220*, 105953. doi:10.1016/j.actatropica.2021.105953 PMID:33979638

Song, Y., Zhang, B., Wang, J., & Kwek, K. (2022). The impact of climate change on China's agricultural green total factor productivity. *Technological Forecasting and Social Change*, *185*, 122054. https://doi.org/10.1016/j.techfore.2022.122054

Soroye, P., Newbold, T., & Kerr, J. (2020). Climate change contributes to widespread declines among bumble bees across continents. *Science*, *367*(6478), 685–688. https://doi.org/10.1126/science.aax8591

Spahn, R., & Lill, J. T. (2022). Higher temperatures reduce the efficacy of a key biocontrol parasitoid. *Biological Control*, *176*, 105079. https://doi.org/10.1016/j.biocontrol.2022.105079

Spooner, F. E. B., Pearson, R. G., & Freeman, R. (2018). Rapid warming is associated with population decline among terrestrial birds and mammals globally. *Global Change Biology*, *24*(10), 4521–4531. https://doi.org/10.1111/gcb.14361

Starks, P. T., & Gilley, D. C. (1999). Heat shielding: A novel method of colonial thermoregulation in honey bees. *Naturwissenschaften*, *86*(9), 438–440. doi:10.1007001140050648 PMID:10501692

Staton, T., Walters, R. J., Smith, J., Breeze, T. D., & Girling, R. D. (2021). Evaluating a trait-based approach to compare natural enemy and pest communities in agroforestry vs. arable systems. *Ecological Applications*, *31*(4), e02294. doi:10.1002/eap.2294 PMID:33427350

Stengaard Hansen, L., Åkerlund, M., Grøntoft, T., Ryhl-Svendsen, M., Schmidt, A. L., Bergh, J.-E., & Vagn Jensen, K.-M. (2012). Future pest status of an insect pest in museums, Attagenus smirnovi: Distribution and food consumption in relation to climate change. *Journal of Cultural Heritage*, *13*(1), 22–27. https://doi.org/10.1016/j.culher.2011.05.005

Storch, D., Šímová, I., Smyčka, J., Bohdalková, E., Toszogyova, A., & Okie, J. G. (2022). Biodiversity dynamics in the Anthropocene: How human activities change equilibria of species richness. *Ecography*, *44*(4), 1–19. doi:10.1111/ecog.05778

Stout, J., & Morales, C. L. (2009). Ecological impacts of invasive alien species on bees. *Apidologie*, *40*(3), 388–409. doi:10.1051/apido/2009023

Strickland, A. H. (1969). Plant Protection and World Crop Production By H. H. Cramer Leverkusen: 'Bayer' Pflanzenschutz (1967), pp. 524. *Experimental Agriculture*, *5*(1), 82–83. https://doi.org/10.1017/S0014479700010036

Suggitt, A. J., Wilson, R. J., Isaac, N. J. B., Beale, C. M., Auffret, A. G., August, T., Bennie, J. J., Crick, H. Q. P., Duffield, S., Fox, R., Hopkins, J. J., Macgregor, N. A., Morecroft, M. D., Walker, K. J., & Maclean, I. M. D. (2018). Extinction risk from climate change is reduced by microclimatic buffering. *Nature Climate Change*, *8*(8), 8. https://doi.org/10.1038/s41558-018-0231-9

Sultan, B., Defrance, D., & Iizumi, T. (2019). Evidence of crop production losses in West Africa due to historical global warming in two crop models. *Scientific Reports*, *9*(1), 1. https://doi.org/10.1038/s41598-019-49167-0

Sutherst, R. (2000). Climate change and invasive species: a conceptual framework. In H. A. Mooney & R. J. Hobbs (Eds.), *Invasive Species in a Changing World* (pp. 211–240). Island Press.

Sutherst, R. W., Constable, F., Finlay, K. J., Harrington, R., Luck, J., & Zalucki, M. P. (2011). Adapting to crop pest and pathogen risks under a changing climate. *Wiley Interdisciplinary Reviews: Climate Change*, *2*(2), 220–237. doi:10.1002/wcc.102

Switanek, M., Crailsheim, K., Truhetz, H., & Brodschneider, R. (2017). Modelling seasonal effects of temperature and precipitation on honey bee winter mortality in a temperate climate. *The Science of the Total Environment*, *579*, 1581–1587. doi:10.1016/j.scitotenv.2016.11.178 PMID:27916302

Talbi, F. Z., El Ouali Lalami, A., Janati Idrissi, A., Sebti, F., & Faraj, C. (2015). Leishmaniasis in central Morocco: Seasonal fluctuations of phlebotomine sand fly in Aichoun locality, from Sefrou province. *Pathology Research International*, *2015*, 1–4. doi:10.1155/2015/438749

Tamura, Y., Osawa, T., Tabuchi, K., Yamasaki, K., Niiyama, T., Sudo, S., Ishigooka, Y., Yoshioka, A., & Takada, M. B. (2022). Estimating plant–insect interactions under climate change with limited data. *Scientific Reports*, *12*(1), 10554. doi:10.103841598-022-14625-9 PMID:35794117

Tang, R. S., Zheng, J. C., Jin, Z. Q., Zhang, D. D., Huang, Y. H., & Chen, L. G. (2008). Possible correlation between high temperature-induced floret sterility and endogenous levels of IAA, GAs and ABA in rice (*Oryza sativa* L.). *Plant Growth Regulation*, *54*(1), 37–43. doi:10.100710725-007-9225-8

Taraz, V. (2018). Can farmers adapt to higher temperatures? Evidence from India. *World Development*, *112*, 205–219. https://doi.org/10.1016/j.worlddev.2018.08.006

Tariq, A. M. (2020). Global warming and climate change: Impact on biodiversity, pest management and food security. *Plant Archives*, *20*(2), 110–115.

Taylor, R. S., & Frisen, V. L. (2017). The role of allochrony in speciation. *Molecular Ecology*, *26*(13), 3330–3342. doi:10.1111/mec.14126 PMID:28370658

Thai, P. H., Huyen, N. T., Van Toan, T., & Jung, C. (2018). *Apis cerana* beekeeping and sacbrood disease management in Vietnam. *Journal of Apiculture*, *33*(4), 269–275. doi:10.17519/apiculture.2018.11.33.4.269

Thomas, C. D., Cameron, A., Green, R. E., Bakkenes, M., Beaumont, L. J., Collingham, Y. C., Erasmus, B. F. N., Fereira de Sigueira, M., & Grainger, A. (2004). Extinction risk from climate change. *Nature*, *427*(6970), 145–148. doi:10.1038/nature02121 PMID:14712274

Thomas, C. D., Franco, A. M. A., & Hill, J. K. (2006). Range retractions and extinction in the face of climate warming. *Trends in Ecology & Evolution*, *21*(8), 415–416. doi:10.1016/j.tree.2006.05.012 PMID:16757062

Thompson, H., & Wilkins, S. (2003). Assessment of the synergy and repellency of pyrethroid/fungicide mixtures. *Bulletin of Insectology*, *56*, 131–134.

Thomson, L. J., Macfadyen, S., & Hoffmann, A. A. (2010). Predicting the effects of climate change on natural enemies of agricultural pests. *Biological Control*, *52*(3), 296–306. https://doi.org/10.1016/j.biocontrol.2009.01.022

Thuiller, W., Lavorel, S., Araújo, M. B., Sykes, M. T., & Prentice, I. C. (2005). Climate change threats to plant diversity in Europe. *Proceedings of the National Academy of Sciences of the United States of America*, *102*(23), 8245–8250. doi:10.1073/pnas.0409902102 PMID:15919825

Tocco, C., Midgley, J. M., & Villet, M. H. (2020). Intermediate disturbance promotes diversity and the conservation of dung beetles (Scarabaeoidea: Scarabaeidae and Aphodiidae) in the Eastern Cape, South Africa. *Basic and Applied Ecology*, *49*, 45–56. doi:10.1016/j.baae.2020.10.005

Todesco, M., Pascual, M. A., Owens, G. L., Ostevik, K. L., Moyers, B. T., Hübner, S., Heredia, S. M., Hahn, M. A., Caseys, C., Bock, D. G., & Rieseberg, L. H. (2016). Hybridization and extinction. *Evolutionary Applications*, *9*(7), 892–908. doi:10.1111/eva.12367 PMID:27468307

Tonnang, H. E. Z., Hervé, B. D. B., Biber-Freudenberger, L., Salifu, D., Subramanian, S., Ngowi, V. B., Guimapi, R. Y. A., Anani, B., Kakmeni, F. M. M., Affognon, H., Niassy, S., Landmann, T., Ndjomatchoua, F. T., Pedro, S. A., Johansson, T., Tanga, C. M., Nana, P., Fiaboe, K. M., Mohamed, S. F., … Borgemeister, C. (2017). Advances in crop insect modelling methods—Towards a whole system approach. *Ecological Modelling, 354*, 88–103. doi:10.1016/j.ecolmodel.2017.03.015

Tonnang, H. E. Z., Sporleder, M., Juarez, H., Carhuapoma, P., & Krosc, J. (2015). Insect Life Cycle Modelling (ILCYM) software—A generic platform for developing insect phenology models, population analysis and risk mapping. *Potato and Sweetpotato in Africa: Transforming the Value Chains for Food and Nutrition Security*, 350–361. doi:10.1079/9781780644202.0350

Tonnang, H. E., Sokame, B. M., Abdel-Rahman, E. M., & Dubois, T. (2022). Measuring and modelling crop yield losses due to invasive insect pests under climate change. *Current Opinion in Insect Science*, *50*, 100873. https://doi.org/10.1016/j.cois.2022.100873

Tremblay, É. D., Duceppe, M.-O., Thurston, G. B., Gagnon, M.-C., Côté, M.-J., & Bilodeau, G. J. (2019). High-resolution biomonitoring of plant pathogens and plant species using metabarcoding of pollen pellet contents collected from a honey bee hive. *Environmental DNA*, *1*(2), 155–175. doi:10.1002/edn3.17

Trubin, A., Mezei, P., Zabihi, K., Surový, P., & Jakuš, R. (2022). Northernmost European spruce bark beetle *Ips typographus* outbreak: Modelling tree mortality using remote sensing and climate data. *Forest Ecology and Management*, *505*, 119829. doi:10.1016/j.foreco.2021.119829

Trudgill, D., Honek, A., Li, D., & van Straalen, N. M. (2005). Thermal time—Concepts and utility. *Annals of Applied Biology*, *146*, 1–14. https://doi.org/10.1111/j.1744-7348.2005.04088.x

Turcotte, M. M., Araki, H., Karp, D. S., Poveda, K., & Whitehead, S. R. (2017). The eco-evolutionary impacts of domestication and agricultural practices on wild species. *Philosophical Transactions of the Royal Society B: Biological Sciences, 372*(1712), 20160033. doi:10.1098/rstb.2016.0033

Turn Down the Heat: Climate Extremes, Regional Impacts, and the Case for Resilience. (n.d.). Retrieved 6 December 2022, from https://elibrary.worldbank.org/doi/abs/10.1596/978-1-4648-0055-9

Ullah, A., Gajger, I. T., Majoros, A., Dar, S. A., Khan, S., Shah, A. H., Khabir, M. N., Hussain, R., Khan, H. U., Hameed, M., & Anjum, S. I. (2021). Viral impacts on honey bee populations: A review. *Saudi Journal of Biological Sciences*, *28*(1), 523–530. doi:10.1016/j.sjbs.2020.10.037 PMID:33424335

Urbaneja, A., Pascual-Ruiz, S., Pina, T., Abad-Moyano, R., Vanaclocha, P., Montón, H., Dembilio, O., Castañera, P., & Jacas, J. A. (2008). Efficacy of five selected acaricides against *Tetranychus urticae* (Acari: Tetranychidae) and their side effects on relevant natural enemies occurring in citrus orchards. *Pest Management Science*, *64*(8), 834–842. doi:10.1002/ps.1572 PMID:18383196

Vadez, V., Berger, J. D., Warkentin, T., Asseng, S., Ratnakumar, P., Poorna Chandra Rao, K., Gaur, P. M., Munier-Jolain, N., Larmure, A., Voisin, A.-S., ... Zaman, M. A. (2012). Adaptation of grain legumes to climate change: A review. *Agronomy for Sustainable Development*, *32*(1), 31–44. doi:10.100713593-011-0020-6

vanEngelsdorp, D., Evans, J. D., Saegerman, C., Mullin, C., Haubruge, E., Nguyen, B. K., Frazier, M., Frazier, J., Cox-Foster, D., Chen, Y., Underwood, R., Tarpy, D. R., & Pettis, J. S. (2009). Colony collapse disorder: A descriptive study. *PLoS One*, *4*(8), e6481. doi:10.1371/journal.pone.0006481 PMID:19649264

Vargas, R. I., Miyashita, D., & Nishida, T. (1984). Life history and demographic parameters of three laboratory-reared tephritids (Diptera: Tephritidae). *Annals of the Entomological Society of America*, *77*(6), 651–656.

Vasiliev, D., & Greenwood, S. (2021). The role of climate change in pollinator decline across the Northern Hemisphere is underestimated. *The Science of the Total Environment*, *775*, 145788. doi:10.1016/j.scitotenv.2021.145788 PMID:33618305

Villalba, A., Maggi, M., Ondarza, P. M., Szawarski, N., & Miglioranza, K. S. B. (2020). Influence of land use on chlorpyrifos and persistent organic pollutant levels in honey bees, bee bread and honey: Beehive exposure assessment. *The Science of the Total Environment*, *713*, 136554. doi:10.1016/j.scitotenv.2020.136554 PMID:31955084

Volney, W. J. A., & Fleming, R. A. (2000). Climate change and impacts of boreal forest insects. *Agriculture, Ecosystems & Environment, 82*(1-3), 283–294. doi:10.1016/S0167-8809(00)00232-2

Walter, J. (2020). Dryness, wetness and temporary flooding reduce floral resources of plant communities with adverse consequences for pollinator attraction. *Journal of Ecology, 108*(4), 1453–1464. doi:10.1111/1365-2745.13364

Wanderley, P. A., Ramalho, F. S., Zanuncio, J. C., & Leite, G. L. D. (2007). Thermal requirements and development ofBracon vulgaris, a parasitoid of the cotton boll weevil. *Phytoparasitica, 35*(4), 336–345. https://doi.org/10.1007/BF02980696

Wang, Y., Yim, S., Yang, Y., & Morin, C. W. (2020). The effect of urbanization and climate change on the mosquito population in the Pearl River Delta region of China. *International Journal of Biometeorology, 64*(3), 501–512. doi:10.100700484-019-01837-4 PMID:31811391

Wania, F., & MacKay, D. (1996). Tracking the Distribution of Persistent Organic Pollutants. *Environmental Science & Technology, 30*(9), 390A–396A. doi:10.1021/es962399q PMID:21649427

Ware, A. (1994). The biology and control of citrus leafminer. *Citrus Journal, 4*(4), 26–28.

Watanabe, M. E. (1994). Pollination worries rise as honey bees decline. *Science, 265*(5176), 1170. doi:10.1126cience.265.5176.1170 PMID:17787573

Welch, K. D., & Harwood, J. D. (2014). Temporal dynamics of natural enemy–pest interactions in a changing environment. *Biological Control, 75*, 18–27. https://doi.org/10.1016/j.biocontrol.2014.01.004

Weldon, C. W., Nyamukondiwa, C., Karsten, M., Chown, S. L., & Terblanche, J. S. (2018). Geographic variation and plasticity in climate stress resistance among southern African populations of Ceratitis capitata (Wiedemann) (Diptera: Tephritidae). *Scientific Reports, 8*(1), 1. https://doi.org/10.1038/s41598-018-28259-3

Whitehead, S. R., Turcotte, M. M., & Poveda, K. (2017). Domestication impacts on plant–herbivore interactions: A meta-analysis. *Philosophical Transactions of the Royal Society B: Biological Sciences, 372*(1712), 20160034. doi:10.1098/rstb.2016.0034

WHO. (2022). *Leishmaniasis*. 8 January 2022. https://www.who.int/news-room/fact-sheets/detail/leishmanias is

Williams, I. H. (1994). The dependence of crop production within the European Union on pollination by honey bees. *Agricultural Zoology Reviews, 6*, 229–257.

Willis, C. G., Ruhfel, B., Primack, R. B., Miller Rushing, A. J., & Davis, C. C. (2008). Phylogenetic patterns of species loss in Thoreau's woods are driven by climate change. *Proceedings of the National Academy of Sciences of the United States of America, 105*(44), 17029–17033. doi:10.1073/pnas.0806446105 PMID:18955707

Wilson, C. G. (1991). Notes on *Phyllocnistis citrella* Stainton (Lepidoptera: Phyllocnistidae) attacking four citrus varieties in Darwin. *Journal of the Australian Entomological Society*, *30*(1), 77–78. doi:10.1111/j.1440-6055.1991.tb02198.x

Winfree, R., Williams, N. M., Gaines, H., Ascher, J. S., & Kremen, C. (2008). Wild bee pollinators provide the majority of crop visitation across land-use gradients in New Jersey and Pennsylvania, USA. *Journal of Applied Ecology*, *45*(3), 793–802. doi:10.1111/j.1365-2664.2007.01418.x

Wood, T. J., Michez, D., Paxton, R. J., Drossart, M., Neumann, P., Gérard, M., Vanderplanck, M., Barraud, A., Martinet, B., Leclercq, N., & Vereecken, N. J. (2020). Managed honey bees as a radar for wild bee decline? *Apidologie*, *51*(6), 1100–1116. doi:10.100713592-020-00788-9

Wu, J. Y., Anelli, C. M., & Sheppard, W. S. (2011). Sub-lethal effects of pesticide residues in brood comb on worker honey bee (*Apis mellifera*) development and longevity. *PLoS One*, *6*(2), e14720. doi:10.1371/journal.pone.0014720 PMID:21373182

Yamamura, K., & Kiritani, K. (1998). A simple method to estimate the potential increase in the number of generations under global warming in temperate zones. *Journal Applied Entomology and Zoology*, *33*(2), 289–298. doi:10.1303/aez.33.289

Yamamura, K., Yokozawa, M., Nishimori, M., Ueda, Y., & Yokosuka, T. (2006). How to analyze long-term insect population dynamics under climate change: 50-year data of three insect pests in paddy fields. *Population Ecology*, *48*(1), 31–48. doi:10.100710144-005-0239-7

Yan, L., & Roy, D. P. (2016). Conterminous United States field size quantification from multitemporal Landsat data. *Remote Sensing of Environment*, *172*, 67–86. doi:10.1016/j.rse.2015.10.034

Yélamos, T. (1992). Nuevos datos sobre los Histéridos de Marruecos. *Zapateri*, *12*, 55–64.

Yoruk, A., & Sahinler, N. (2013). Potential effects of global warming on the honey bee. *Uludag Bee Journal*, *13*, 79–87.

Zampieri, M., Ceglar, A., Dentener, F., & Toreti, A. (2017). Wheat yield loss attributable to heat waves, drought and water excess at the global, national and subnational scales. *Environmental Research Letters*, *12*(6), 064008. https://doi.org/10.1088/1748-9326/aa723b

Zaragoza-Trello, C., Vilà, M., Botías, C., & Bartomeus, I. (2021). Interactions among global change pressures act in a non-additive way on bumblebee individuals and colonies. *Functional Ecology*, *35*(2), 420–434. doi:10.1111/1365-2435.13703

Zarić, N. M., Ilijević, K., Stanisavljević, L., & Gržetić, I. (2017). Use of honeybees (*Apis mellifera* L.) as bioindicators for assessment and source appointment of metal pollution. *Environmental Science and Pollution Research International*, *24*(33), 25828–25838. doi:10.100711356-017-0196-7 PMID:28936680

Zaveri, E., & Lobell, B. D. (2019). The role of irrigation in changing wheat yields and heat sensitivity in India. *Nature Communications, 10*(1), Article 1. doi:10.1038/s41467-019-12183-9

Zhang, X., He, S. Y., Evans, J. D., Pettis, J. S., Yin, G. F., & Chen, Y. P. (2012). New evidence that deformed wing virus and black queen cell virus are multi-host pathogens. *Journal Invertebrate Pathology, 109*, 156–159. doi: .09.010 doi:10.1016/j.jip.2011

Zvereva, E. L., & Kozlov, M. V. (2006). Consequences of simultaneous elevation of carbon dioxide and temperature for plant–herbivore interactions: A metaanalysis. *Global Change Biology*, *12*(1), 27–41. https://doi.org/10.1111/j.1365-2486.2005.01086.x

Zvereva, E. L., & Kozlov, M. V. (2010). Responses of terrestrial arthropods to air pollution: A metaanalysis. *Environmental Science and Pollution Research International*, *17*(2), 297–311. doi:10.100711356-009-0138-0 PMID:19319587

Related References

To continue our tradition of advancing academic research, we have compiled a list of recommended IGI Global readings. These references will provide additional information and guidance to further enrich your knowledge and assist you with your own research and future publications.

Ajiboye, O. E., & Yusuff, O. S. (2017). Foreign Land Acquisition: Food Security and Food Chains – The Nigerian Experience. In I. Management Association (Ed.), Natural Resources Management: Concepts, Methodologies, Tools, and Applications (pp. 1524-1545). Hershey, PA: IGI Global. https://doi.org/ doi:10.4018/978-1-5225-0803-8.ch072

Alapiki, H. E., & Amadi, L. A. (2018). Sustainable Food Consumption in the Neoliberal Order: Challenges and Policy Implications. In A. Obayelu (Ed.), *Food Systems Sustainability and Environmental Policies in Modern Economies* (pp. 90–123). Hershey, PA: IGI Global. doi:10.4018/978-1-5225-3631-4.ch005

Altaş, A. (2018). Geographical Information System Applications Utilized in Museums in Turkey Within the Scope of the Cultural Heritage Tourism: A Case Study of Mobile Application of Müze Asist. In S. Chaudhuri & N. Ray (Eds.), *GIS Applications in the Tourism and Hospitality Industry* (pp. 42–60). Hershey, PA: IGI Global. doi:10.4018/978-1-5225-5088-4.ch002

Andreea, I. R. (2018). Beyond Macroeconomics of Food and Nutrition Security. *International Journal of Sustainable Economies Management*, *7*(1), 13–22. doi:10.4018/IJSEM.2018010102

Anwar, J. (2017). Reproductive and Mental Health during Natural Disaster: Implications and Issues for Women in Developing Nations – A Case Example. In I. Management Association (Ed.), Gaming and Technology Addiction: Breakthroughs in Research and Practice (pp. 446-472). Hershey, PA: IGI Global. https://doi.org/ doi:10.4018/978-1-5225-0778-9.ch021

Awadh, H., Aksissou, M., Benhardouze, W., Darasi, F., & Snaiki, J. (2018). Socioeconomic Status of Artisanal Fishers in the West Part of Moroccan Mediterranean. *International Journal of Social Ecology and Sustainable Development*, *9*(1), 40–52. doi:10.4018/IJSESD.2018010104

Aye, G. C., & Haruna, R. F. (2018). Effect of Climate Change on Crop Productivity and Prices in Benue State, Nigeria: Implications for Food Security. In V. Erokhin (Ed.), *Establishing Food Security and Alternatives to International Trade in Emerging Economies* (pp. 244–268). Hershey, PA: IGI Global. doi:10.4018/978-1-5225-2733-6.ch012

Azizan, S. A., & Suki, N. M. (2017). Consumers' Intentions to Purchase Organic Food Products. In T. Esakki (Ed.), *Green Marketing and Environmental Responsibility in Modern Corporations* (pp. 68–82). Hershey, PA: IGI Global. doi:10.4018/978-1-5225-2331-4.ch005

Barakabitze, A. A., Fue, K. G., Kitindi, E. J., & Sanga, C. A. (2017). Developing a Framework for Next Generation Integrated Agro Food-Advisory Systems in Developing Countries. In I. Management Association (Ed.), Agri-Food Supply Chain Management: Breakthroughs in Research and Practice (pp. 47-67). Hershey, PA: IGI Global. doi:10.4018/978-1-5225-1629-3.ch004

Beachcroft-Shaw, H., & Ellis, D. (2018). Using Successful Cases to Promote Environmental Sustainability: A Social Marketing Approach. In I. Management Association (Ed.), Sustainable Development: Concepts, Methodologies, Tools, and Applications (pp. 936-953). Hershey, PA: IGI Global. doi:10.4018/978-1-5225-3817-2.ch042

Behnassi, M., Kahime, K., Boussaa, S., Boumezzough, A., & Messouli, M. (2017). Infectious Diseases and Climate Vulnerability in Morocco: Governance and Adaptation Options. In I. Management Association (Ed.), Public Health and Welfare: Concepts, Methodologies, Tools, and Applications (pp. 91-109). Hershey, PA: IGI Global. https://doi.org/ doi:10.4018/978-1-5225-1674-3.ch005

Bekele, F., & Bekele, I. (2017). Social and Environmental Impacts on Agricultural Development. In W. Ganpat, R. Dyer, & W. Isaac (Eds.), *Agricultural Development and Food Security in Developing Nations* (pp. 21–56). Hershey, PA: IGI Global. doi:10.4018/978-1-5225-0942-4.ch002

Benaouda, A., & García-Peñalvo, F. J. (2018). Towards an Intelligent System for the Territorial Planning: Agricultural Case. In F. García-Peñalvo (Ed.), *Global Implications of Emerging Technology Trends* (pp. 158–178). Hershey, PA: IGI Global. doi:10.4018/978-1-5225-4944-4.ch010

Bhaskar, A., Rao, G. B., & Vencatesan, J. (2017). Characterization and Management Concerns of Water Resources around Pallikaranai Marsh, South Chennai. In P. Rao & Y. Patil (Eds.), *Reconsidering the Impact of Climate Change on Global Water Supply, Use, and Management* (pp. 102–121). Hershey, PA: IGI Global. doi:10.4018/978-1-5225-1046-8.ch007

Bhyan, P., Shrivastava, B., & Kumar, N. (2022). Requisite Sustainable Development Contemplating Buildings: Economic and Environmental Sustainability. In A. Hussain, K. Tiwari, & A. Gupta (Eds.), *Addressing Environmental Challenges Through Spatial Planning* (pp. 269–288). IGI Global. https://doi.org/10.4018/978-1-7998-8331-9.ch014

Bogataj, D., & Drobne, D. (2017). Control of Perishable Goods in Cold Logistic Chains by Bionanosensors. In S. Joo (Ed.), *Applying Nanotechnology for Environmental Sustainability* (pp. 376–402). Hershey, PA: IGI Global. doi:10.4018/978-1-5225-0585-3.ch016

Bogataj, D., & Drobne, D. (2017). Control of Perishable Goods in Cold Logistic Chains by Bionanosensors. In I. Management Association (Ed.), Materials Science and Engineering: Concepts, Methodologies, Tools, and Applications (pp. 471-497). Hershey, PA: IGI Global. https://doi.org/ doi:10.4018/978-1-5225-1798-6.ch019

Bogueva, D., & Marinova, D. (2018). What Is More Important: Perception of Masculinity or Personal Health and the Environment? In D. Bogueva, D. Marinova, & T. Raphaely (Eds.), *Handbook of Research on Social Marketing and Its Influence on Animal Origin Food Product Consumption* (pp. 148–162). Hershey, PA: IGI Global. doi:10.4018/978-1-5225-4757-0.ch010

Bouzid, M. (2017). Waterborne Diseases and Climate Change: Impact and Implications. In M. Bouzid (Ed.), *Examining the Role of Environmental Change on Emerging Infectious Diseases and Pandemics* (pp. 89–108). Hershey, PA: IGI Global. doi:10.4018/978-1-5225-0553-2.ch004

Bowles, D. C. (2017). Climate Change-Associated Conflict and Infectious Disease. In M. Bouzid (Ed.), *Examining the Role of Environmental Change on Emerging Infectious Diseases and Pandemics* (pp. 68–88). Hershey, PA: IGI Global. doi:10.4018/978-1-5225-0553-2.ch003

Buck, J. J., & Lowry, R. K. (2017). Oceanographic Data Management: Quills and Free Text to the Digital Age and "Big Data". In P. Diviacco, A. Leadbetter, & H. Glaves (Eds.), *Oceanographic and Marine Cross-Domain Data Management for Sustainable Development* (pp. 1–22). Hershey, PA: IGI Global. doi:10.4018/978-1-5225-0700-0.ch001

Buse, C. G. (2017). Are Climate Change Adaptation Policies a Game Changer?: A Case Study of Perspectives from Public Health Officials in Ontario, Canada. In M. Bouzid (Ed.), *Examining the Role of Environmental Change on Emerging Infectious Diseases and Pandemics* (pp. 230–257). Hershey, PA: IGI Global. doi:10.4018/978-1-5225-0553-2.ch010

Calderon, F. A., Giolo, E. G., Frau, C. D., Rengel, M. G., Rodriguez, H., Tornello, M., ... Gallucci, R. (2018). Seismic Microzonation and Site Effects Detection Through Microtremors Measures: A Review. In N. Ceryan (Ed.), *Handbook of Research on Trends and Digital Advances in Engineering Geology* (pp. 326–349). Hershey, PA: IGI Global. doi:10.4018/978-1-5225-2709-1.ch009

Carfì, D., Donato, A., & Panuccio, D. (2018). A Game Theory Coopetitive Perspective for Sustainability of Global Feeding: Agreements Among Vegan and Non-Vegan Food Firms. In I. Management Association (Ed.), Game Theory: Breakthroughs in Research and Practice (pp. 71-104). Hershey, PA: IGI Global. https://doi.org/doi:10.4018/978-1-5225-2594-3.ch004

Castagnolo, V. (2018). Analyzing, Classifying, Safeguarding: Drawing for the Borgo Murattiano Neighbourhood of Bari. In G. Carlone, N. Martinelli, & F. Rotondo (Eds.), *Designing Grid Cities for Optimized Urban Development and Planning* (pp. 93–108). Hershey, PA: IGI Global. doi:10.4018/978-1-5225-3613-0.ch006

Chekima, B. (2018). The Dilemma of Purchase Intention: A Conceptual Framework for Understanding Actual Consumption of Organic Food. *International Journal of Sustainable Economies Management, 7*(2), 1–13. doi:10.4018/IJSEM.2018040101

Chen, Y. (2017). Sustainable Supply Chains and International Soft Landings: A Case of Wetland Entrepreneurship. In B. Christiansen & F. Kasarcı (Eds.), *Corporate Espionage, Geopolitics, and Diplomacy Issues in International Business* (pp. 232–247). Hershey, PA: IGI Global. doi:10.4018/978-1-5225-1031-4.ch013

Çıtak, L., Akel, V., & Ersoy, E. (2018). Investors' Reactions to the Announcement of New Constituents of BIST Sustainability Index: An Analysis by Event Study and Mean-Median Tests. In M. Risso & S. Testarmata (Eds.), *Value Sharing for Sustainable and Inclusive Development* (pp. 270–289). Hershey, PA: IGI Global. doi:10.4018/978-1-5225-3147-0.ch012

D'Aleo, V., D'Aleo, F., & Bonanno, R. (2018). New Food Industries Toward a New Level of Sustainable Supply: Success Stories, Business Models, and Strategies. In V. Erokhin (Ed.), *Establishing Food Security and Alternatives to International Trade in Emerging Economies* (pp. 74–97). Hershey, PA: IGI Global. doi:10.4018/978-1-5225-2733-6.ch004

Dagevos, H., & Reinders, M. J. (2018). Flexitarianism and Social Marketing: Reflections on Eating Meat in Moderation. In D. Bogueva, D. Marinova, & T. Raphaely (Eds.), *Handbook of Research on Social Marketing and Its Influence on Animal Origin Food Product Consumption* (pp. 105–120). Hershey, PA: IGI Global. doi:10.4018/978-1-5225-4757-0.ch007

Danisman, G. O. (2022). What Drives Eco-Design Innovations in European SMEs? In U. Akkucuk (Ed.), *Disruptive Technologies and Eco-Innovation for Sustainable Development* (pp. 191–206). IGI Global. https://doi.org/10.4018/978-1-7998-8900-7.ch011

Deenapanray, P. N., & Ramma, I. (2017). Adaptations to Climate Change and Climate Variability in the Agriculture Sector in Mauritius: Lessons from a Technical Needs Assessment. In I. Management Association (Ed.), Natural Resources Management: Concepts, Methodologies, Tools, and Applications (pp. 655-680). Hershey, PA: IGI Global. https://doi.org/ doi:10.4018/978-1-5225-0803-8.ch030

Deenapanray, P. N., & Ramma, I. (2017). Adaptations to Climate Change and Climate Variability in the Agriculture Sector in Mauritius: Lessons from a Technical Needs Assessment. In I. Management Association (Ed.), Natural Resources Management: Concepts, Methodologies, Tools, and Applications (pp. 655-680). Hershey, PA: IGI Global. https://doi.org/ doi:10.4018/978-1-5225-0803-8.ch030

Deshpande, S., Basu, S. K., Li, X., & Chen, X. (2017). Smart, Innovative and Intelligent Technologies Used in Drug Designing. In I. Management Association (Ed.), Pharmaceutical Sciences: Breakthroughs in Research and Practice (pp. 1175-1191). Hershey, PA: IGI Global. https://doi.org/ doi:10.4018/978-1-5225-1762-7.ch045

Dlamini, P. N. (2017). Use of Information and Communication Technologies Tools to Capture, Store, and Disseminate Indigenous Knowledge: A Literature Review. In P. Ngulube (Ed.), *Handbook of Research on Theoretical Perspectives on Indigenous Knowledge Systems in Developing Countries* (pp. 225–247). Hershey, PA: IGI Global. doi:10.4018/978-1-5225-0833-5.ch010

Dolejsova, M., & Kera, D. (2017). The Fermentation GutHub Project and the Internet of Microbes. In S. Konomi & G. Roussos (Eds.), *Enriching Urban Spaces with Ambient Computing, the Internet of Things, and Smart City Design* (pp. 25–46). Hershey, PA: IGI Global. doi:10.4018/978-1-5225-0827-4.ch002

Dolunay, O. (2018). A Paradigm Shift: Empowering Farmers to Eliminate the Waste in the Form of Fresh Water and Energy Through the Implementation of 4R+T. In I. Management Association (Ed.), Sustainable Development: Concepts, Methodologies, Tools, and Applications (pp. 882-892). Hershey, PA: IGI Global. https://doi.org/doi:10.4018/978-1-5225-3817-2.ch039

Dube, P., Heijman, W. J., Ihle, R., & Ochieng, J. (2018). The Potential of Traditional Leafy Vegetables for Improving Food Security in Africa. In V. Erokhin (Ed.), *Establishing Food Security and Alternatives to International Trade in Emerging Economies* (pp. 220–243). Hershey, PA: IGI Global. doi:10.4018/978-1-5225-2733-6.ch011

Duruji, M. M., & Urenma, D. F. (2017). The Environmentalism and Politics of Climate Change: A Study of the Process of Global Convergence through UNFCCC Conferences. In I. Management Association (Ed.), *Natural Resources Management: Concepts, Methodologies, Tools, and Applications* (pp. 77-108). Hershey, PA: IGI Global. 10.4018/978-1-5225-0803-8.ch004

Dutta, U. (2017). Agro-Geoinformatics, Potato Cultivation, and Climate Change. In S. Londhe (Ed.), *Sustainable Potato Production and the Impact of Climate Change* (pp. 247–271). Hershey, PA: IGI Global. doi:10.4018/978-1-5225-1715-3.ch012

Edirisinghe, R., Stranieri, A., & Wickramasinghe, N. (2017). A Taxonomy for mHealth. In N. Wickramasinghe (Ed.), *Handbook of Research on Healthcare Administration and Management* (pp. 596–615). Hershey, PA: IGI Global. doi:10.4018/978-1-5225-0920-2.ch036

Ekpeni, N. M., & Ayeni, A. O. (2018). Global Natural Hazard and Disaster Vulnerability Management. In A. Eneanya (Ed.), *Handbook of Research on Environmental Policies for Emergency Management and Public Safety* (pp. 83–104). Hershey, PA: IGI Global. doi:10.4018/978-1-5225-3194-4.ch005

Ene, C., Voica, M. C., & Panait, M. (2017). Green Investments and Food Security: Opportunities and Future Directions in the Context of Sustainable Development. In M. Mieila (Ed.), *Measuring Sustainable Development and Green Investments in Contemporary Economies* (pp. 163–200). Hershey, PA: IGI Global. doi:10.4018/978-1-5225-2081-8.ch007

Escamilla, I., Ruíz, M. T., Ibarra, M. M., Soto, V. L., Quintero, R., & Guzmán, G. (2018). Geocoding Tweets Based on Semantic Web and Ontologies. In M. Lytras, N. Aljohani, E. Damiani, & K. Chui (Eds.), *Innovations, Developments, and Applications of Semantic Web and Information Systems* (pp. 372–392). Hershey, PA: IGI Global. doi:10.4018/978-1-5225-5042-6.ch014

Escribano, A. J. (2018). Marketing Strategies for Trendy Animal Products: Sustainability as a Core. In F. Quoquab, R. Thurasamy, & J. Mohammad (Eds.), *Driving Green Consumerism Through Strategic Sustainability Marketing* (pp. 169–203). Hershey, PA: IGI Global. doi:10.4018/978-1-5225-2912-5.ch010

Eudoxie, G., & Roopnarine, R. (2017). Climate Change Adaptation and Disaster Risk Management in the Caribbean. In W. Ganpat & W. Isaac (Eds.), *Environmental Sustainability and Climate Change Adaptation Strategies* (pp. 97–125). Hershey, PA: IGI Global. doi:10.4018/978-1-5225-1607-1.ch004

Farmer, L. S. (2017). Data Analytics for Strategic Management: Getting the Right Data. In V. Wang (Ed.), *Encyclopedia of Strategic Leadership and Management* (pp. 810–822). Hershey, PA: IGI Global. doi:10.4018/978-1-5225-1049-9.ch056

Fattal, L. R. (2017). Catastrophe: An Uncanny Catalyst for Creativity. In R. Shin (Ed.), *Convergence of Contemporary Art, Visual Culture, and Global Civic Engagement* (pp. 244–262). Hershey, PA: IGI Global. doi:10.4018/978-1-5225-1665-1.ch014

Forti, I. (2017). A Cross Reading of Landscape through Digital Landscape Models: The Case of Southern Garda. In A. Ippolito (Ed.), *Handbook of Research on Emerging Technologies for Architectural and Archaeological Heritage* (pp. 532–561). Hershey, PA: IGI Global. doi:10.4018/978-1-5225-0675-1.ch018

Gharbi, A., De Runz, C., & Akdag, H. (2017). Urban Development Modelling: A Survey. In S. Faiz & K. Mahmoudi (Eds.), *Handbook of Research on Geographic Information Systems Applications and Advancements* (pp. 96–124). Hershey, PA: IGI Global. doi:10.4018/978-1-5225-0937-0.ch004

Ghosh, I., & Ghoshal, I. (2018). Implications of Trade Liberalization for Food Security Under the ASEAN-India Strategic Partnership: A Gravity Model Approach. In V. Erokhin (Ed.), *Establishing Food Security and Alternatives to International Trade in Emerging Economies* (pp. 98–118). Hershey, PA: IGI Global. doi:10.4018/978-1-5225-2733-6.ch005

Glaves, H. M. (2017). Developing a Common Global Framework for Marine Data Management. In P. Diviacco, A. Leadbetter, & H. Glaves (Eds.), *Oceanographic and Marine Cross-Domain Data Management for Sustainable Development* (pp. 47–68). Hershey, PA: IGI Global. doi:10.4018/978-1-5225-0700-0.ch003

Godulla, A., & Wolf, C. (2018). Future of Food: Transmedia Strategies of National Geographic. In R. Gambarato & G. Alzamora (Eds.), *Exploring Transmedia Journalism in the Digital Age* (pp. 162–182). Hershey, PA: IGI Global. doi:10.4018/978-1-5225-3781-6.ch010

Gomes, P. P. (2018). Food and Environment: A Review on the Sustainability of Six Different Dietary Patterns. In A. Obayelu (Ed.), *Food Systems Sustainability and Environmental Policies in Modern Economies* (pp. 15–31). Hershey, PA: IGI Global. doi:10.4018/978-1-5225-3631-4.ch002

Gonzalez-Feliu, J. (2018). Sustainability Evaluation of Green Urban Logistics Systems: Literature Overview and Proposed Framework. In A. Paul, D. Bhattacharyya, & S. Anand (Eds.), *Green Initiatives for Business Sustainability and Value Creation* (pp. 103–134). Hershey, PA: IGI Global. doi:10.4018/978-1-5225-2662-9.ch005

Goodland, R. (2017). A Fresh Look at Livestock Greenhouse Gas Emissions and Mitigation. In I. Management Association (Ed.), Natural Resources Management: Concepts, Methodologies, Tools, and Applications (pp. 124-139). Hershey, PA: IGI Global. doi:10.4018/978-1-5225-0803-8.ch006

Goundar, S., & Appana, S. (2018). Mainstreaming Development Policies for Climate Change in Fiji: A Policy Gap Analysis and the Role of ICTs. In I. Management Association (Ed.), Sustainable Development: Concepts, Methodologies, Tools, and Applications (pp. 402-432). Hershey, PA: IGI Global. doi:10.4018/978-1-5225-3817-2.ch020

Granell-Canut, C., & Aguilar-Moreno, E. (2018). Geospatial Influence in Science Mapping. In M. Khosrow-Pour, D.B.A. (Ed.), Encyclopedia of Information Science and Technology, Fourth Edition (pp. 3473-3483). Hershey, PA: IGI Global. doi:10.4018/978-1-5225-2255-3.ch302

Grigelis, A., Blažauskas, N., Gelumbauskaitė, L. Ž., Gulbinskas, S., Suzdalev, S., & Ferrarin, C. (2017). Marine Environment Data Management Related to the Human Activity in the South-Eastern Baltic Sea (The Lithuanian Sector). In P. Diviacco, A. Leadbetter, & H. Glaves (Eds.), *Oceanographic and Marine Cross-Domain Data Management for Sustainable Development* (pp. 282–302). Hershey, PA: IGI Global. doi:10.4018/978-1-5225-0700-0.ch012

Guma, I. P., Rwashana, A. S., & Oyo, B. (2018). Food Security Policy Analysis Using System Dynamics: The Case of Uganda. *International Journal of Information Technologies and Systems Approach*, *11*(1), 72–90. doi:10.4018/IJITSA.2018010104

Guma, I. P., Rwashana, A. S., & Oyo, B. (2018). Food Security Indicators for Subsistence Farmers Sustainability: A System Dynamics Approach. *International Journal of System Dynamics Applications*, *7*(1), 45–64. doi:10.4018/IJSDA.2018010103

Gupta, P., & Goyal, S. (2017). Wildlife Habitat Evaluation. In A. Santra & S. Mitra (Eds.), *Remote Sensing Techniques and GIS Applications in Earth and Environmental Studies* (pp. 258–264). Hershey, PA: IGI Global. doi:10.4018/978-1-5225-1814-3.ch013

Hanson, T., & Hildebrand, E. (2018). GPS Travel Diaries in Rural Transportation Research: A Focus on Older Drivers. In I. Management Association (Ed.), Intelligent Transportation and Planning: Breakthroughs in Research and Practice (pp. 609-625). Hershey, PA: IGI Global. doi:10.4018/978-1-5225-5210-9.ch027

Hartman, M. B. (2017). Research-Based Climate Change Public Education Programs. In I. Management Association (Ed.), Natural Resources Management: Concepts, Methodologies, Tools, and Applications (pp. 992-1003). Hershey, PA: IGI Global. doi:10.4018/978-1-5225-0803-8.ch046

Hashim, N. (2017). Zanzibari Seaweed: Global Climate Change and the Promise of Adaptation. In I. Management Association (Ed.), Natural Resources Management: Concepts, Methodologies, Tools, and Applications (pp. 365-391). Hershey, PA: IGI Global. https://doi.org/ doi:10.4018/978-1-5225-0803-8.ch019

Herrera, J. E., Argüello, L. V., Gonzalez-Feliu, J., & Jaimes, W. A. (2017). Decision Support System Design Requirements, Information Management, and Urban Logistics Efficiency: Case Study of Bogotá, Colombia. In G. Jamil, A. Soares, & C. Pessoa (Eds.), *Handbook of Research on Information Management for Effective Logistics and Supply Chains* (pp. 223–238). Hershey, PA: IGI Global. doi:10.4018/978-1-5225-0973-8.ch012

Huizinga, T., Ayanso, A., Smoor, M., & Wronski, T. (2017). Exploring Insurance and Natural Disaster Tweets Using Text Analytics. *International Journal of Business Analytics*, *4*(1), 1–17. doi:10.4018/IJBAN.2017010101

Hung, K., Kalantari, M., & Rajabifard, A. (2017). An Integrated Method for Assessing the Text Content Quality of Volunteered Geographic Information in Disaster Management. *International Journal of Information Systems for Crisis Response and Management*, *9*(2), 1–17. doi:10.4018/IJISCRAM.2017040101

Husnain, A., & Avdic, A. (2018). Identifying the Contemporary Status of E-Service Sustainability Research. In I. Management Association (Ed.), Sustainable Development: Concepts, Methodologies, Tools, and Applications (pp. 467-485). Hershey, PA: IGI Global. https://doi.org/ doi:10.4018/978-1-5225-3817-2.ch022

Iarossi, M. P., & Ferro, L. (2017). "The Past is Never Dead. It's Not Even Past": Virtual Archaeological Promenade. In A. Ippolito & M. Cigola (Eds.), *Handbook of Research on Emerging Technologies for Digital Preservation and Information Modeling* (pp. 228–255). Hershey, PA: IGI Global. doi:10.4018/978-1-5225-0680-5.ch010

Ignjatijević, S., & Cvijanović, D. (2018). Analysis of Serbian Honey Production and Exports. In *Exploring the Global Competitiveness of Agri-Food Sectors and Serbia's Dominant Presence: Emerging Research and Opportunities* (pp. 109–139). Hershey, PA: IGI Global. doi:10.4018/978-1-5225-2762-6.ch005

Jana, S. K., & Karmakar, A. K. (2017). Globalization, Governance, and Food Security: The Case of BRICS. In I. Management Association (Ed.), Natural Resources Management: Concepts, Methodologies, Tools, and Applications (pp. 692-712). Hershey, PA: IGI Global. https://doi.org/ doi:10.4018/978-1-5225-0803-8.ch032

Jana, S. K., & Karmakar, A. K. (2017). Food Security in Asia: Is There Convergence? In I. Management Association (Ed.), Natural Resources Management: Concepts, Methodologies, Tools, and Applications (pp. 109-123). Hershey, PA: IGI Global. https://doi.org/ doi:10.4018/978-1-5225-0803-8.ch005

John, J., & Kumar, S. (2018). A Locational Decision Making Framework for Shipbreaking Under Multiple Criteria. In I. Management Association (Ed.), Operations and Service Management: Concepts, Methodologies, Tools, and Applications (pp. 504-527). Hershey, PA: IGI Global. doi:10.4018/978-1-5225-3909-4.ch024

John, J., & Srivastava, R. K. (2018). Decision Insights for Shipbreaking using Environmental Impact Assessment: Review and Perspectives. *International Journal of Strategic Decision Sciences*, *9*(1), 45–62. doi:10.4018/IJSDS.2018010104

Joshi, Y., & Rahman, Z. (2018). Determinants of Sustainable Consumption Behaviour: Review and Conceptual Framework. In A. Paul, D. Bhattacharyya, & S. Anand (Eds.), *Green Initiatives for Business Sustainability and Value Creation* (pp. 239–262). Hershey, PA: IGI Global. doi:10.4018/978-1-5225-2662-9.ch011

Juma, D. W., Reuben, M., Wang, H., & Li, F. (2018). Adaptive Coevolution: Realigning the Water Governance Regime to the Changing Climate. In I. Management Association (Ed.), Hydrology and Water Resource Management: Breakthroughs in Research and Practice (pp. 346-357). Hershey, PA: IGI Global. https://doi.org/ doi:10.4018/978-1-5225-3427-3.ch014

K., S., & Tripathy, B. K. (2018). Neighborhood Rough-Sets-Based Spatial Data Analytics. In M. Khosrow-Pour, D.B.A. (Ed.), *Encyclopedia of Information Science and Technology, Fourth Edition* (pp. 1835-1844). Hershey, PA: IGI Global. https:// doi.org/ doi:10.4018/978-1-5225-2255-3.ch160

Kabir, F. (2018). Towards a More Gender-Inclusive Climate Change Policy. In N. Mahtab, T. Haque, I. Khan, M. Islam, & I. Wahid (Eds.), *Handbook of Research on Women's Issues and Rights in the Developing World* (pp. 354–369). Hershey, PA: IGI Global. doi:10.4018/978-1-5225-3018-3.ch022

Kabir, F. (2018). Towards a More Gender-Inclusive Climate Change Policy. In I. Management Association (Ed.), Climate Change and Environmental Concerns: Breakthroughs in Research and Practice (pp. 525-540). Hershey, PA: IGI Global. https://doi.org/ doi:10.4018/978-1-5225-5487-5.ch027

Kanyamuka, J. S., Jumbe, C. B., & Ricker-Gilbert, J. (2018). Making Agricultural Input Subsidies More Effective and Profitable in Africa: The Role of Complementary Interventions. In A. Obayelu (Ed.), *Food Systems Sustainability and Environmental Policies in Modern Economies* (pp. 172–187). Hershey, PA: IGI Global. doi:10.4018/978-1-5225-3631-4.ch008

Karimi, H., & Gholamrezafahimi, F. (2017). Study of Integrated Coastal Zone Management and Its Environmental Effects: A Case of Iran. In R. Singh, A. Singh, & V. Srivastava (Eds.), *Environmental Issues Surrounding Human Overpopulation* (pp. 64–88). Hershey, PA: IGI Global. doi:10.4018/978-1-5225-1683-5.ch004

Kaya, I. R., Hutabarat, J., & Bambang, A. N. (2018). "Sasi": A New Path to Sustain Seaweed Farming From Up-Stream to Down-Stream in Kotania Bay, Molucass. *International Journal of Social Ecology and Sustainable Development*, *9*(2), 28–36. doi:10.4018/IJSESD.2018040103

Khader, V. (2018). Technologies for Food, Health, Livelihood, and Nutrition Security. In I. Management Association (Ed.), Food Science and Nutrition: Breakthroughs in Research and Practice (pp. 94-112). Hershey, PA: IGI Global. https://doi.org/doi:10.4018/978-1-5225-5207-9.ch005

Kocadağlı, A. Y. (2017). The Temporal and Spatial Development of Organic Agriculture in Turkey. In W. Ganpat, R. Dyer, & W. Isaac (Eds.), *Agricultural Development and Food Security in Developing Nations* (pp. 130–156). Hershey, PA: IGI Global. doi:10.4018/978-1-5225-0942-4.ch006

Koundouri, P., Giannouli, A., & Souliotis, I. (2017). An Integrated Approach for Sustainable Environmental and Socio-Economic Development Using Offshore Infrastructure. In I. Management Association (Ed.), Renewable and Alternative Energy: Concepts, Methodologies, Tools, and Applications (pp. 1581-1601). Hershey, PA: IGI Global. doi:10.4018/978-1-5225-1671-2.ch056

Kumar, A., & Dash, M. K. (2017). Sustainability and Future Generation Infrastructure on Digital Platform: A Study of Generation Y. In N. Ray (Ed.), *Business Infrastructure for Sustainability in Developing Economies* (pp. 124–142). Hershey, PA: IGI Global. doi:10.4018/978-1-5225-2041-2.ch007

Kumar, A., Mukherjee, A. B., & Krishna, A. P. (2017). Application of Conventional Data Mining Techniques and Web Mining to Aid Disaster Management. In A. Kumar (Ed.), *Web Usage Mining Techniques and Applications Across Industries* (pp. 138–167). Hershey, PA: IGI Global. doi:10.4018/978-1-5225-0613-3.ch006

Kumar, C. P. (2017). Impact of Climate Change on Groundwater Resources. In I. Management Association (Ed.), Natural Resources Management: Concepts, Methodologies, Tools, and Applications (pp. 1094-1120). Hershey, PA: IGI Global. doi:10.4018/978-1-5225-0803-8.ch052

Kumari, S., & Patil, Y. (2017). Achieving Climate Smart Agriculture with a Sustainable Use of Water: A Conceptual Framework for Sustaining the Use of Water for Agriculture in the Era of Climate Change. In P. Rao & Y. Patil (Eds.), *Reconsidering the Impact of Climate Change on Global Water Supply, Use, and Management* (pp. 122–143). Hershey, PA: IGI Global. doi:10.4018/978-1-5225-1046-8.ch008

Kumari, S., & Patil, Y. (2018). Achieving Climate Smart Agriculture With a Sustainable Use of Water: A Conceptual Framework for Sustaining the Use of Water for Agriculture in the Era of Climate Change. In I. Management Association (Ed.), Climate Change and Environmental Concerns: Breakthroughs in Research and Practice (pp. 111-133). Hershey, PA: IGI Global. doi:10.4018/978-1-5225-5487-5.ch006

Kursah, M. B. (2017). Least-Cost Pipeline using Geographic Information System: The Limit to Technicalities. *International Journal of Applied Geospatial Research, 8*(3), 1–15. doi:10.4018/ijagr.2017070101

Lahiri, S., Ghosh, D., & Bhakta, J. N. (2017). Role of Microbes in Eco-Remediation of Perturbed Aquatic Ecosystem. In J. Bhakta (Ed.), *Handbook of Research on Inventive Bioremediation Techniques* (pp. 70–107). Hershey, PA: IGI Global. doi:10.4018/978-1-5225-2325-3.ch004

Lallo, C. H., Smalling, S., Facey, A., & Hughes, M. (2018). The Impact of Climate Change on Small Ruminant Performance in Caribbean Communities. In I. Management Association (Ed.), Climate Change and Environmental Concerns: Breakthroughs in Research and Practice (pp. 193-218). Hershey, PA: IGI Global. doi:10.4018/978-1-5225-5487-5.ch010

Laurini, R. (2017). Nature of Geographic Knowledge Bases. In S. Faiz & K. Mahmoudi (Eds.), *Handbook of Research on Geographic Information Systems Applications and Advancements* (pp. 29–60). Hershey, PA: IGI Global. doi:10.4018/978-1-5225-0937-0.ch002

Lawrence, J., Simpson, L., & Piggott, A. (2017). Protected Agriculture: A Climate Change Adaptation for Food and Nutrition Security. In I. Management Association (Ed.), Natural Resources Management: Concepts, Methodologies, Tools, and Applications (pp. 140-158). Hershey, PA: IGI Global. doi:10.4018/978-1-5225-0803-8.ch007

Leadbetter, A., Cheatham, M., Shepherd, A., & Thomas, R. (2017). Linked Ocean Data 2.0. In P. Diviacco, A. Leadbetter, & H. Glaves (Eds.), *Oceanographic and Marine Cross-Domain Data Management for Sustainable Development* (pp. 69–99). Hershey, PA: IGI Global. doi:10.4018/978-1-5225-0700-0.ch004

Lucas, M. R., Rego, C., Vieira, C., & Vieira, I. (2017). Proximity and Cooperation for Innovative Regional Development: The Case of the Science and Technology Park of Alentejo. In L. Carvalho (Ed.), *Handbook of Research on Entrepreneurial Development and Innovation Within Smart Cities* (pp. 199–228). Hershey, PA: IGI Global. doi:10.4018/978-1-5225-1978-2.ch010

Ma, X., Beaulieu, S. E., Fu, L., Fox, P., Di Stefano, M., & West, P. (2017). Documenting Provenance for Reproducible Marine Ecosystem Assessment in Open Science. In P. Diviacco, A. Leadbetter, & H. Glaves (Eds.), *Oceanographic and Marine Cross-Domain Data Management for Sustainable Development* (pp. 100–126). Hershey, PA: IGI Global. doi:10.4018/978-1-5225-0700-0.ch005

Mabe, L. K., & Oladele, O. I. (2017). Application of Information Communication Technologies for Agricultural Development through Extension Services: A Review. In T. Tossy (Ed.), *Information Technology Integration for Socio-Economic Development* (pp. 52–101). Hershey, PA: IGI Global. doi:10.4018/978-1-5225-0539-6.ch003

Malomo, B. I. (2018). A Review of Psychological Resilience as a Response to Natural Hazards in Nigeria. In A. Eneanya (Ed.), *Handbook of Research on Environmental Policies for Emergency Management and Public Safety* (pp. 147–165). Hershey, PA: IGI Global. doi:10.4018/978-1-5225-3194-4.ch008

Manchiraju, S. (2018). Predicting Behavioral Intentions Toward Sustainable Fashion Consumption: A Comparison of Attitude-Behavior and Value-Behavior Consistency Models. In I. Management Association (Ed.), Fashion and Textiles: Breakthroughs in Research and Practice (pp. 1-21). Hershey, PA: IGI Global. doi:10.4018/978-1-5225-3432-7.ch001

Manzella, G. M., Bartolini, R., Bustaffa, F., D'Angelo, P., De Mattei, M., Frontini, F., ... Spada, A. (2017). Semantic Search Engine for Data Management and Sustainable Development: Marine Planning Service Platform. In P. Diviacco, A. Leadbetter, & H. Glaves (Eds.), *Oceanographic and Marine Cross-Domain Data Management for Sustainable Development* (pp. 127–154). Hershey, PA: IGI Global. doi:10.4018/978-1-5225-0700-0.ch006

Mbonigaba, J. (2018). Comparing the Effects of Unsustainable Production and Consumption of Food on Health and Policy Across Developed and Less Developed Countries. In A. Obayelu (Ed.), *Food Systems Sustainability and Environmental Policies in Modern Economies* (pp. 124–158). Hershey, PA: IGI Global. doi:10.4018/978-1-5225-3631-4.ch006

McKeown, A. E. (2017). Nurses, Healthcare, and Environmental Pollution and Solutions: Breaking the Cycle of Harm. In I. Management Association (Ed.), Natural Resources Management: Concepts, Methodologies, Tools, and Applications (pp. 392-415). Hershey, PA: IGI Global. https://doi.org/ doi:10.4018/978-1-5225-0803-8.ch020

Mili, B., Barua, A., & Katyaini, S. (2017). Climate Change and Adaptation through the Lens of Capability Approach: A Case Study from Darjeeling, Eastern Himalaya. In I. Management Association (Ed.), Natural Resources Management: Concepts, Methodologies, Tools, and Applications (pp. 1351-1365). Hershey, PA: IGI Global. https://doi.org/ doi:10.4018/978-1-5225-0803-8.ch064

Mir, S. A., Shah, M. A., Mir, M. M., & Iqbal, U. (2017). New Horizons of Nanotechnology in Agriculture and Food Processing Industry. In B. Nayak, A. Nanda, & M. Bhat (Eds.), *Integrating Biologically-Inspired Nanotechnology into Medical Practice* (pp. 230–258). Hershey, PA: IGI Global. doi:10.4018/978-1-5225-0610-2.ch009

Moallem, M., Sterrett, W. L., Gordon, C. R., Sukhera, S. M., Mahmood, A., & Bashir, A. (2021). An Investigation of the Effects of Integrating Computing and Project- or Problem-Based Learning in the Context of Environmental Sciences: A Case of Pakistani STEM Teachers. In S. Schroth & J. Daniels (Eds.), *Building STEM Skills Through Environmental Education* (pp. 49–89). IGI Global. https://doi.org/10.4018/978-1-7998-2711-5.ch003

Mujere, N., & Moyce, W. (2017). Climate Change Impacts on Surface Water Quality. In W. Ganpat & W. Isaac (Eds.), *Environmental Sustainability and Climate Change Adaptation Strategies* (pp. 322–340). Hershey, PA: IGI Global. doi:10.4018/978-1-5225-1607-1.ch012

Mukherjee, A. B., Krishna, A. P., & Patel, N. (2018). Geospatial Technology for Urban Sciences. In *Geospatial Technologies in Urban System Development: Emerging Research and Opportunities* (pp. 99–120). Hershey, PA: IGI Global. doi:10.4018/978-1-5225-3683-3.ch005

Nagarajan, S. K., & Sangaiah, A. K. (2017). Vegetation Index: Ideas, Methods, Influences, and Trends. In N. Kumar, A. Sangaiah, M. Arun, & S. Anand (Eds.), *Advanced Image Processing Techniques and Applications* (pp. 347–386). Hershey, PA: IGI Global. doi:10.4018/978-1-5225-2053-5.ch016

Naraine, L., & Meehan, K. (2017). Strengthening Food Security with Sustainable Practices by Smallholder Farmers in Lesser Developed Economies. In W. Ganpat, R. Dyer, & W. Isaac (Eds.), *Agricultural Development and Food Security in Developing Nations* (pp. 57–81). Hershey, PA: IGI Global. doi:10.4018/978-1-5225-0942-4.ch003

Naraine, L., & Meehan, K. (2017). Strengthening Food Security with Sustainable Practices by Smallholder Farmers in Lesser Developed Economies. In W. Ganpat, R. Dyer, & W. Isaac (Eds.), *Agricultural Development and Food Security in Developing Nations* (pp. 57–81). Hershey, PA: IGI Global. doi:10.4018/978-1-5225-0942-4.ch003

Naraine, L., & Meehan, K. (2017). Strengthening Food Security with Sustainable Practices by Smallholder Farmers in Lesser Developed Economies. In W. Ganpat, R. Dyer, & W. Isaac (Eds.), *Agricultural Development and Food Security in Developing Nations* (pp. 57–81). Hershey, PA: IGI Global. doi:10.4018/978-1-5225-0942-4.ch003

Nikolaou, K., Tsakiridou, E., Anastasiadis, F., & Mattas, K. (2017). Exploring Alternative Distribution Channels of Agricultural Products. *International Journal of Food and Beverage Manufacturing and Business Models*, *2*(2), 36–66. doi:10.4018/IJFBMBM.2017070103

Nishat, K. J., & Rahman, M. S. (2018). Disaster, Vulnerability, and Violence Against Women: Global Findings and a Research Agenda for Bangladesh. In N. Mahtab, T. Haque, I. Khan, M. Islam, & I. Wahid (Eds.), *Handbook of Research on Women's Issues and Rights in the Developing World* (pp. 235–250). Hershey, PA: IGI Global. doi:10.4018/978-1-5225-3018-3.ch014

O'Hara, S., Jones, D., & Trobman, H. B. (2018). Building an Urban Food System Through UDC Food Hubs. In A. Burtin, J. Fleming, & P. Hampton-Garland (Eds.), *Changing Urban Landscapes Through Public Higher Education* (pp. 116–143). Hershey, PA: IGI Global. doi:10.4018/978-1-5225-3454-9.ch006

Obayelu, A. E. (2018). Integrating Environment, Food Systems, and Sustainability in Feeding the Growing Population in Developing Countries. In A. Obayelu (Ed.), *Food Systems Sustainability and Environmental Policies in Modern Economies* (pp. 1–14). Hershey, PA: IGI Global. doi:10.4018/978-1-5225-3631-4.ch001

Othman, R., Nath, N., & Laswad, F. (2018). Environmental Reporting and Accounting: Sustainability Hybridisation. In G. Azevedo, J. da Silva Oliveira, R. Marques, & A. Ferreira (Eds.), *Handbook of Research on Modernization and Accountability in Public Sector Management* (pp. 130–158). Hershey, PA: IGI Global. doi:10.4018/978-1-5225-3731-1.ch007

Ouadi, A., & Zitouni, A. (2021). Phasor Measurement Improvement Using Digital Filter in a Smart Grid. In A. Recioui & H. Bentarzi (Eds.), *Optimizing and Measuring Smart Grid Operation and Control* (pp. 100–117). IGI Global. https://doi.org/10.4018/978-1-7998-4027-5.ch005

Padigala, B. S. (2018). Traditional Water Management System for Climate Change Adaptation in Mountain Ecosystems. In I. Management Association (Ed.), Climate Change and Environmental Concerns: Breakthroughs in Research and Practice (pp. 630-655). Hershey, PA: IGI Global. doi:10.4018/978-1-5225-5487-5.ch033

Panda, C. K. (2018). Mobile Phone Usage in Agricultural Extension in India: The Current and Future Perspective. In F. Mtenzi, G. Oreku, D. Lupiana, & J. Yonazi (Eds.), *Mobile Technologies and Socio-Economic Development in Emerging Nations* (pp. 1–21). Hershey, PA: IGI Global. doi:10.4018/978-1-5225-4029-8.ch001

Pandian, S. L., Yarrakula, K., & Chaudhury, P. (2018). GIS-Based Decision Support System for Village Level: A Case Study in Andhra Pradesh. In S. Chaudhuri & N. Ray (Eds.), *GIS Applications in the Tourism and Hospitality Industry* (pp. 275–295). Hershey, PA: IGI Global. doi:10.4018/978-1-5225-5088-4.ch012

Quaranta, G., & Salvia, R. (2017). Social-Based Product Innovation and Governance in The Milk Sector: The Case of Carciocacio and Innonatura. In T. Tarnanidis, M. Vlachopoulou, & J. Papathanasiou (Eds.), *Driving Agribusiness With Technology Innovations* (pp. 293–310). Hershey, PA: IGI Global. doi:10.4018/978-1-5225-2107-5.ch015

Rahman, M. K., Schmidlin, T. W., Munro-Stasiuk, M. J., & Curtis, A. (2017). Geospatial Analysis of Land Loss, Land Cover Change, and Landuse Patterns of Kutubdia Island, Bangladesh. *International Journal of Applied Geospatial Research*, *8*(2), 45–60. doi:10.4018/IJAGR.2017040104

Rajack-Talley, T. A. (2017). Agriculture, Trade Liberalization and Poverty in the ACP Countries. In W. Ganpat, R. Dyer, & W. Isaac (Eds.), *Agricultural Development and Food Security in Developing Nations* (pp. 1–20). Hershey, PA: IGI Global. doi:10.4018/978-1-5225-0942-4.ch001

Rajamanickam, S. (2018). Exploring Landscapes in Regional Convergence: Environment and Sustainable Development in South Asia. In I. Management Association (Ed.), Sustainable Development: Concepts, Methodologies, Tools, and Applications (pp. 1051-1087). Hershey, PA: IGI Global. https://doi.org/doi:10.4018/978-1-5225-3817-2.ch047

Rizvi, S. M., & Dearden, A. (2018). KHETI: ICT Solution for Agriculture Extension and Its Replication in Open and Distance Learning. In U. Pandey & V. Indrakanti (Eds.), *Open and Distance Learning Initiatives for Sustainable Development* (pp. 163–174). Hershey, PA: IGI Global. doi:10.4018/978-1-5225-2621-6.ch008

Roşu, L., & Macarov, L. I. (2017). Management of Drought and Floods in the Dobrogea Region. In I. Management Association (Ed.), Agri-Food Supply Chain Management: Breakthroughs in Research and Practice (pp. 372-403). Hershey, PA: IGI Global. https://doi.org/ doi:10.4018/978-1-5225-1629-3.ch016

Rouzbehani, K., & Rouzbehani, S. (2018). Mapping Women's World: GIS and the Case of Breast Cancer in the US. *International Journal of Public Health Management and Ethics*, *3*(1), 14–25. doi:10.4018/IJPHME.2018010102

Roy, D. (2018). Success Factors of Adoption of Mobile Applications in Rural India: Effect of Service Characteristics on Conceptual Model. In M. Khosrow-Pour, D.B.A. (Ed.), Green Computing Strategies for Competitive Advantage and Business Sustainability (pp. 211-238). Hershey, PA: IGI Global. https://doi.org/ doi:10.4018/978-1-5225-5017-4.ch010

Sajeva, M., Lemon, M., & Sahota, P. S. (2017). Governance for Food Security: A Framework for Social Learning and Scenario Building. *International Journal of Food and Beverage Manufacturing and Business Models*, *2*(2), 67–84. doi:10.4018/IJFBMBM.2017070104

Sambhanthan, A., & Potdar, V. (2017). A Study of the Parameters Impacting Sustainability in Information Technology Organizations. *International Journal of Knowledge-Based Organizations*, *7*(3), 27–39. doi:10.4018/IJKBO.2017070103

Sanga, C., Kalungwizi, V. J., & Msuya, C. P. (2017). Bridging Gender Gaps in Provision of Agricultural Extension Service Using ICT: Experiences from Sokoine University of Agriculture (SUA) Farmer Voice Radio (FVR) Project in Tanzania. In I. Management Association (Ed.), Discrimination and Diversity: Concepts, Methodologies, Tools, and Applications (pp. 682-697). Hershey, PA: IGI Global. https://doi.org/ doi:10.4018/978-1-5225-1933-1.ch031

Santra, A., & Mitra, D. (2017). Role of Remote Sensing in Potential Fishing Zone Forecast. In A. Santra & S. Mitra (Eds.), *Remote Sensing Techniques and GIS Applications in Earth and Environmental Studies* (pp. 243–257). Hershey, PA: IGI Global. doi:10.4018/978-1-5225-1814-3.ch012

Schaap, D. (2017). SeaDataNet: Towards a Pan-European Infrastructure for Marine and Ocean Data Management. In P. Diviacco, A. Leadbetter, & H. Glaves (Eds.), *Oceanographic and Marine Cross-Domain Data Management for Sustainable Development* (pp. 155–177). Hershey, PA: IGI Global. doi:10.4018/978-1-5225-0700-0.ch007

Seckin-Celik, T. (2017). Sustainability Reporting and Sustainability in the Turkish Business Context. In U. Akkucuk (Ed.), *Ethics and Sustainability in Global Supply Chain Management* (pp. 115–132). Hershey, PA: IGI Global. doi:10.4018/978-1-5225-2036-8.ch006

Segbefia, A. Y., Barnes, V. R., Akpalu, L. A., & Mensah, M. (2018). Environmental Location Assessment for Seaweed Cultivation in Ghana: A Spatial Multi-Criteria Approach. *International Journal of Applied Geospatial Research*, *9*(1), 51–64. doi:10.4018/IJAGR.2018010104

Sen, Y. (2018). How to Manage Sustainability: A Framework for Corporate Sustainability Tools. In I. Management Association (Ed.), Sustainable Development: Concepts, Methodologies, Tools, and Applications (pp. 568-589). Hershey, PA: IGI Global. doi:10.4018/978-1-5225-3817-2.ch026

Shamshiry, E., Abdulai, A. M., Mokhtar, M. B., & Komoo, I. (2015). Regional Landfill Site Selection with GIS and Analytical Hierarchy Process Techniques: A Case Study of Langkawi Island, Malaysia. In P. Thomas, M. Srihari, & S. Kaur (Eds.), *Handbook of Research on Cultural and Economic Impacts of the Information Society* (pp. 248–282). Hershey, PA: IGI Global. doi:10.4018/978-1-4666-8598-7.ch011

Sharma, Y. K., Mangla, S. K., Patil, P. P., & Uniyal, S. (2018). Analyzing Sustainable Food Supply Chain Management Challenges in India. In M. Ram & J. Davim (Eds.), *Soft Computing Techniques and Applications in Mechanical Engineering* (pp. 162–180). Hershey, PA: IGI Global. doi:10.4018/978-1-5225-3035-0.ch008

Silvestrelli, P. (2018). The Impact of Events: To Which Extent Are They Sustainable for Tourist Destinations? Some Evidences From Expo Milano 2015. In M. Risso & S. Testarmata (Eds.), *Value Sharing for Sustainable and Inclusive Development* (pp. 185–204). Hershey, PA: IGI Global. doi:10.4018/978-1-5225-3147-0.ch008

Silvius, G. (2018). Sustainability Evaluation of IT/IS Projects. In I. Management Association (Ed.), Sustainable Development: Concepts, Methodologies, Tools, and Applications (pp. 26-40). Hershey, PA: IGI Global. https://doi.org/ doi:10.4018/978-1-5225-3817-2.ch002

Singh, R., Srivastava, P., Singh, P., Upadhyay, S., & Raghubanshi, A. S. (2017). Human Overpopulation and Food Security: Challenges for the Agriculture Sustainability. In R. Singh, A. Singh, & V. Srivastava (Eds.), *Environmental Issues Surrounding Human Overpopulation* (pp. 12–39). Hershey, PA: IGI Global. doi:10.4018/978-1-5225-1683-5.ch002

Singh, R., Srivastava, P., Singh, P., Upadhyay, S., & Raghubanshi, A. S. (2017). Human Overpopulation and Food Security: Challenges for the Agriculture Sustainability. In R. Singh, A. Singh, & V. Srivastava (Eds.), *Environmental Issues Surrounding Human Overpopulation* (pp. 12–39). Hershey, PA: IGI Global. doi:10.4018/978-1-5225-1683-5.ch002

Srivastava, N. (2017). Climate Change Mitigation: Collective Efforts and Responsibly. In I. Management Association (Ed.), Natural Resources Management: Concepts, Methodologies, Tools, and Applications (pp. 64-76). Hershey, PA: IGI Global. https://doi.org/ doi:10.4018/978-1-5225-0803-8.ch003

Stanganelli, M., & Gerundo, C. (2017). Understanding the Role of Urban Morphology and Green Areas Configuration During Heat Waves. *International Journal of Agricultural and Environmental Information Systems*, *8*(2), 50–64. doi:10.4018/IJAEIS.2017040104

Stewart, M. K., Hagood, D., & Ching, C. C. (2017). Virtual Games and Real-World Communities: Environments that Constrain and Enable Physical Activity in Games for Health. *International Journal of Game-Based Learning*, *7*(1), 1–19. doi:10.4018/IJGBL.2017010101

Stone, R. J. (2017). Modelling the Frequency of Tropical Cyclones in the Lower Caribbean Region. In W. Ganpat & W. Isaac (Eds.), *Environmental Sustainability and Climate Change Adaptation Strategies* (pp. 341–349). Hershey, PA: IGI Global. doi:10.4018/978-1-5225-1607-1.ch013

Syed, A., & Jabeen, U. A. (2018). Climate Change Impact on Agriculture and Food Security. In A. Eneanya (Ed.), *Handbook of Research on Environmental Policies for Emergency Management and Public Safety* (pp. 223–237). Hershey, PA: IGI Global. doi:10.4018/978-1-5225-3194-4.ch012

Tam, G. C. (2017). The Global View of Sustainability. In *Managerial Strategies and Green Solutions for Project Sustainability* (pp. 1–24). Hershey, PA: IGI Global. doi:10.4018/978-1-5225-2371-0.ch001

Tam, G. C. (2017). Understanding Project Sustainability. In *Managerial Strategies and Green Solutions for Project Sustainability* (pp. 110–139). Hershey, PA: IGI Global. doi:10.4018/978-1-5225-2371-0.ch005

Tam, G. C. (2017). Perspectives on Sustainability. In *Managerial Strategies and Green Solutions for Project Sustainability* (pp. 53–76). Hershey, PA: IGI Global. doi:10.4018/978-1-5225-2371-0.ch003

Tang, M., & Karunanithi, A. T. (2018). Visual Logic Maps (vLms). In *Advanced Concept Maps in STEM Education: Emerging Research and Opportunities* (pp. 108–149). Hershey, PA: IGI Global. doi:10.4018/978-1-5225-2184-6.ch005

Taşçıoğlu, M., & Yener, D. (2018). The Value and Scope of GIS in Marketing and Tourism Management. In S. Chaudhuri & N. Ray (Eds.), *GIS Applications in the Tourism and Hospitality Industry* (pp. 189–211). Hershey, PA: IGI Global. doi:10.4018/978-1-5225-5088-4.ch009

Thanh Tung, B. (2021). Pharmacology and Therapeutic Applications of Resveratrol. In A. Hussain & S. Behl (Eds.), *Treating Endocrine and Metabolic Disorders With Herbal Medicines* (pp. 321–333). IGI Global. https://doi.org/10.4018/978-1-7998-4808-0.ch014

Tianming, G. (2018). Food Security and Rural Development on Emerging Markets of Northeast Asia: Cases of Chinese North and Russian Far East. In V. Erokhin (Ed.), *Establishing Food Security and Alternatives to International Trade in Emerging Economies* (pp. 155–176). Hershey, PA: IGI Global. doi:10.4018/978-1-5225-2733-6.ch008

Tiftikçigil, B. Y., Yaşgül, Y. S., & Güriş, B. (2017). Sustainability of Foreign Trade Deficit in Energy: The Case of Turkey. In N. Ray (Ed.), *Business Infrastructure for Sustainability in Developing Economies* (pp. 94–109). Hershey, PA: IGI Global. doi:10.4018/978-1-5225-2041-2.ch005

Tiwari, S., Vaish, B., & Singh, P. (2018). Population and Global Food Security: Issues Related to Climate Change. In I. Management Association (Ed.), Climate Change and Environmental Concerns: Breakthroughs in Research and Practice (pp. 41-64). Hershey, PA: IGI Global. https://doi.org/ doi:10.4018/978-1-5225-5487-5.ch003

Toujani, A., & Achour, H. (2018). A Data Mining Framework for Forest Fire Mapping. In I. Management Association (Ed.), Information Retrieval and Management: Concepts, Methodologies, Tools, and Applications (pp. 771-794). Hershey, PA: IGI Global. doi:10.4018/978-1-5225-5191-1.ch033

Tripathy, B., & K., S. B. (2017). Rough Fuzzy Set Theory and Neighbourhood Approximation Based Modelling for Spatial Epidemiology. In I. Management Association (Ed.), *Public Health and Welfare: Concepts, Methodologies, Tools, and Applications* (pp. 1257-1268). Hershey, PA: IGI Global. https://doi.org/ doi:10.4018/978-1-5225-1674-3.ch058

Trukhachev, A. (2017). New Approaches to Regional Branding through Green Production and Utilization of Existing Natural Advantages. In I. Management Association (Ed.), Advertising and Branding: Concepts, Methodologies, Tools, and Applications (pp. 1758-1778). Hershey, PA: IGI Global. doi:10.4018/978-1-5225-1793-1.ch081

Tsobanoglou, G. O., & Vlachopoulou, E. I. (2017). Social-Ecological Systems in Local Fisheries Communities. In G. Korres, E. Kourliouros, & M. Michailidis (Eds.), *Handbook of Research on Policies and Practices for Sustainable Economic Growth and Regional Development* (pp. 306–316). Hershey, PA: IGI Global. doi:10.4018/978-1-5225-2458-8.ch026

Tuydes-Yaman, H., & Karatas, P. (2018). Evaluation of Walkability and Pedestrian Level of Service. In I. Management Association (Ed.), Intelligent Transportation and Planning: Breakthroughs in Research and Practice (pp. 264-291). Hershey, PA: IGI Global. doi:10.4018/978-1-5225-5210-9.ch012

Uddin, S., Chakravorty, S., Ray, A., & Sherpa, K. S. (2018). Optimal Location of Sub-Station Using Q-GIS and Multi-Criteria Decision Making Approach. *International Journal of Decision Support System Technology, 10*(2), 65–79. doi:10.4018/IJDSST.2018040104

Uddin, S., Chakravorty, S., Sherpa, K. S., & Ray, A. (2018). Power Distribution System Planning Using Q-GIS. *International Journal of Energy Optimization and Engineering*, *7*(2), 61–75. doi:10.4018/IJEOE.2018040103

Uzun, F. V. (2018). Natural Resources Management. In A. Eneanya (Ed.), *Handbook of Research on Environmental Policies for Emergency Management and Public Safety* (pp. 1–21). Hershey, PA: IGI Global. doi:10.4018/978-1-5225-3194-4.ch001

V, M., Agrawal, R., Sharma, V., & T.N., K. (2018). Supply Chain Social Sustainability and Manufacturing. In I. Management Association (Ed.), *Technology Adoption and Social Issues: Concepts, Methodologies, Tools, and Applications* (pp. 226-252). Hershey, PA: IGI Global. https://doi.org/ doi:10.4018/978-1-5225-5201-7.ch011

van der Vliet-Bakker, J. M. (2017). Environmentally Forced Migration and Human Rights. In C. Akrivopoulou (Ed.), *Defending Human Rights and Democracy in the Era of Globalization* (pp. 146–180). Hershey, PA: IGI Global. doi:10.4018/978-1-5225-0723-9.ch007

Vargas-Hernández, J. G. (2022). A Comprehensive Entrepreneurship Model for the Internationalization of Green Innovation Businesses. In U. Akkucuk (Ed.), *Disruptive Technologies and Eco-Innovation for Sustainable Development* (pp. 131–149). IGI Global. https://doi.org/10.4018/978-1-7998-8900-7.ch008

Vaskov, A. G., Lin, Z. Y., Tyagunov, M. G., Shestopalova, T. A., & Deryugina, G. V. (2018). Design of Renewable Sources GIS for ASEAN Countries. In V. Kharchenko & P. Vasant (Eds.), *Handbook of Research on Renewable Energy and Electric Resources for Sustainable Rural Development* (pp. 1–25). Hershey, PA: IGI Global. doi:10.4018/978-1-5225-3867-7.ch001

Vázquez, D. G., & Gil, M. T. (2017). Sustainability in Smart Cities: The Case of Vitoria-Gasteiz (Spain) – A Commitment to a New Urban Paradigm. In L. Carvalho (Ed.), *Handbook of Research on Entrepreneurial Development and Innovation Within Smart Cities* (pp. 248–268). Hershey, PA: IGI Global. doi:10.4018/978-1-5225-1978-2.ch012

Wahab, I. N., & Soonthodu, S. (2018). Geographical Information System in Eco-Tourism. In S. Chaudhuri & N. Ray (Eds.), *GIS Applications in the Tourism and Hospitality Industry* (pp. 61–75). Hershey, PA: IGI Global. doi:10.4018/978-1-5225-5088-4.ch003

Weiss-Randall, D. (2018). Cultivating Environmental Justice. In *Utilizing Innovative Technologies to Address the Public Health Impact of Climate Change: Emerging Research and Opportunities* (pp. 110–143). Hershey, PA: IGI Global. doi:10.4018/978-1-5225-3414-3.ch004

Weiss-Randall, D. (2018). Cultivating Resilience. In *Utilizing Innovative Technologies to Address the Public Health Impact of Climate Change: Emerging Research and Opportunities* (pp. 204–235). Hershey, PA: IGI Global. doi:10.4018/978-1-5225-3414-3.ch007

Weiss-Randall, D. (2018). Climate Change Solutions: Where Do We Go From Here? In *Utilizing Innovative Technologies to Address the Public Health Impact of Climate Change: Emerging Research and Opportunities* (pp. 236–268). Hershey, PA: IGI Global. doi:10.4018/978-1-5225-3414-3.ch008

Whyte, K. P., List, M., Stone, J. V., Grooms, D., Gasteyer, S., Thompson, P. B., . . . Bouri, H. (2018). Uberveillance, Standards, and Anticipation: A Case Study on Nanobiosensors in U.S. Cattle. In I. Management Association (Ed.), Biomedical Engineering: Concepts, Methodologies, Tools, and Applications (pp. 577-596). Hershey, PA: IGI Global. https://doi.org/ doi:10.4018/978-1-5225-3158-6.ch025

Wulff, E. (2017). Data and Operational Oceanography: A Review in Support of Responsible Fisheries and Aquaculture. In P. Diviacco, A. Leadbetter, & H. Glaves (Eds.), *Oceanographic and Marine Cross-Domain Data Management for Sustainable Development* (pp. 303–324). Hershey, PA: IGI Global. doi:10.4018/978-1-5225-0700-0.ch013

Yener, D. (2017). Geographic Information Systems and Its Applications in Marketing Literature. In S. Faiz & K. Mahmoudi (Eds.), *Handbook of Research on Geographic Information Systems Applications and Advancements* (pp. 158–172). Hershey, PA: IGI Global. doi:10.4018/978-1-5225-0937-0.ch006

Yu, K., Liu, Y., & Sharma, A. (2021). Analyze the Effectiveness of the Algorithm for Agricultural Product Delivery Vehicle Routing Problem Based on Mathematical Model. *International Journal of Agricultural and Environmental Information Systems, 12*(3), 26–38. https://doi.org/10.4018/IJAEIS.2021070103

Related References

Zhou, X., Sharma, A., & Mohindru, V. (2021). Research on Linear Programming Algorithm for Mathematical Model of Agricultural Machinery Allocation. *International Journal of Agricultural and Environmental Information Systems*, *12*(3), 1–12. https://doi.org/10.4018/IJAEIS.2021070101

About the Contributors

Ahmed Karmaoui started his academic career as Research Fellow at Cadi Ayyad University and awarded Ph.D. from the Faculty of Semlalia (UCA); graduated from Publons Academy, and awarded Top 1% of international Reviewers (2018) by Publons. He has published various peer-reviewed international journal articles, book chapters, and books on a range of environmental topics. He is currently the president of the MoroccanCenter for Culture and Science, Morocco, and an assistant professor at FSTE (UMI). Broad areas of his research interests include ecosystem services assessment.

* * *

Hassan Alahyane is a biology doctor within the Cadi Ayyad University, specialty: animal ecology and microbiology. Member researcher of the laboratory molecular modeling and ecophysiology.

Debraj Biswal is currently employed as Assistant Professor of Zoology in a Government college in West Bengal, India. His teaching career spans for more than 13 years. His research activities for over 14 years mainly include areas in parasitology, helminthology, vector biology and microbiology. He has published a number of articles related to these topics. Additionally, he has keen interests in ecology, environmental biology and eco-toxicology with a few publications in these fields too.

Majidi Lhou was awarded a PhD degree in Applied Organic Synthesis in 1996. He then joined the Sciences and Technologies Faculty Errachidia, University as Professor, as Director of the Laboratory "Ingénierie des Matériaux pour l'Environnement & Ressources Naturelles", and as head of Doctoral Program in Chemistry and Environment at Chemical Department. He is currently the Dean of the Polydisciplinary Faculty of Errachidia, Morocco and Coordinator of national and international research projects. His research interest covers Characterization of plants by the detailed study of the chemical composition of extractable by conventional chromatographic methods (GC/FID, GC/MS, LC/MS/MS) and spectral methods for identification (1D and 2D NMR 1H and 13C); Valuation by chemical, electrochemical and biological medicinal of aromatic plants. Study of different methods of obtaining essential oils of WFP and their influence on the yield and chemical composition of the essential oil obtained; Synthesis, semi-synthesis and chemical transformation of the main components of essential oils and extracts of aromatic and medicinal plants. He is editorial Board and a reviewer at several journals. He is author of two books in organic chemistry, two Patents and over 141 scientific research papers and he has given many scientific conferences at several international meetings.

Mohamed Ouknin has a PhD in Organic, Analytical and Phytochemistry from the Faculty of Science and Techniques-Errachidia, Moulay Ismail University. Member researcher of the Laboratory of Natural Substances & Synthesis and Molecular Dynamics, Faculty of Sciences and Techniques. Our research team works on the valorization of natural products and plants in the agricultural, agri-food and pharmaceutical fields.

Index

I

L

M

N

O

P

T

V

www.ingramcontent.com/pod-product-compliance
Lightning Source LLC
LaVergne TN
LVHW080856300926
849162LV00003B/441
* 9 7 8 1 6 6 8 4 4 8 2 4 3 *